Electric Vehicle Charge Point Infrastructure

Other related titles:

You may also like

- PSFS002P | Guide to Electric Vehicle Charging Installations at Filling Stations, 2nd Edition | 2025
- PSEV005P | Code of Practice for Electric Vehicle Charging Equipment Installation, 5th Edition | 2023
- PBTR039 | Electric Vehicle Components and Charging Technologies: Design, modeling, simulation and control | Sanjeev Singh, Sanjay Gairola, Sanjeet Dwivedi | 2023
- PBTR032 | Cable Based and Wireless Charging Systems for Electric Vehicles: Technology and control, management and grid integration | Rajiv Singh, Sanjeevikumar Padmanaban, Sanjeet Dwivedi, Marta Molinas, Frede Blaabjerg | 2021

We also publish a wide range of books on the following topics:
Computing and Networks
Control, Robotics and Sensors
Electrical Regulations
Electromagnetics and Radar
Energy Engineering
Healthcare Technologies
History and Management of Technology
IET Codes and Guidance
Materials, Circuits and Devices
Model Forms
Nanomaterials and Nanotechnologies
Optics, Photonics and Lasers
Production, Design and Manufacturing
Security
Telecommunications
Transportation

All books are available in print via https://shop.theiet.org or as eBooks via our Digital Library https://digital-library.theiet.org.

IET TRANSPORTATION SERIES 48

Electric Vehicle Charge Point Infrastructure

From planning to operation

Edited by
David Trousdale

The Institution of Engineering and Technology

About the IET

This book is published by the Institution of Engineering and Technology (The IET).

We inspire, inform and influence the global engineering community to engineer a better world. As a diverse home across engineering and technology, we share knowledge that helps make better sense of the world, to accelerate innovation and solve the global challenges that matter.

The IET is a not-for-profit organisation. The surplus we make from our books is used to support activities and products for the engineering community and promote the positive role of science, engineering and technology in the world. This includes education resources and outreach, scholarships and awards, events and courses, publications, professional development and mentoring, and advocacy to governments.

To discover more about the IET, please visit https://www.theiet.org/.

About IET books

The IET publishes books across many engineering and technology disciplines. Our authors and editors offer fresh perspectives from universities and industry. Within our subject areas, we have several book series steered by editorial boards made up of leading subject experts.

We peer review each book at the proposal stage to ensure the quality and relevance of our publications.

Get involved

If you are interested in becoming an author, editor, series advisor, or peer reviewer, please visit https://www.theiet.org/publishing/publishing-with-iet-books/ or contact author_support@theiet.org.

Discovering our electronic content

All of our books are available online via the IET's Digital Library. Our Digital Library is the home of technical documents, eBooks, conference publications, real-life case studies and journal articles. To find out more, please visit https://digital-library.theiet.org.

In collaboration with the United Nations and the International Publishers Association, the IET is a Signatory member of the SDG Publishers Compact. The Compact aims to accelerate progress to achieve the Sustainable Development Goals (SDGs) by 2030. Signatories aspire to develop sustainable practices and act as champions of the SDGs during the Decade of Action (2020–30), publishing books and journals that will help inform, develop, and inspire action in that direction.

In line with our sustainable goals, our UK printing partner has FSC accreditation, which is reducing our environmental impact to the planet. We use a print-on-demand model to further reduce our carbon footprint.

Published by The Institution of Engineering and Technology, London, United Kingdom

The Institution of Engineering and Technology (the "**Publisher**") is registered as a Charity in England & Wales (no. 211014) and Scotland (no. SC038698).

First published 2025

The Institution of Engineering and Technology
Futures Place,
Kings Way,
Stevenage,
Herts, SG1 2UA
United Kingdom

www.theiet.org

British Library Cataloguing in Publication Data
A catalogue record for this product is available from the British Library

ISBN 978-1-83724-119-4 (hardback)
ISBN 978-1-83724-120-0 (PDF)
ISBN 978-1-80705-120-4 (EPUB3)

Typeset in India by MPS Limited

Cover image: SolStock/E+ via Getty Images

Contents

About the editor

David Trousdale is a specialist in sustainability with 6 years of experience in transport decarbonisation. He has developed an in-depth knowledge of charge point infrastructure having worked for two international consultancies. He currently works as a principal advisor at Newcastle City Council delivering the city's ambitious climate change action plan. He happily lives in York with his wife and two children.

Chapter 1
Introduction

David Trousdale[1]

To date, the vast majority of written material regarding electric vehicles (EV) and charge point infrastructure has been written for technical audiences, with recent examples [1,2] focusing primarily on emerging charge point standards and new technologies. This book does not follow this well-trodden path and views charge point infrastructure from a more 'end-to-end' perspective, providing an overview ranging from the process to determine where to deploy charge points all the way through to decommissioning. Throughout this, we analyse the different policies and strategies guiding the direction of travel and dive into different stakeholder perspectives. Furthermore, it places EV charge points within the broader context of the drive towards transport decarbonisation and net zero, recognising that this infrastructure is part of a transport evolution (or revolution depending on your perspective) which started with the rise of the motor car powered by an internal combustion engine (ICE) in the late nineteenth century. I have yet to find a book that covers the whole story. This is a key reason why I chose to write it.

I also wanted to put pen to paper because I believe passionately in delivering net zero as this is the only way to undo the damage we have done to our planet over the past centuries. This is not just for my generation, but for future generations, including both my sons, the eldest who has recently turned 6, and the youngest, 3. To them, the climate they are experiencing now, with all of its erratic and unpredictable weather patterns, is all that they will know, unless of course, we can undo over the next 100 years the damage we have done by taking action now. The consequences of previous generations emitting excessive greenhouse gases will take decades to be reversed. Reversing this can only be achieved by making far-reaching and radical changes to the way we live our lives. In 2020, the United Nations published its development goals to 'protect nature, tackle climate change, and take care of our planet'. Of the 12 goals, three specifically focus on transport. These are that we should all walk, bike or take public transport, switch to an EV and consider our travel in terms of whether to fly to a destination or holiday at home [3]. My eldest took the second to heart at an early age. When he was 2, he got the car keys, ventured to the car, pulled out

[1]Principal Advisor - Climate Change and Sustainability, Newcastle City Council, United Kingdom

the power cable then took it to the garden and tried to plug in his Little Tykes ride on car. Apart from possibly demonstrating irresponsible parenting by leaving car keys in reach, this shows that from an early age children learn from their parents and surroundings.

I've owned or leased an EV for over 5 years and it is clear that my son has learnt that driving and charging one is entirely normal. He has not inherited any of the baggage of the ICE era or poisonous divisive narrative which has somehow found EVs at its heart [4]. To him EVs are not one option to get around by car; they are the only option. He has also started to understand the reasons for their adoption, saying cars with exhausts are 'dirty' and damage the environment with 'smelly fumes'. He is indeed correct that EVs have zero tailpipe emissions and, if owned for over 3–4 years after buying new, reach carbon neutrality, though this is of course determined by miles driven [5]. After that point and every subsequent year, their carbon dioxide impact on the environment diminishes when compared with an ICE vehicle which will continue to produce tailpipe emissions throughout its lifespan. This is not to say that EVs do not produce any emissions. That is not true. EVs, like their ICE counterparts, produce particulate matter (measured as fine particles, known as PM2.5) from tyre and brake wear and tear, though the research is not clear as to how much greater or lesser this is than an ICE vehicle [6]. Recognising all of this, it is clear that from a net zero perspective, EVs are core to decarbonising private car travel, as is the charge point infrastructure that powers the vehicles.

I learned early in my transport career that placing charge point infrastructure is not as simple as dropping a pin on a map. There are various steps that need to be taken which must be achieved to ensure that the charge points are deployed in the right place, are fed with the right amount of power and are used frequently enough to justify the investment. This is a step change from planning fuel station infrastructure for ICE vehicles as the only constraints in installing those is the size of the land, the vicinity of a nearby road to feed traffic and anticipated demand. This knowledge I gained through several years working at two international consultancies. In both organisations, I was a technical director, focused on transport decarbonisation. Following this, I then went to work for a local authority in the North of England in a principal advisor role focused on net zero and climate change. I worked on several projects for public and private sector clients ranging from developing EV strategies for local authorities to providing technical due diligence as part of an investment in a charge point operator (CPO). Across all of these projects I have looked at installing infrastructure from an end-to-end perspective, starting with planning, going through design and deployment, all the way through to operations, maintenance and decommissioning. Throughout the full lifecycle, I've engaged with a range of stakeholders, all with different perspectives. This experience, combined with my passion to deliver net zero, has meant that I am able to write a comprehensive, holistic narrative of EV charge point infrastructure from start to finish. What it has also enabled me to do is dismiss the ideas and concepts proposed in transport as alternatives to EVs, namely synthetic fuels including electrofuels (e-fuels) and hydrogen. One final point to note is that what is written throughout the various chapters is the opinions of each of the authors, based on their own experiences and does not represent the position of the organisations they represent.

1.1 Alternatives to EV batteries

1.1.1 Hydrogen

Hydrogen has recently been touted by various news agencies as the solution to all of our future transport problems [7]. And what is not to like. Zero tailpipe emissions (bar water) and comparable refuelling rates when compared with petrol and diesel, however these articles and news stories seem to ignore the major challenges. Firstly, hydrogen as a fuel is incredibly energy intensive to produce. A recent paper from the European Parliament stated that it takes between 50 and 55 kWh to produce 1 kg of hydrogen [8]. The source of the energy to produce hydrogen is also a key area of challenge. Hydrogen is known as grey, blue, turquoise or green based on how it is produced, with the first three based on fossil fuels. Grey hydrogen is produced via gasification or steam methane reforming. Blue hydrogen is produced using the same process but with carbon capture of around 85–95% of the carbon produced through the process. Turquoise hydrogen, an emerging form, is produced by pyrolysis but again is reliant on a fossil fuel in the form of methane. Green hydrogen is produced via electrolysis using renewable energy and is the most sustainable; however, it is also one of the most energy intensive and is reliant on spare power capacity. Green hydrogen only currently makes up 0.1% of overall hydrogen production but is anticipated to grow as a solution to storing excess renewable energy produced during off-peak hours (e.g. when high winds are experienced at wind farms overnight) [9]. The complexity of hydrogen production and the challenge of most current production methods being reliant on fossil fuels are key reasons why it cannot be considered as a net zero fuel.

Secondly, when burnt as a fuel, it is estimated that the same 1 kg would release around 39 kWh of energy [10], losing around 11–16 kWh during the process. This is inefficient when compared with an EV where anything between 12% and 36% can be lost in the process of refuelling a car. Using the same power input of 50–55 kWh, the vehicle might lose between 6 and 20 kWh of energy, though the higher figure only applies in the absolute worst charging scenarios where the vehicle has a very low state of charge (SoC), is very cold and charging is alternating current (AC) to direct current (DC) at very slow speeds [11]. Furthermore, when considering the full end-to-end cycle of source to vehicle hydrogen is very inefficient. 33% of the energy required to produce hydrogen at source and get it to the wheel is used to propel the vehicle. This compares poorly to 77% for EVs, though is more favourable than the 16% for e-fuel petrol [12]. On efficiency grounds, it makes little sense to adopt hydrogen over battery-powered vehicles.

Thirdly, and perhaps the most important of all, most countries lack the required infrastructure to support mass hydrogen as a fuel and are unwilling to invest the vast sums of money to upgrade existing infrastructure and install new fuel networks to get hydrogen where it needs to be. In the United Kingdom, as an example it is estimated that the cost to build a single hydrogen fuel station is in the range of £1,000,000 to £2,000,000, when compared with the average cost for an EV charging station of between £60,000 and £120,000 [13]. Replicating a United Kingdom wide fuel network akin to that of the petrol station network, where there are roughly 7,000 in England and Wales alone [14], would cost just short of £10 billion. To install a pipe network connecting production facilities could cost hundreds of billion, based on recent estimates for a 400 mile pipeline between the United Kingdom and Germany, which is estimated to cost £2 billion [15]. Recent studies across Europe have proposed that €19 billion to €25 billion of investment is needed for hydrogen infrastructure up to 2050 [16]. Existing gas lines have been proposed as a possible way to move hydrogen [17]; however, investment is needed to make this transition. Further work is required to determine the funds needed in the United Kingdom and Europe to enable this at scale. An additional challenge that makes hydrogen infrastructure expensive is that for green hydrogen to be produced, the most suitable locations for production facilities is at or near power generation. In most cases, this means at coastal locations near where wind farm connections meet land or at hydroelectric or geothermal power plants. The locations tend to be the furthest from where the demand for hydrogen as a fuel is highest. As a result, additional infrastructure costs are required to connect these to a new or gas converted pipeline network. And even if the infrastructure was there, there are only two hydrogen fuel cell vehicles on sale in the United Kingdom [18].

It is for these reasons that I dismiss hydrogen as a viable fuel for private cars and vans. EV charge point infrastructure is less complex to install, has broad industry buy-in and relies on existing power networks (though upgrades will be required) that are integrating renewable energy capacity as it continues to expand globally. This is not to say that there is no case for hydrogen as a fuel for buses, heavy goods vehicles, trains and other heavy vehicles like refuse collection vehicles and diggers. In these use cases, hydrogen fuelling infrastructure is required on a smaller, more concentrated scale with fuelling stations able to serve multiple vehicle types at depots and strategic locations. Hydrogen is also better suited to these vehicles as lithium-ion-based batteries currently struggle under load [19]. But seeing as the focus of this book is on private cars and light goods vehicles, hydrogen can be discounted. The same can also be said for synthetic fuels including e-fuels.

1.1.2 Synthetic fuels and e-fuels

Synthetic fuels are classed as 'a liquid or gaseous fuel derived from a source such as coal, shale oil, tar sands or biomass, used as a substitute for oil or natural gas' [20]. E-fuels are classed as a derivative of synthetic fuels and are manufactured through capturing carbon dioxide or carbon monoxide. E-fuels can also be made sourcing these compounds from industrial processes. This is currently the most common way to ascertain carbon dioxide and monoxide. These compounds are then combined with hydrogen, derived through the aforementioned green hydrogen production process to create e-fuel [21]. This is a drop-in fuel meaning it can be refined to match current fuels like gasoline and kerosene and then used in existing combustion engines to power the vehicle. Synthetic fuels are seen by some in the industry as a straightforward way to decarbonise transport, mainly because the consumer does not need to change the process of refuelling or their behaviour. They simply drive to a fuel station as they currently do and fill up their tank as normal. In Europe, e-fuels are currently being promoted by Porsche and Ferrari through investment in e-fuel manufacturers. Both vehicle manufacturers offer limited EVs on the market. Others, like BMW, are also hedging their bets, developing a range of EVs whilst investing in an e-fuel startup, Prometheus Fuels [22].

Although the process to produce e-fuels often uses carbon capture, the vehicle still emits the same amount of carbon dioxide, nitric oxide/nitrogen dioxide and other polluting compounds as oil-derived petrol. If all vehicles were to switch to using e-fuels, the outcome would effectively be an endorsement of the atmospheric status quo in that there would be no reduction in carbon dioxide as what is extracted via carbon capture is fed straight back in through combustion. And even if it is accepted that an e-petrol-powered car produces 70% less emissions than a conventional petrol car when well to wheel emissions are included, an EV emits 90% less based on the same comparison. And unless fully carbon neutral, an e-fuel car produces around five times as much carbon dioxide as an EV [23]. Furthermore, taking an earlier quoted figure, e-fuels are less efficient when combusted when compared with hydrogen and battery electric as only 16% of the energy is useable following combustion [12]. So e-fuels, and synthetic fuels more broadly are not only less efficient; they will also have minimal impact on achieving our climate goals. The cynic in me would suggest that e-fuels are being pushed by parts of the industry not ready for the EV revolution and see this as an opportunity to continue manufacturing ICE vehicles with little change to their existing processes. Neither hydrogen nor synthetic fuels are the future. EVs are, and these are underpinned by charge point infrastructure.

1.2 Plug-in charge point infrastructure

When powering EVs, there are multiple technological solutions out there, including wireless and pantograph charging. The former involves a wireless pad placed under the vehicle to transfer energy, whereas the latter, primarily used by buses and heavy goods vehicles, involves a connecting arm on the roof that connects to an overhead power pad or connector. There is also an emerging market for battery swapping, with Nio becoming one of the leaders in this space [24]. Each of these technologies will have a role to play in the future of keeping vehicles charged up; however, they are niche solutions when compared with the primary charging infrastructure being deployed globally, plug-in charge points. Deploying this infrastructure forms the basis of this book. I have discounted these niche charging solutions because it is only plug-in charge point infrastructure which is and will continue to dominate charging for the foreseeable future. This is because plug-in charge points can be deployed at scale and utilise well-developed technologies that continue to advance in charging rates and convenience. Furthermore, they are well supported by a mature manufacturing base and are purchased globally in their hundreds of thousands by CPOs who have placed plug-in charge points at the centre of their business models. Every EV sold supports one of the main charging standards when it rolls off the production line. Other charging solutions require adaptation to existing vehicles to make them work. I cannot see how this is likely to change.

Not only is this book focused on plug-in charge point infrastructure. It concentrates primarily on EV cars and vans, outlining the full end-to-end process determining how charge points should be deployed to support the transition to EV. The latter are often referred to as light goods vehicles and weigh less than 3,500 kg though for the purposes of this book they will be referred to as vans. This is not to say that a similar transition, underpinned by different charging solutions, is not currently happening for heavy goods vehicles. Recent developments in battery technologies for this type of vehicle are resulting in an advancement in charge point technologies to support the large 400 kW plus sized batteries that need to be rapidly charged [25]. Advancements are also happening with electric buses. Similarly, in the world of micro-mobility, e-scooters and e-bikes are now increasingly common on our streets, and at sea we are seeing electric-powered boats, albeit in their very early stages. But across these modes, there is yet to be recognised a single charging technology or approach to deploying the infrastructure required at scale to make the transition to EV easier and achievable. In the heavy goods vehicle space for example, there are a myriad of charging solutions ranging from pantographs to plug-in charge points. When charge point infrastructure is installed in motorway truck stations, the power drain from several of these charge points operating together can be the same as a small town. As a result, and unlike charge point infrastructure for cars and vans which can be installed at scale and in many cases, within existing power constraints, these charging stations require major planning, often at a national scale. As their deployment is an area in its infancy this would benefit as the focus for articles, periodicals and research papers and is not the focus of this book. Likewise, electric flight and electric ships are areas where it is not yet clear what charging infrastructure will be proposed to power vehicles. It is also not clear whether current battery technology is even suitable to meet the demanding needs of these vehicles.

Plug-in hybrids in their various forms are also not the focus of this book. This is because the technology, though a useful solution in enabling drivers to switch away from ICE vehicles, will not achieve meaningful reductions in emissions due to their underlying reliance on a conventional engine, though their short distance travelled using battery propulsion has some impact. Recent studies have indicated that plug-in hybrids produce significantly more emission than quoted in their standard laboratory tests [26]. Furthermore, the range under which they can operate purely on batteries alone, limited by battery size, means that they invariably start to produce tailpipe emissions after around 30–40 miles once their batteries are drained and their ICE kicks in, assuming batteries are charged to 100%. The small size of their batteries (around 7–10 kW) means that the need for charging infrastructure is greatly reduced, and that infrastructure only needs to provide charging rates of 7–11 kW to fully charge a battery within 1 hour. Vehicles are often restricted to charging rates of this level as there is little case for on-board systems capable of rapid (50 kW plus) and ultra-rapid (100 kW plus) charging. This is primarily because the charging speed benefit outweighs the additional system upgrades to handle these speeds. Most plug-in hybrids will also charge a battery using energy generated from regenerative breaking, negating the need for infrastructure and, ultimately, if the battery runs out, the vehicle can always switch to a traditional ICE. In summary, plug-in hybrids may be a stop gap on the road to private vehicle decarbonisation but they do not achieve meaningful emission reductions or require major infrastructure investment equivalent to that required for EVs.

When defining what is meant by charge point infrastructure, there are multiple terms used across the industry that may confuse the reader. Throughout this book, we have opted for the term charge point to describe the device which an EV is connected to for charging. A charge point may have multiple sockets, enabling it to charge more than one vehicle at a time or it may only have one socket. A charge point may be tethered meaning it has a cable which is permanently attached and is used to connect to the charge port on the vehicle, or it may be untethered, meaning the driver must provide the cable to connect the charge port to the socket on the charge point. Some charge points offer a tethered cable for 50 kW charging and above alongside a socket for charging at speeds of around 11–22 kW. Like the definition for charge point and socket, there are also multiple terms used for different rated charge point speeds. Throughout this book we will refer to level 1, 2 and 3 charge points. Level 1 (also known as slow) charge points are AC and rated anywhere between 2.3 and 6 kW. They tend to be used in a residential setting, with rates at the bottom end of the spectrum similar to charging via a home power outlet or socket. Some refer to these charge points as 'standard' though we will not use that term in this book. Level 2 charge points (also known as fast) are AC and rated at between 7 and 22 kW, with the former being single-phase and the latter three-phase. 11 kW charge points are also three-phase. Level 2 charge points tend to be used in a destination setting, though single-phase level 2 charge points tend to be installed in residential off-street and a number of on-street use cases. Level 3 charge points are DC and are rated at between 50 kW (known as rapid) and 350 kW (anything over 100 kW is classed as ultra-rapid. The upper end figure continues to increase). Throughout this book we will refer to level 3 as level 3 DC to differentiate from level 1 and 2 AC charge points. Level 3 charge points are installed in en route use cases where a driver may only wait for 30 minutes or so before continuing on with their journey [27]. The focus of this book is primarily on level 2 and level 3 DC charge points as these are mainly installed in a public setting.

1.3 Geographic focus

The adoption of EVs is occurring globally in all developed and most developing countries. The top five countries where EVs have made up a majority of sales, based on 2022 data, are Norway (80% of vehicle sales for that year), Iceland (41%), Sweden (32%), the Netherlands (24%) and China (22%). When looking regionally, the European Union and United States are starting from lower EV market shares (12% and 6%, respectively); however, both are seeing increasing sales. In the United Kingdom, 16.8% of all car and van sales are EV [28].

Variances in geographical adoption rates are down to a number of reasons. In countries like Norway, government incentives to encourage adoption have played an important part, whereas in the United States and Australia, lower adoption rates are primarily down to the lack of infrastructure and greater average distances travelled meaning that range anxiety is putting people off the transition [29,30]. This book is not focused on the reasons why adoption rates differ but rather the role EVs play in the drive to decarbonise and the broader process and steps which must be followed to rollout infrastructure effectively. As this process is largely driven by governance systems and societal structures, the approach in the West to deploying infrastructure greatly differs to that in the East [31]. As my experience of EV infrastructure deployment has developed based on working in the United Kingdom and European market, the geographic focus of this book is on the United Kingdom, Europe, North America and Australasia. I understand the governance and societal structures across these geographies enough to outline the process and key considerations underpinning EV deployment there. I recognise that there are some key difference across those four geographies but the broad consistencies in approach mean that trends in how to deploy infrastructure can be outlined. This is not to say that elements of this book might not apply in other geographies outside the focus; however, care should always be taken when elements are applied outside the United Kingdom, Europe, North America and Australasia. For example, other geographies might follow similar processes in terms of determining demand for EV charge points, but the stakeholders operating in these areas and the local decision-makers involved in granting approval may differ greatly, thus impacting on the ability to get charge points installed. Ultimately, case studies and learnings derived from the United Kingdom and Europe have shaped my understanding of charge point infrastructure. This book should be seen through that lens. Throughout the relevant chapters of this book, I outline the commonalities and differences across these four geographies, whilst demonstrating how the broader approach to deploying infrastructure is roughly the same. This might involve slightly different terminologies for the same thing and, where this is the case, I call out clearly which applies to what geography.

1.4 The structure of this book

This book introduces the reader to EV charge points and how they have developed from a historical perspective as part of the transition to decarbonised transport. It then takes the reader through the broad steps required to install charge points from the initial strategy all the way through to deployment, operation and decommission. It is not intended to be a detailed guide or handbook for charge point deployment but rather it takes the reader through the broader themes and considerations which should be understood, covering both technical and non-technical aspects. Case studies are used to 'bring to life' examples of deployment whilst highlighting the perspectives of different stakeholders in deployment ranging from national and local governments all the way through to power companies (known as distribution network operators (DNOs) in some geographies) and CPOs. The structure set out below provides an overview of what each chapter covers:

- **The drive to decarbonise, a historical perspective and where EV infrastructure plays a part**: this chapter outlines the climate crisis from a historical perspective, initial attempts to decarbonise and the role transport will play in achieving future emission reductions. It sets out the emergence of EVs and their role in transport decarbonisation, and the need to deploy EV infrastructure to support the transition to net zero.
- **Deploying EV infrastructure: the overall process and key stakeholders**: this chapter sets out the overall end-to-end process for installing EV infrastructure. It considers the different charging use cases (destination, en route and residential), the public and private sector perspective and the need for a strategy. It then outlines the overall process at a high level, including scoping, concept and detailed design, tendering (in a public context), deployment, operations and maintenance including upgrade and decommissioning. It notes key stakeholders and the role they play.
- **The strategy and policy environment**: this chapter discusses the current strategy and policy environment in the Western world, specifically covering the United Kingdom, Europe, North America and Australasia. This includes which EV charging infrastructure is being rolled out, target dates, technology requirements and how these link with broader transport and industry decarbonisation targets.
- **Demand and supply modelling including predicting utilisation**: this chapter covers why demand and supply modelling is key to charge point deployment, the tools available to do this and how analysis should be conducted in the real world. It highlights the differences in approaches for AC and DC modelling which is largely driven by the charging use case. It also provides an overview of tools and techniques used to assess power capacity and likely demand and how this is fed into the overarching model.
- **Charge point technologies**: this chapter focuses on current charge point technologies, what contexts they are being deployed in and technological developments over the next 5–10 years. This includes the impact of likely vehicle battery and charging advancements.
- **Planning for deployment**: this chapter outlines what needs to be done to deploy charge points on the ground including concept and site designs, alignment with standards, inclusivity and practical aspects like traffic management (including road traffic orders) and enforcement planning. This also includes the power aspect of planning such as power capacity and implications on cost.

- **Deploying charging infrastructure**: this chapter covers how charge points are installed including practical real world learnings and observations from delivered projects.
- **Operating, maintaining and decommissioning charge points**: this chapter outlines the key elements of operating and maintaining charge points ranging from back office platforms and payments systems to maintaining charge points over the duration of their operation. The final part of the chapter covers what needs to be done to decommission a charge point.
- **Stakeholder perspectives**: this chapter provides an overview of the different stakeholders in the EV charging space including summaries of their role, and the opportunities and challenges they face.
- **Looking to the future: where is EV infrastructure heading**: this chapter summarises the emerging trends in EV charge points and sets out their likely impacts. This includes a view on new technologies, anticipated future charging patterns and commercial models. This includes emerging trends in smart charging, vehicle to grid and also whether wireless charging will take off. It touches on the broader implications of how the advancements in private cars and associated infrastructure is influencing freight for example. It concludes by highlighting areas which should be the focus of further research to understand their impact on charging in the future.
- **Conclusions**: this provides short summaries for each chapter and draws together key observations from across them.

1.5 Summary

I hope that in reading this book, you learn more about the role that charge points are playing now, and will continue to play in the future, in decarbonising cars and vans. I also hope that it outlines a clearer structure, set within the context of why the infrastructure is important, of how to deploy charge points in the real world and the steps that should be followed. If this book results in the rollout of additional charging infrastructure, I will have achieved my goal of enabling the transition from ICE to EVs. Without this transition, I fear transport emissions will continue to increase rapidly, resulting in irreversible damage to our climate, more extreme weather events and the destruction of life as we know it.

References

[1] Vahidinasab V., and Mohammadi-Ivatloo B. (eds), *Electric Vehicle Integration via Smart Charging*. Switzerland: Springer Cham; 2022.

[2] Palanisamy S., Chenniappan S., and Padmanaban S. (eds), *Fast-Charging Infrastructure for Electric and Hybrid Electric Vehicles: Methods for Large-Scale Penetration into Electric Distribution Networks*. Hoboken: John Wiley and Sons; 2023.

[3] United Nations, *Actions for a healthy planet [online]*. 2025. Available from https://www.un.org/en/actnow/ten-actions [Accessed 07 July 2024].

[4] Jolly J., *How electric cars became a battleground in the culture wars [online]*. 2023. Available from https://www.theguardian.com/business/2023/aug/04/electric-cars-culture-wars-evs#:~:text=Since%20at%20least%202014%20rolling,to%20net%20zero%20carbon%20emissions [Accessed 07 July 2024].

[5] Ritchie H., *Electric cars are better for the climate than petrol or diesel [online]*. 2023. Available from https://www.sustainabilitybynumbers.com/p/ev-fossil-cars-climate [Accessed 07 July 2024].

[6] Brogan C., *Prioritise tackling toxic emissions from tyres, urge Imperial experts [online]*. 2023. Available from https://www.imperial.ac.uk/news/243333/prioritise-tackling-toxic-emissions-from-tyres/ [Accessed 07 July 2024].

[7] Bleakley D., *Fossil fuel backers still arguing hydrogen has advantage over batteries for transport [online]*. 2024. Available from https://thedriven.io/2024/04/14/fossil-fuel-backers-still-arguing-hydrogen-has-advantage-over-batteries-for-transport/ [Accessed 07 July 2024].

[8] Kurrer C., *The potential of hydrogen for decarbonising steel production [online]*. 2020. Available from https://www.europarl.europa.eu/RegData/etudes/BRIE/2020/641552/EPRS_BRI(2020)641552_EN.pdf [Accessed 07 July 2024].

[9] Marchant N., *Grey, blue, green – why are there so many colours of hydrogen? [online]*. 2021. Available from https://www.weforum.org/agenda/2021/07/clean-energy-green-hydrogen/ [Accessed 07 July 2024].

[10] University of Oxford, *Heat, energy, and chemical bonds [online]*. 2005. Available from https://vrchemistry.chem.ox.ac.uk/energy/Page_13.htm [Accessed 07 July 2024].

[11] Apostolaki-Iosifidou E., Codani P., and Kempton W., *Measurement of power loss during electric vehicle charging and discharging [online]*. 2017. Available from https://www.sciencedirect.com/science/article/pii/S0360544217303730 [Accessed 07 July 2024].

[12] Jolly J., *Will hydrogen overtake batteries in the race for zero-emission cars? [online]*. 2024. Available from https://www.theguardian.com/business/2024/feb/13/will-hydrogen-overtake-batteries-in-the-race-for-zero-emission-cars [Accessed 07 July 2024].

[13] Cranfield University, *How a UK hydrogen car industry could cut fuel costs and carbon emissions [online]*. 2023. Available from https://www.cranfield.ac.uk/press/news-2023/how-a-uk-hydrogen-car-industry-could-cut-fuel-costs-and-carbon-emissions [Accessed 07 July 2024].

[14] The Valuation Office Agency, *How we value: petrol stations [online]*. 2022. Available from https://valuationoffice.blog.gov.uk/2022/03/25/how-we-value-petrol-stations/ [Accessed 07 July 2024].

[15] Barton C., *UK and Germany agree to explore a hydrogen pipeline between countries [online]*. 2023. https://www.hydrogeninsight.com/sponsored-content/dbt/uk-and-germany-agree-to-explore-a-hydrogen-pipeline-between-countries/ [Accessed 07 July 2024].

[16] Kotek P., Takácsné Tóth B., and Selei A., *Designing a future-proof gas and hydrogen infrastructure for Europe – a modelling-based approach [online]*. 2023. Available from https://www.sciencedirect.com/science/article/pii/S0301421523002264 [Accessed 07 July 2024].

[17] Lipiäinen S., Lipiäinen K., Ahola A., and Vakkilainen E., *Use of existing gas infrastructure in European hydrogen economy [online]*. 2023. Available from https://www.sciencedirect.com/science/article/pii/S0360319923021134 [Accessed 07 July 2024].

[18] Warrick J., and Scullion M., *Everything you need to know about hydrogen fuel cell vehicles [online]*. 2024. Available from https://www.autocar.co.uk/car-news/technology/hydrogen-cars [Accessed 07 July 2024].

[19] Nykvist B., *The feasibility of heavy battery electric trucks [online]*. 2021. Available from https://www.sciencedirect.com/science/article/pii/S2542435121001306 [Accessed 07 July 2024].

[20] Cleveland C. J., and Morris C. (eds), *Dictionary of energy [online]*. 2015. p. 583. Available from https://www.sciencedirect.com/topics/engineering/synthetic-fuel [Accessed 09 July 2024].

[21] Tesfaye M., *Priming the e-pump: Explaining e-fuels [online*]. 2024. Available from https://bipartisanpolicy.org/blog/e-fuel-basics [Accessed 09 July 2024].

[22] Waldersee V., and Abnett K., *Explainer: What are e-fuels, and can they help make cars CO2-free? [online]*. 2023. Available from https://www.reuters.com/business/autos-transportation/what-are-e-fuels-can-they-help-make-cars-co2-free-2023-03-07/ [Accessed 09 July 2024].

[23] Transport and Environment, *E-fuel cars are not zero emission [online]*.2023. Available from https://www.transportenvironment.org/articles/e-fuel-cars-are-not-zero-emission [Accessed 09 July 2024].

[24] NIO, *Nio power [online]*. 2024. Available from https://www.nio.com/nio-power?&noredirect= [Accessed 09 July 2024].

[25] Innovate UK, *Demonstrating the transition to electric HGVs [online]*. 2024. Available from https://www.ukri.org/who-we-are/how-we-are-doing/research-outcomes-and-impact/innovate-uk/demonstrating-the-transition-to-electric-hgvs/ [Accessed 09 July 2024].

[26] Jolly J., *Major plug-in hybrid cars pollute more than official measures suggest [online]*. 2023. Available from https://www.theguardian.com/environment/2023/feb/08/major-plug-in-hybrid-cars-pollute-more-than-official-measures-suggest [Accessed 09 July 2024].

[27] Zapmap, *EV charging connector types [online]*. 2024. Available from https://www.zap-map.com/ev-guides/connector-types [Accessed 12 May 2025].

[28] Jaeger J., *These countries are adopting electric vehicles the fastest [online]*. 2023. Available from https://www.wri.org/insights/countries-adopting-electric-vehicles-fastest#:~:text=There%20are%20already%20a%20number,%25)%2C%20according%20to%20our%20analysis. [Accessed 09 July 2024].

[29] Randall T., *Americans insist on 300 miles of EV range. They're right [online]*. 2023. Available from https://www.bloomberg.com/news/articles/2023-05-04/americans-insist-on-300-miles-of-ev-range-they-re-right [Accessed 09 July 2024].

[30] Philip T., Whitehead J., and Prato C. G., *Adoption of electric vehicles in a laggard, car-dependent nation: Investigating the potential influence of V2G and broader energy benefits on adoption [online]*. 2023. Available from https://www.sciencedirect.com/science/article/abs/pii/S0965856422002981 [Accessed 09 July 2024].

[31] International Energy Agency, *Trends in charging infrastructure [online]*. 2022. Available from https://www.iea.org/reports/global-ev-outlook-2022/trends-in-charging-infrastructure [Accessed 09 July 2024].

Chapter 2

The drive to decarbonise, a historical perspective and where EV infrastructure plays a part

David Trousdale[1]

This chapter sets out an overview of the history of electric vehicles (EVs) as a key transport intervention to tackle climate change and the need for charge point infrastructure to enable the transition. For us to understand, this requires a deep dive into the origins of the scientific study of climate change and the ways to tackle it through decarbonisation and net zero. This helps us then understand what is driving policy, strategy and delivery of interventions to reduce emissions. This requires an analysis of when the terms emerged historically and when societies and governments began to take climate change, and climate change action, seriously. It also requires us to understand which sectors of the economy net zero and decarbonisation were applied to first and how there gradually was a focus on transport as a key greenhouse emitter. We also must understand the origins of the international movement to tackle climate change on a global scale and the policies developed across various sectors to reduce emissions. An increasing focus was placed on the role that emissions generated by transport were having on the climate and the interventions required across different transport modes to deliver emissions reductions. This chapter will help the reader gain an understanding of all of these areas.

Furthermore, this chapter will outline the emergence of different transport modes over the past two centuries, their contribution to emissions and the different interventions focused on decarbonisation. Understanding this enables us then to understand the history of the motor car, its role in contributing to emissions increases and the history of EVs as a solution to the decarbonisation of road based transport, with a specific focus on EV cars and vans. Going hand in hand with this is the historical emergence of infrastructure enabling vehicles to charge on the go and the development of this technology over time. We must also historically track the upwards trajectory of EV and charge point infrastructure numbers, adoption rates and how this growth aligns with targets set in the past and what has been

[1]Principal Advisor - Climate Change and Sustainability, Newcastle City Council, United Kingdom

achieved to date. This helps us to understand where the projections are taking us into the future. This chapter will address these points, whilst also demonstrating, through recent historical examples, the critical importance of properly planning charge point deployment as core to the transition. It will also introduce to the reader the role that charge point infrastructure will play in the future, based on the trends of the past.

2.1 Climate change, net zero and decarbonisation: the origins of the concept

2.1.1 *The nineteenth century: the foundations of the theory*

The concept of climate change originated in the nineteenth century following various scientific discoveries that eventually resulted in scientific evidence and theory taking shape in the twentieth century regarding man-made atmospheric changes which cause the climate to change. It was the French mathematician and physicist Joseph Fourier who in 1824 observed the basics of the greenhouse effect. In his work, he identified that gases in the Earth's atmosphere prevented heat escaping, trapping it, thereby making it warmer than it would otherwise be. This discovery was further advanced in 1856 when Eunice Newton Foote, an amateur American scientist, identified that water vapour and carbon dioxide were the key gases responsible for trapping heat in the atmosphere. These discoveries only became of interest because there started at the time to be a focus on the ice ages, what caused the melting of the great ice sheets and the extinction of creatures like the woolly mammoth and sabre-toothed tiger. A further scientific discovery from the Irish physicist John Tyndall was that burning heating coal produced gases, primarily carbon dioxide, methane and volatile hydrocarbons, which absorbed large amounts of energy. Tyndall linked his discovery to Earth's changing atmospheric composition which contributed to the prehistoric ice age.

In 1896, Swedish physicist Svante Arrhenius built on Tyndall's findings and developed a climate change model that demonstrated how different concentrations of atmospheric carbon dioxide influenced global temperatures. It is at this point we start to see an understanding that Earth's atmospheric composition, and, ultimately, climate, had not remained constant. Arrhenius, like Tyndall, started to develop theories as to what atmospheric compositions might have resulted in Earth's ice ages and started to formulate a link between human activities as part of the industrial revolution, namely the burning of fossil fuels, and the potential increases in averages temperatures that this could cause. He predicted that it would take 3,000 years for atmospheric carbon dioxide levels to double, leading to an increase of 5–6°C [1].

2.1.2 *The twentieth century: the theory takes shape*

As the twentieth century progressed, the various pieces of the climate change jigsaw started to fall into place as scientists across a range of fields made discoveries that demonstrated a clear evidential link between man-made greenhouse gas emissions and global temperature increase. By the end of the century, the picture was complete enough to spur on national and international action to reduce emissions; however, this also began to sow the seeds of a backlash against the concept, its impact and our role as humans contributing to it.

Guy Callendar, an amateur scientist, is credited with developing the concept of global warming. In 1938, Callendar began to gather data from 147 weather stations across the world and in doing so discovered that global temperatures had risen 0.3°C over the previous 50 years. Drawing on the previous work of scientists including Tyndall and Arrhenius he argued that carbon dioxide derived from man-made activities including industrial emissions were responsible for global warming. His work, however, was ignored by scientists, primarily because they did not believe humans could exert such an impact on global climate.

By 1958, there was a realisation that the climate was changing and that humans were to blame. The geochemist, Charles David Keeling developed an instrument that was capable of measuring carbon dioxide levels in the atmosphere. By setting this up at a weather station on the top of Mauna Loa, a volcano in Hawaii, and taking detailed measurements every day over a period of 5 years, Doctor Keeling was able to evidence that carbon dioxide levels in the atmosphere were increasing. Before this, it had been assumed to be the case; however, no concrete evidence existed to prove it. The understanding of how the Earth's climate functioned took a major step forward in 1966 with the development of a climate model. Researchers Syukuro Manabe and Richard Wetherald created what is recognised as the world's first accurate computer model of Earth's climate [2]. Various adjustable parameters were built into the model, including carbon dioxide levels, which could then be tweaked to understand the impact on global temperatures. Manabe and Wetherald wrote in 1966 that 'according to our estimate, a doubling of the CO_2 content in the atmosphere has the effect of raising the temperature of the atmosphere (whose relative humidity is fixed) by about 2°C' [3]. This statement has stood the test of time as our current understanding that atmospheric carbon dioxide levels increased over time aligns closely with temperature increases we have seen since scientific measurement began in the 1880s. Over this period, carbon dioxide levels have increased by 50% and global temperatures by 1.1°C [2].

Following the work completed by Manabe and Wetherald, Doctor John Mercer completed a detailed analysis of the impact of global warming on Antarctic ice sheets in 1968, discovering that a glacier in West Antarctica was once a freshwater lake. In discovering this and looking at the analysis of the data, Doctor Mercer identified that not only could glaciers form, they could smelt away in their entirety and re-form again. This was considered impossible prior to his study. Alongside his work, advancements in satellite technology were providing data to scientists that transformed their understanding of the Earth and its climate. In 1969, the National Aeronautics and Space

Administration launched Nimbus III, a satellite that could accurately measure global atmospheric temperature. The satellite provided accurate temperature data that enabled scientists to understand how various factors influence this, and also, how temperature changes impact the Earth. From that point onwards, and as more satellites have been launched into orbit, our understanding of the atmosphere and its warming effects have grown exponentially, enabling scientists to adapt the basics of Manabe and Wetherald's model to run ever more complex scenarios showing how carbon dioxide emissions increase global temperature.

As scientists were using satellites in the skies above to understand climate change in the present, others began to seek answers to historical trends by drilling below to extract ice cores containing frozen bubbles of long lost Paleolithic climates literally frozen in time. In 1985, a team of French and former Soviet scientists began drilling into Antarctic ice, extracting ice cores over 2,000 m in length. These cores represented a period of around 150,000 years, with the base of the core (the part furthest away from the surface) representing the furthest back in time. Analysis of the frozen gas bubbles in slices of the core along different depths demonstrated changes in the atmospheric composition of gas and showed that over time carbon dioxide and methane levels have increased rapidly. This analysis, combined with more recent cores representing a period stretching back 800,000 years demonstrates that up until the early nineteenth century, carbon dioxide and methane levels remained relatively constant. After that point, concentrations began to rise and are now around 50% higher than what is recorded before the industrial revolution. By the late 1980s, scientists had amassed enough data from various sources to clearly state that climate change is happening. Man-made activity since the industrial revolution has accelerated and continues to accelerate this as a result of burning fossil fuels ranging from peat and coal to oil and gas. It is also at this time where we start to see both the term and definition for climate change becoming established.

2.1.3 *Settling on the definitions of climate change, decarbonisation and net zero*

As the scientific realisation dawned that climate change was real, so did the desire at an international level to understand the pace of change and its impact. In 1988, the Intergovernmental Panel on Climate Change (IPCC) was established by the United Nations Environment Programme (UNEP) and the World Meteorological Organisation (WMO) to fulfil that role, with the specific remit for assessing the science behind climate change. This is achieved through the organisation producing scientific reports that analyse climate change worldwide. Since its establishment, five reports have been produced and since 1990, these have highlighted repeatedly that the Earth is experiencing global warming and that this is caused by human activity [2]. The IPCC established the definition for climate change which is now the basis from which all scientific study in this field is derived.

Climate change

A change in the state of the climate that can be identified (e.g., by using statistical tests) by changes in the mean and/or the variability of its properties and that persists for an extended period, typically decades or longer. Climate change may be due to natural internal processes or external forces, or to persistent anthropogenic changes in the composition of the atmosphere or in land use.

Source: IPCC, 2012: Glossary of terms. In: *Managing the Risks of Extreme Events and Disasters to Advance Climate Change Adaptation* [Field, C.B., V. Barros, T.F. Stocker, D. Qin, D.J. Dokken, K.L. Ebi, M.D. Mastrandrea, K.J. Mach, G.-K. Plattner, S.K. Allen, M. Tignor, and P.M. Midgley (eds.)]. A Special Report of Working Groups I and II of the Intergovernmental Panel on Climate Change (IPCC). Cambridge University Press, Cambridge, UK, and New York, NY, USA, pp. 555–564.

The structured approach the IPCC adopted in publishing its first assessment report in 1990 enabled governments to understand the size and nature of the challenge facing the planet and also set in motion different approaches to mitigating the impact of human activity. It is through this activity that the terms net zero and decarbonisation were developed.

Net zero emissions

Net zero emissions are achieved when anthropogenic emissions of greenhouse gases to the atmosphere are balanced by anthropogenic removals over a specified period. Where multiple greenhouse gases are involved, the quantification of net zero emissions depends on the climate metric chosen to compare emissions of different gases (such as global warming potential, global temperature change potential, and others, as well as the chosen time horizon).

Source: IPCC, 2018: Annex I: Glossary [Matthews, J.B.R. (ed.)]. In: *Global Warming of 1.5°C. An IPCC Special Report on the impacts of global warming of 1.5°C above pre-industrial levels and related global greenhouse gas emission pathways, in the context of strengthening the global response to the threat of climate change, sustainable development, and efforts to eradicate poverty* [Masson-Delmotte, V., P. Zhai, H.-O. Pörtner, D. Roberts, J. Skea, P.R. Shukla, A. Pirani, W. Moufouma-Okia, C. Péan, R. Pidcock, S. Connors, J.B.R. Matthews, Y. Chen, X. Zhou, M.I. Gomis, E. Lonnoy, T. Maycock, M. Tignor, and T. Waterfield (eds.)]. Cambridge University Press, Cambridge, UK and New York, NY, USA, pp. 541–562. https://doi.org/10.1017/9781009157940.008.

The origins of the term net zero began in 2009 as a result of the activities of various scientists, including Myles Allen and Dave Frame, who published a paper which noted that the eventual extent of global warming is mainly determined by cumulative emissions of carbon dioxide. In the same year, Susan Solomon and other scientists demonstrated through their modelling work that temperatures do not decline for a number of centuries after a complete cessation of carbon dioxide emissions [4]. These works, when combined, demonstrated that there is a delayed feedback loop between carbon dioxide emissions and global temperatures and that this loop is cumulative in that as carbon dioxide levels increase feedback manifests in the form of increased temperatures, which further impacts emissions. This feedback is not immediate. Further to this Damon Matthew, alongside other scientists, proposed in 2012 that 'cumulative carbon emissions represent an alternative framework that is applicable both as a tool for climate mitigation and for the assessment of potential climate impacts'.

Decarbonisation

The process by which countries, individuals or other entities aim to achieve zero fossil carbon existence. This typically refers to a reduction of the carbon emissions associated with electricity, industry and transport.

Source: IPCC, 2018: Annex I: Glossary [Matthews, J.B.R. (ed.)]. In: *Global Warming of 1.5°C. An IPCC Special Report on the impacts of global warming of 1.5°C above pre-industrial levels and related global greenhouse gas emission pathways, in the context of strengthening the global response to the threat of climate change, sustainable development, and efforts to eradicate poverty* [Masson-Delmotte, V., P. Zhai, H.-O. Pörtner, D. Roberts, J. Skea, P.R. Shukla, A. Pirani, W. Moufouma-Okia, C. Péan, R. Pidcock, S. Connors, J.B.R. Matthews, Y. Chen, X. Zhou, M.I. Gomis, E. Lonnoy, T. Maycock, M. Tignor, and T. Waterfield (eds.)]. Cambridge University Press, Cambridge, UK and New York, NY, USA, pp. 541–562. https://doi.org/10.1017/9781009157940.008.

In 2013, the IPCC produced its Fifth Assessment Report which stated that limiting global temperature change means limiting the cumulative (or stock) of carbon dioxide emissions in the atmosphere. To eventually stop global warming, net anthropogenic additions of carbon dioxide into the atmosphere have to reach zero. The report also set out a clear definition of net zero and decarbonisation to guide policymakers in shaping the direction of travel. In 2014, the President of the World Bank, Jim Yong Kim was the first head of a major commercial organisation to recognise that net zero must play a part in the global response to climate change, stating that a climate agreement should 'provide a clear pathway to zero net emissions before 2100' [4]. This recognition, alongside developments in scientific understanding, prompted the push to establish international climate agreements and a governance regime to monitor progress. This process started with the IPCC and continues to this day.

2.2 The rise of climate change governance: when did governments and societies start taking climate change seriously?

Prior to the establishment of the IPCC in 1988, the basis for international cooperation on environmental matters began in 1972 at the United Nations Scientific Conference held in Stockholm. The conference adopted a declaration that 'set out principles for the preservation and enhancement of the human environment, and an action plan containing recommendations for international environmental action'. The declaration raised the issue of climate change for the first time at an intergovernmental level, warning governments that their activities might lead to climate change and that they need to assess and evaluate the likelihood and magnitude of climatic effects. The conference also established the United Nations Environmental Programme (UNEP) Governing Council. Following the conference, focus on climate change and its impacts increased, leading in 1979 to the UNEP setting out to monitor and evaluate the long-range transport of air pollutants. As a result, the first international instrument on climate, known as the Convention on Long-Range Transboundary Air Pollution, was then adopted. In 1980, the UNEP began to focus its efforts on the ozone layer, recommending measures to limit the production and use of chlorofluorocarbons F-11 and F-12. This culminated in 1985 in the Vienna Convention for the Protection of the Ozone Layer, following on from the 1979 Transboundary Air Pollution Convention, which aimed at reducing sulphur emissions by 30%.

In 1987, the United Nations General Assembly placed environmental issues front and centre when it adopted 'the Environmental Perspective to the Year 2000 and beyond'. This, according to Peter Jackson, Chief Editor of the Yearbook Unit of the United Nations Department of Public Information and a font of information regarding the international climate movement, established 'a framework to guide national action and international cooperation on policies and programmes aimed at achieving environmentally sound development'. By 1988, global warming and the growing hole in the ozone layer captured the public imagination. Sensing the international public debate and political agenda, the UNEP organised in January of that year an internal seminar to identify environmental sectors that might be sensitive to climate change, and in November of the same year, the IPCC was established as a forum for the examination of greenhouse warming and global climate change. Following this and according to Jackson, '1989 was a watershed year for climate change, as the first significant global efforts were taken'. The General Assembly, in resolution 44/207, endorsed the UNEP Governing Council's request to begin preparations with the WMO for negotiations on a framework convention on climate change. In the same year, the Montreal Protocol was adopted to tackle emissions of substances causing the formation of the hole in the ozone layer.

In 1990, a world climate conference was held which stated that climate change was a global problem of unique character for which a global response was required. In 1992, the General Assembly decided to convene the United Nations Conference on Environment and Development in Rio de Janeiro, Brazil. This came to be known as the Earth Summit and resulted in the establishment of a new framework for seeking international agreements to protect the integrity of the global environment. Most importantly, the Summit was the first time where a link was made between specific human activities and climate change, establishing a connection between science, sustainable development, energy development and consumption, transportation, industrial development, stratospheric ozone depletion and transboundary atmospheric pollution. Jackson notes that the 'most significant event during the Conference was the opening for signature of the United Nations Framework Convention on Climate Change (UNFCCC); by the end of 1992, 158 States had signed it'. In 1994, it came into force. In 1995, the Conference of the Parties (COP) to the Convention adopted the Berlin Mandate, launching talks on a protocol containing stronger commitments for developed countries and those in transition. The COP is the supreme decision-making body of the Convention [5]. Following this, the IPCC's Second Assessment Report was published in 1995 which led in 1997 to the Kyoto Protocol [2], signed by 192 parties. According to the United Nations, 'the Kyoto Protocol operationalises the United Nations Framework Convention on Climate Change by committing industrialised countries and economies in transition to limit and reduce greenhouse gases emissions in accordance with agreed individual targets. The Convention itself only asks those countries to adopt policies and measures on mitigation and to report periodically'. Progress against the protocol is tracked against an annex-based structure, with Annex B setting binding emissions reduction targets for 37 industrialised countries and economies in transition including the European Union as a nation block. These targets add up to an average 5% emission reduction compared to 1990 levels over the 5-year period 2008–2012 (the first commitment period) [6]. A notable omission is the United States which indicated its intention not to ratify the Kyoto Protocol. Those countries that did ratify it agreed to reduce the six main greenhouse gases which are carbon dioxide, methane, nitrous oxide, hydrofluorocarbons, perfluorocarbons and sulphur hexafluoride. The Annexes do not specify how countries are meant to achieve their targets.

The United States chose to remove itself from the Kyoto process in 2001, and those who remained in it witnessed the protocol become international law in 2005. In the same year, the United Kingdom began its first steps on the road to establishing a long-term legally binding framework to cut carbon emissions, the first country to do so [7]. These steps were first taken by Prime Minister Tony Blair who selected climate change as the key priority for his terms as chair of the G8 and president of the European Union. Over the following 3 years, climate change began to play a more central role in the United Kingdom, culminating in the Climate Change Act. The Act set a target requiring emissions reductions of at least 100% by 2050. It also established the Committee on Climate Change (CCC), an organisation that sets out carbon budgets every 5 years and tracks progress against this, holding the United Kingdom Government to account [8]. Each budget is set to enable the government to prepare, plan, make laws and secure investment to deliver the activities required to hit the budget. To date, there have been three completed budgets, the first (2008–2012), second (2013–2017) and the third (2018–2022) have all been met. The fourth (2023–2027) is currently being delivered against [9].

The United Kingdom may well have started to lead the way; however, the international community continued to drive forward with its focus on climate action. Intergovernmental policy making continued to progress as the United Kingdom Government was shaping the Climate Act. In 2007, the IPCC's Fourth Assessment Report concluded that it is more than 90% likely that humanity's emissions of greenhouse gases are responsible for modern-day climate change. In the same year, significant effort was made by the United Nations to secure a 'Bali Roadmap' aimed at establishing a new treaty by the end of 2009. However, by the time 2009 came around, a new global achievement could not be agreed. What came instead was the Copenhagen Accord, a political agreement which placed renewed emphasis on the need to combat climate change [7]. Major progress came 6 years later when the Paris Agreement was signed by 196 countries at COP 21 (the 21st session of the convention, the first being in Berlin in 1995). This agreed to limit global warming to less than 2°C above pre-industrial levels. The agreement is a legally binding international treaty and came into force in November 2016 [2]. Since COP 21, a further seven COPs have taken place with further international climate agreements being made at each, the most significant being Glasgow in November 2021 (COP 26) and Dubai in November 2023 (COP 28) [10].

At a domestic policy level, various countries have seen the United Kingdom's Climate Change Act as a template to meet the legal requirements of the Paris Agreement and have followed suit, adapting a similar approach to achieve similar aims. For example, both Sweden and New Zealand have based their equivalent acts on the Climate Change Act [11]. The European Union has taken a slightly different approach and has set a European climate law. European Union Member States must cut greenhouse gas emissions by at least 55% by 2030 with the overarching goal to make the European Union climate neutral by 2050 [12]. Australia has based its approach on the Paris Agreement, committing itself to reducing greenhouse gas emissions by 43% below 2005 levels by 2030 and achieving net zero emissions by 2050 [13]. Similarly, Canada has set a 2030 Emission Reduction Plan, providing a roadmap for the Canadian economy to achieve 40–45% emissions reductions below 2005 levels by 2030 and, to achieve net zero by 2050 [14]. As the United States is not a signatory of the Paris Agreement or Kyoto Protocol, its goals are not legally binding and are susceptible to political horse-trading symptomatic of the United States political system. The election of Donald Trump in November 2024 put an end to efforts of the Biden-Harris presidency to align the United States with international climate agreements.

2.3 Early attempts to introduce interventions to achieve net zero

In the United Kingdom, the Climate Change Act not only determined a system of carbon budgets, it established the CCC. The CCC is an independent body that provides evidence-based advice to the UK Government and Parliament on the mandatory carbon budgets [15]. The CCC provides policy direction to the UK Government, devolved governments and local authorities. Its first report, 'Building a low-carbon economy – the United Kingdom's contribution to tackling climate change' was published on 1 December 2008 and contained the CCC's recommendations on the 2050 emissions reduction target and advised on the levels of the United Kingdom's first three legally binding carbon budgets for 2008–2022. The report set out three core approaches to meeting the carbon budget targets; decarbonising electricity generation; reducing energy use in buildings and industry; and reducing domestic transport emissions. It also provided views on decarbonising international aviation and shipping. Furthermore, the report considered wider economic and social considerations, including competitiveness challenges and opportunities, economic costs and fiscal implications, fuel poverty implications and energy security of supply [16].

It is fair to say that this report helped spur the rapid investment in renewable energy generation in the United Kingdom, with the percentage of electricity generated by renewables increasing from 2.8% in 2000 to 40% in 2022. This increase has largely resulted from the construction of large-scale wind farms in the North and Irish seas. Reducing energy usage in buildings and industry focused on the delivery of four key action areas. Of these four, the first is those which reduce electricity demand, whether via energy improvements or lifestyle changes. The second focused on actions that reduce direct fossil fuel use, whether via energy efficiency improvements or lifestyle change. The third covered actions to produce heat more carbon efficiently, whether via the development of renewable heat sources, or the use of electricity to generate low carbon heat. The fourth and final action area was the microgeneration of electricity at locally distributed level [17].

According to the CCC, reducing domestic transport emissions in the United Kingdom centred on abatement opportunities from surface transport and the reductions achieved through using policy levers. Regarding the former, proposed carbon reductions could only be achieved by improving the fuel efficiency of cars, vans and heavy goods vehicles, eco-driving (smoother acceleration and braking), modal shift and better journey planning. Furthermore, an initial mention of the introduction of EVs and plug-in hybrids, albeit not as central to emission reductions, was proposed alongside effective speed limit enforcement to achieve the carbon budget. The policy levers focused on using European Union legislation to force vehicle manufacturers to improve fuel efficiency. This was to be achieved through using a range of levers including better information and driver training to encourage behaviour change, modal shift and better journey planning [16]. For aviation and shipping, the Committee proposed that emissions from both of these transport areas should be tracked so that future interventions could be proposed.

The United Kingdom was not the only country to introduce policy levers to try to reduce emissions. Norway introduced a National Strategy for Sustainable Development in 2002 and a National Action Plan for Sustainable Development in 2003. A revised strategy was adopted in 2007. Key parts of this strategy included climate, ozone layer and transboundary atmospheric pollution, and natural resources [18]. These parts were developed as distinct policy areas. Early areas of focus included decarbonising energy production through promoting wind farms and hydro power and exploring approaches to decarbonising transport. Numerous other countries have now enshrined net zero in law (including France and Sweden) and have adopted policy levers to achieve these targets; however, their efforts to do this came roughly 10 years after the United Kingdom [19]. In many ways, the United Kingdom continues to lead globally in achieving net zero, having cut its greenhouse gas emissions by 50% between 1990 and 2022, while growing the economy by nearly 80%. The United Kingdom is the first major economy to achieve this [20].

2.4 Understanding net zero and decarbonisation: a sectoral approach

Western governments recognised that declaring net zero targets in political manifestos or enshrining these targets in law would not actually achieve net zero unless this was accompanied with policy interventions, set out in net zero strategies. As previously mentioned, the United Kingdom was early to establish an approach to delivering net zero through the CCC's publication of its first report entitled 'Building a low-carbon economy – the United Kingdom's contribution to tackling climate change' [16]. Other governments followed similar approaches. The European Union invited all parties to communicate, by 2020, their mid-century, long-term low greenhouse gas emission development strategies. Alongside this, the European Union developed a European wide net zero strategy. As part of these strategies, emission reductions in individual sectors were proposed. The sectors of focus included electricity, industry, transport, the heating and cooling of buildings, agriculture, waste and land use, and forestry [21]. Similarly, the Australian Government set out in its strategy, launched in 2023, practical actions required to reduce emissions, including from electricity, buildings and energy efficiency, procurement, vehicle fleet, business travel, computing and data systems and driving change through people, culture and behaviour [22]. Across all of these strategies, it is clear that a sectoral approach to delivering net zero is key as this makes carbon emission reduction achievable by assigning actions and activities, underpinned by legislation if required, to key stakeholders in each sector.

The United Kingdom's Net Zero Strategy: Build Back Greener, produced in 2021, established a straightforward sectoral approach, breaking down emission reduction plans into seven areas, details of which are highlighted below:

1. **Power**: this focuses primarily on action relating to promoting low cost renewable energy generation, investing in nuclear power, delivering a smart and flexible energy system and reforming planning to support the deployment of low carbon energy infrastructure;
2. **Fuel supply and hydrogen**: this set out plans to increase hydrogen production, establish a low carbon fuel strategy for transport and establish regulations to minimise greenhouse gas emissions in the oil and gas sector;

3. **Industry**: this established a range of interventions including industrial carbon capture, encouraging hydrogen industry clustering, enabling energy efficiency measures in industrial processes and incentivising industry to switch to low and zero carbon production methods (in iron and steel specifically);
4. **Heat and buildings**: interventions focused on upskilling the workforce to take advantage of green skilled jobs (e.g., heat pump installers), making the transition to low carbon buildings affordable and achievable, helping households and business reduce energy bills, promoting heat networks and exploring hydrogen as a home heating solution replacing gas;
5. **Transport**: the policy interventions focused on ending the sale of petrol and diesel vehicles from 2030, encouraging the rollout of charging infrastructure, ending the sale of all non-zero emission road vehicles by 2040, leading by example through ensuring all government car and van fleet vehicles are zero emission by 2027, encouraging the use of low carbon fuels where vehicles cannot decarbonise, spending £12 billion in local transport through the current parliament (which ran until July 2024), investing £3 billion in a national bus strategy, electrifying more railway lines, kick starting sustainable aviation fuel and investing £2 billion in cycling and walking infrastructure;
6. **Natural resources and waste**: this set out plans for farmers to adopt low carbon practices, investment in net zero agriculture and horticulture, treble woodland creation rates by the end of parliament (July 2024), restoring at least 35,000 hectares of peatland by 2025 and working with local authorities to roll out food waste collections for all households from 2025;
7. **Greenhouse gas removals**: policy interventions included investment in direct air carbon capture and storage and encouraging the development of markets and incentives for investment in greenhouse gas removal methods [23].

Our focus is primarily on the role that transport decarbonisation interventions, namely the adoption of EVs and EV infrastructure, can play in achieving net zero. In Section 2.5, we explore in more detail what role transport decarbonisation can play in achieving net zero.

2.5 Transport, its contribution to emissions and interventions to decarbonise

Across the Western world, transport plays a key role in contributing to greenhouse gas emissions. This is primarily due to the fact that vehicles are powered by internal combustion engines (ICE) that rely on petrol, diesel or, in a rare number of cases, liquefied petroleum gas. When combusted these fuels primarily release carbon dioxide alongside other gases (e.g., nitrogen oxide). Prior to the combustion engine, the burning of coal in steam engines to propel trains and traction engines also released carbon dioxide alongside sulphur dioxide, nitrogen oxides and other gases. Ever since the steam engine powered the first train in the United Kingdom in the 1820s and ICE the first car in Germany in the 1880s, carbon dioxide derived from transport has added 2,500 billion tonnes to the atmosphere since 1850 (up until 2021) [24].

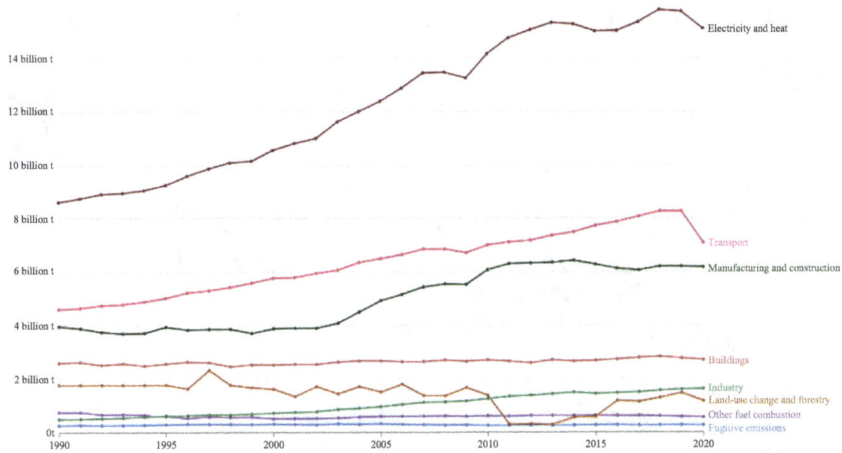

Figure 2.1 *Global carbon dioxide emissions by sector. [Source: Hannah Ritchie, Pablo Rosado and Max Roser (2023) – 'CO_2 and Greenhouse Gas Emissions' Published online at OurWorldinData.org. Retrieved from: https://ourworldindata.org/co2-and-greenhouse-gas-emissions [Online Resource] [Accessed 28 August 2024]].*

A detailed global breakdown of carbon dioxide emissions in Figure 2.1 notes that in 1990 (the first date where a sectoral breakdown of emissions is available), transport contributed to 4.61 billion tonnes, rising to 7.10 billion tonnes in 2020, having dipped slightly from a peak of 8.19 billion tonnes in 2019. This fall was directly the result of the coronavirus epidemic that forced countrywide lockdowns across the globe, impacting people movement [25]. These transport figures are for all modes, including aviation and maritime; however, according to Hannah Ritchie writing in Our World in Data, 'Road travel accounts for three-quarters of transport emissions. Most of this comes from passenger vehicles – cars and buses – contributing 45.1%. The other 29.4% comes from trucks carrying freight' [26]. As a whole, transport accounts for about 21% of total emissions. Road transport accounts for about 15% of total emissions. Figure 2.2 shows the breakdown of global carbon dioxide emissions using actual data up until 2020 and then projecting emissions into the future, up until 2070 [27]. As can be clearly seen, road-based transport makes up the bulk of transport emissions and does so up until 2065 when it starts to fall under 50%.

The modal split of emissions differs slightly by Western geography though broadly speaking road-based transport makes up the bulk of emissions regardless of location. Where there is a slight difference, this is primarily the result of different geographies choosing to fly, thereby increasing emissions, or travelling by rail, reducing emissions. Flight-based emissions per capita on average are highest in the United States followed by Australia, New Zealand and Canada. This is primarily due to affluence (people are richer so will fly to save time even if it costs more) and the fact that these counties tend to be larger or have spread out geographies with poor public transport connectivity [28]. Western countries with the largest railway

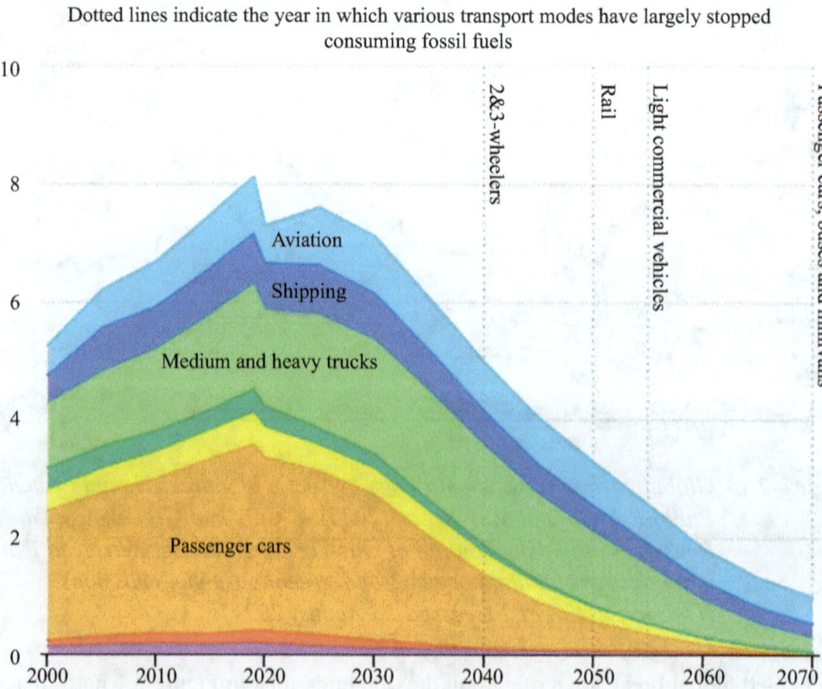

Figure 2.2 Global CO_2 emissions in transport by mode in the Sustainable Development Scenario, 2000–2070 [Source: IEA 2020; Global CO_2 emissions in transport by mode in the Sustainable Development Scenario, 2000–2070, https://www.iea.org/data-and-statistics/charts/global-co2-emissions-in-transport-by-mode-in-the-sustainable-development-scenario-2000-2070, License: CC BY 4.0 [Accessed 28 August 2024]].

networks, in contrast, tend to have lower domestic flight emission. Germany, for example, is the fourth highest country by rail usage globally, seeing roughly 1.2 billion passenger trips over 46 billion km in 2024. France is fifth, seeing 900 million passenger trips over 75 billion km [29]. Germany ranks 40th in per capita carbon dioxide emissions from domestic aviation in 2018, and France ninth [28].

Recognising that most emissions are derived from road-based transport, a majority of interventions target a reduction in emissions from road-based vehicles, though there is also focus on electrifying railways, developing sustainable aviation fuel and decarbonising maritime via ammonia as a fuel.

2.6 Transport decarbonisation from a road-based perspective

The CCC in the United Kingdom was the first to set out in 2008 a range of intervention to decarbonise road-based transport focusing primarily on cars, vans and heavy goods vehicles. Many of these interventions were technological solutions and included:

- **Improvements in carbon efficiency without a change in energy source**: this applied to all road-based vehicles and focused on the promotion of more fuel efficient diesel engine vehicles (misguided at the time), the improvement in the efficiency of petrol and diesel combustion vehicles, a wider use of hybrid engine technologies that capture energy otherwise lost when decelerating and breaking, and non-powertrain measures including aerodynamics and weight reduction.
- **The use of electricity as an energy source**: the CCC identified that EVs (specifically cars and vans) could play a key role in decarbonising transport as battery technology improves. Heavy goods vehicles were less suitable for electrical powertrains; however, small heavy goods vehicles might benefit from some battery technologies.
- **Biofuels as an energy source**: these were promoted across cars, vans and heavy goods vehicles but were deemed to have a bigger impact for heavier vehicles due to the difficulties in electrifying these.

Hydrogen was dismissed at the time as not being a viable alternative because low carbon sources of hydrogen and the associated hydrogen supply infrastructure did not exist.

The CCC also noted that demand interventions should also play a part in transport decarbonisation. These included the following:

- **Changes in driver behaviour including things like eco-driving**: a smoother driving style with less aggressive acceleration and braking increases fuel efficiency by up to 10% thereby reducing emissions.
- **Speed limit enforcement and possible reduction:** the CCC identified that 'a typical petrol fuelled car driven at 70 mph emits about 19% more grams of carbon dioxide per km than when driven at 50 mph'. Large-scale enforcement of speed limits and lower limits would have a positive impact in terms of reducing emissions.
- **Modal shift to less carbon intensive transport**: travelling by rail (electric where the power generation is renewable), walking or cycling produces zero emissions when compared with ICE vehicle travel. This should be encouraged where possible, especially in urban areas [16].

The interventions proposed in 2008 are similar to those proposed by the International Energy Agency (IEA) in 2022 with some caveats. A recent study of progress in tackling road-based emissions noted that the electrification of road transport will play a key role in the Western world, and one greater than anticipated in 2008. This applies not only to cars and vans but also to heavy goods vehicle where powering infrastructure including pantograph, inductive and ultra–rapid charging, combined with larger more efficient batteries will make it increasingly possible for freight to move by electric on roads. Hydrogen is also noted as a key fuel, having been discounted in 2008 as not being developed enough. The report notes that hydrogen could play a key role in fuelling heavy goods vehicles, alongside alternative fuels and hybrid engine technologies to reduce emissions. Modal shift is still noted as key to reducing emissions alongside first mile – last mile solutions for urban freight. The report, when compared with 2008, does not focus heavily on improvements in vehicle efficiency and no longer calls out diesel as a fuel following this being discredited during the 2010s due to its polluting nature. The report strongly emphasises that when it comes to decarbonising road-based transport, electric is king, noting that 'total spending on electric cars increased by 50% over 2021 to reach $425 billion. Most of this was spent by private or corporate consumers on electric vehicles' and that 'rapid and continuing progress in the electrification of road transport has been one of the bright spots of the clean energy transition' [30].

To understand the importance of EVs as key to the future of decarbonising cars and vans requires us to understand how EVs have developed and how rapid advancements over the past decade mean that the various barriers to adoption have fallen one by one.

2.7 The history of EVs

The first EV can be traced to the 1830s when British inventor Robert Anderson developed a basic carriage powered by electric; however, the vehicle lacked a functioning battery to make it useable. In 1859, French physicist Gaston Planté invented the first lead-acid battery that enabled EVs to be reliably powered. This battery technology was further developed in 1881 by French scientist Camille Alphonse, improving the design and capacity of the battery. In 1884, the British inventor, Thomas Parker built the first production electric car as shown in Figure 2.3. This was soon followed by Andreas Flocken in 1888 who created a more advanced version. A larger six passenger EV soon followed, invented by American chemist William Morrison, in the United States in 1890. This was capable of speeds of 14 mph, sparking interest in EVs across America. Further advancements were made in the late 1890s with the development of a fleet of battery-powered taxis by Walter Bersey in London. By 1900, EVs accounted for a third of all vehicles on the road in the United States.

The early 1900s and 1910s saw attempts to further the advancement of EVs when Henry Ford and Thomas Edison partnered to develop a low-cost EV; however, Ford decided to invest his time and money in the petrol engine, launching the Model T in 1908, transforming the automotive industry and setting the direction of travel for roughly the next 100 years. By 1935 EVs, had all but disappeared from roads across the world. There was renewed interest in EVs in the 1960s and 1970s during the oil crises caused by tensions in the Middle East; however, they were unviable due to limited performance and range when compared with gasoline-powered vehicles. It was only in the 1990s where the foundations of the modern EV were laid. In the early 1990s, California's Clean Air Agency started to advocate for zero emissions vehicles (ZEVs), prompting car manufacturers to start the development of EV models.

Figure 2.3 An EV developed by Thomas Parker, 1895 [Source: Wikimedia Commons. Thomas Parker Electric Car [online]. 1895. Available from https://commons.wikimedia.org/wiki/File:Thomas_Parker_Electric_car.jpg [Accessed 31 July 2024]].

In 1991, Sony and Asahi Kasei commercialised the lithium-ion battery which began to unlock longer vehicle ranges and better performance; however, it took a further decade for this technology to make its way into the manufacturing process.

In 2004, Tesla Motors was established and launched its first EV, the roadster in 2008. Built around the lithium-ion battery, the Tesla Roadster was the first production EV to travel 200 miles on a single charge; however, this cost just under $100,000, meaning it was not a mass market vehicle. The Nissan Leaf, launched in 2010 in the United States and Japan, is considered the first modern family hatchback EV produced for the mass market by a recognised manufacturer. From that point onwards, the growth of EVs began to gain pace as can be seen in Figure 2.4. By 2016, EVs on world roads reached just over 2 million. During the same year, Norway saw 100,000 EVs registered [31]. In 2020, when EVs on global roads reached over 10 million, Tesla became the first car manufacturer to produce 1 million electric cars. Tesla's lead looked solid even though other manufacturers were all beginning to bring mass-produced EVs to the market. It is fair to say however that none have yet surpassed the Tesla Model Y, the world's most popular EV that has cumulatively sold 2 million units worldwide up until the end of 2023 [32]. At the time of writing (2024), Tesla is however being challenged by emerging

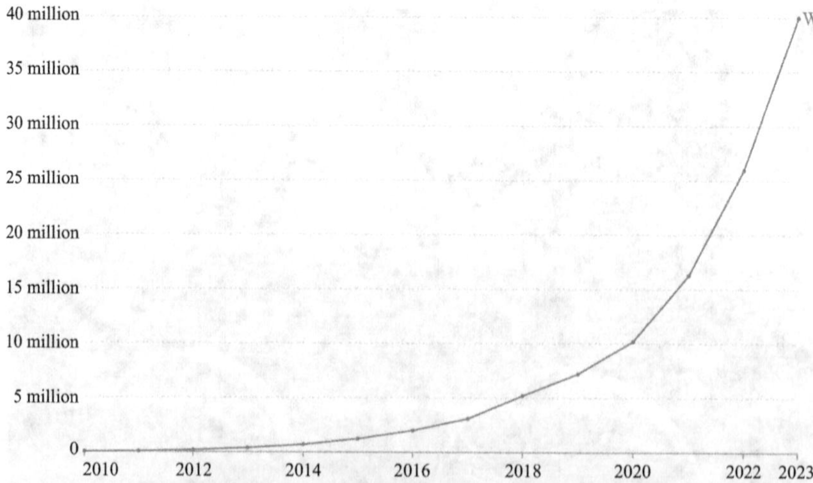

Figure 2.4 Electric car stocks, 2010–2023. Car stocks represent the number of cars that are in use. It is the balance of cumulative sales over time and the number of cars that have been retired or taken off the road. Electric cars include fully battery-electric vehicles and plug-in hybrids [Source: International Energy Agency. Global EV Outlook 2024. – processed by Our World in Data. 'Electric car stocks' [dataset]. International Energy Agency. Global EV Outlook 2024, 'electric_cars_iea' [original data] [Accessed 28 August 2024]].

EV manufacturers from China, such as BYD, that are able to manufacture EVs at much lower costs than Western rivals.

2.8 The history of charge point infrastructure

As EVs developed, so did the charging technologies required to power them. The history of charge points is much less well documented when compared with EVs; however, there are useful online summaries that set out key changes through time. The initial EVs of the early to mid-nineteenth century relied on batteries which when discharged could not be recharged. It was only with the development of the lead-acid battery during the mid to late nineteenth century that the need to charge an EV's battery became a key requirement. These batteries relied on a direct current (DC) power source to charge them. As EVs took off in the early twentieth century, so did the need for basic charging infrastructure. In the United States, EVs made up 38% of all automobiles, only 2% behind steam-powered vehicles. ICE vehicles made up 22% [33]. Charging of EVs at this time involved removing the battery and transferring it to a facility known as a battery room, and then re-installing this back into the vehicle. Battery rooms were not the only way to charge an EV. One of first dedicated charge points, the Pope Manufacturing Company charger was developed in 1912, using ground-breaking technology that converted alternating current (AC) to DC [34]. The Pope Manufacturing Company were not the only manufactures of EV charge points. In 1914, General Electric introduced the first public DC charging station known as the Electrant that could bill users for the power they used. General Electric hoped that these chargers would be installed across the United States; however, the decline of EVs during the 1920s and 1930s meant that the demand for charge points fell through the floor.

Further advances in EVs over the next 70 years meant that when new more technologically advanced EVs came to the market following the turn of the twenty-first century, charge point technologies had also advanced. A key development was the emergence of on-board chargers (OBC) that could make the conversion from AC to DC. Prior to the 1950s, the basics of OBCs were too cumbersome for vehicles at that time and it was only with the invention of solid-state rectifiers by Russel Ohl in 1940 that OBCs could be made smaller and more cost-effective. Advancements in semiconductors in the 1970s, combined with advancements in OBCs, resulted in more compact and efficient units that eventually made their way into two very early EVs, the General Motors EV1, launched in 1996, and the Toyota RAV4 EV, launched in 1997 [33]. From that point onwards, OBCs became a standard part of an EV. The next step in the journey to develop modern charge points was the emergence of DC fast chargers and AC chargers in public places.

The introduction of OBCs meant that vehicles could charge at home using AC from their household electricity sockets, removing the need to charge whilst out and about, providing a vehicle could make it home on a single charge. However, the limitations of lithium-ion batteries in terms of range versus an ICE vehicle meant that depending on the journey, EVs would need to top up their batteries to continue onwards. From this charging landscape emerged the concept of level 1, level 2 and level 3 charging. Level 1 charging is the method using a standard 120 V AC household socket to charge at home, albeit at slow charging speeds (around 1 to 1.2 kW). Level 2 charging also uses AC but doubles the voltage and results in faster charging speeds of anywhere between 3.7 and 49 kW. Level 2 chargers are either single phase, delivering speeds of 3.7 to 7.4 kW, or three phase, delivering speeds of between 11 and 49 kW. Level 3 rapid and ultra-rapid chargers double or quadruple the voltage further depending on the rate of charge and can deliver charging speeds of anywhere between 50 and 350 kW [35]. Level 1 chargers developed alongside early EVs, enabling them to charge at home. Level 2 and 3 chargers developed slowly at first, but at pace during the 2010s and 2020s alongside the increased adoption rates of EVs. Level 3 chargers, and specifically those at the upper end of charging rates, are developing rapidly and have made major advancements in 2020s. This has occurred alongside developments in 400 and 800 V EV battery architectures, all focused on reducing the time a driver has to stay stationary to achieve an 80% charge. Chapter 6 provides further detail.

EV charging standards developed alongside Level 2 and 3 charge points. This ensured interoperability across vehicles and charge points in terms of the plug and socket to charge the vehicle. The ecosystem is still developing and consolidation around charge point standards continues. Major progress has been made around this, with the process having started in the late 1990s. The below list highlights the standards and how they have developed over the past 30 years:

- **AC standard: Magne Charge (in use 1990s–early 2000s)**: an inductive charging solution, developed in accordance with the standard organisation SAE's standard, J1773. This was used in early EVs like the GM EV1; however, it became obsolete by 2003 when no manufacturers supported it anymore.
- **AC standard: Avcon Connector (in use 2006–2010):** developed by the California Air Resources Board (CARB), this solution differed from Magne Charge in terms of opting for conductive coupling over inductive technology. The solution adopted the SAE J1772-2001 standard in 2001. The connector, mainly used in the United States, was rectangular, could supply 6.6 kW but was discontinued due to the adoption of the J1772 Yazaki Connector.
- **AC standard: J1772 Yazaki Connector (in use 2010–present)**: developed by Yazaki to satisfy CARB's higher current delivery requirements, the J1772 connector was developed to deliver up to 19.2 kW. This connector is mainly used in the United States and a few other countries.
- **AC standard: Mennekes Type 2 Connectors (in use 2013–present)**: developed for the European market to meet European standards, this connector supports single-phase and three-phase level 2 AC charging. This connector is used in Europe, Australia and a number of other regions worldwide.
- **AC standard: EC 62196 Type 3 (in use 1999–2013)**: this connector, primarily in use in France and Italy, supported single-phase and three-phase level 2 charging; however, the European Union mandated that this needed to be replaced by the Mennekes Type 2 standard.
- **AC standard: GB/T AC (GB/T 20234.2) (in use 2011–present)**: this connector is primarily used in China but is gaining traction worldwide. It supports single-phase and three-phase level 2 AC charging.
- **DC standard: CHAdeMO (in use 2010–present)**: developed in Japan, the CHAdeMO DC standard gained global recognition and, for a period, was the only level 3 DC standard available. An early version only delivered charging speeds of 62.5 kW; however, second generations deliver speeds of up to 400 kW. The standard is primarily used in Japan and, although still in use, is beginning to lose traction against other standards.
- **DC standard: GB/T DC (GB/T 20234.3) (in use 2011–present)**: developed in China, this is primarily used there and can deliver level 3 speeds of up to 250 kW.
- **AC and DC standard: North American Charging Standard (NACS) (in use 2012–present)**: NACS was developed by Tesla in 2012 and then subsequently opened up to other manufacturers in 2022. The connector supports single-phase and high power level 3 DC charging.
- **AC and DC standard: Combined Charging System 1 (CCS1) (in use 2013–present)**: CCS1 incorporates pins for AC and DC charging. Level 3 DC charging is capable of 350 kW speeds. This standard is primarily used in North America.

- **AC and DC standard: Combined Charging System 2 (CCS2) (in use 2013–present):** CCS2 is similar to CCS1 but is primarily used in Europe.
- **DC standard: ChaoJi (in development)**: developed by Japan and China in partnership, this level 3 DC standard can deliver charging rates of up to 900 kW [33].

Now that we have gained an understanding of the history of EVs and charge points, it is important to summarise the targets we have set over the past 15 years in terms of EV adoption and charge point deployment. These targets are key in driving forward transport decarbonisation.

2.9 EVs and charge points: the history of targets and progress to date

Both Norway and California are considered early leaders in setting EV adoption targets; however, both adopted different approaches, with the former primarily using incentives and the latter regulations. In 1990, California, through CARB, set a ZEV mandate which included the requirement that of all vehicles sold in the State, 2% would be battery EVs from 1998, 5% from 2001 and 10% from 2003. The Board stated it would hold biennial reviews of the program to track progress and make adjustments to the policy direction where required. The mandate was underpinned by a ZEV credits system where CARB releases 'annual credit bank balances each year, as well as the total number of vehicles produced for that model year and the total number of zero-emission vehicles and plug-in hybrids'. In 1996, at a biennial review, the intermediary years were removed and the focus placed specifically on the 10% target from 2003. In 1998, the ZEV mandate was further adjusted to support low emissions vehicles that were not EVs (e.g., plug-in hybrids). Further changes to the ZEV mandate, partly driven by litigation, resulted in changes to regulations regarding the sell by date of ZEVs. This came into force in 2004. In 2009, as part of a ZEV Program Review, CARB staff began investigating infrastructure requirements. Between 2010 and 2012, CARB adopted 'regulatory changes to the ZEV program in 2012 that substantially increased and simplified requirements for 2018 and subsequent model years. California's low-emission vehicle (LEV) regulations were amended, known as LEV III, to increase the stringency of tailpipe and greenhouse gas emission standards for new passenger vehicles. The 2012 modifications combined the control of smog-causing pollutants and greenhouse gas emissions into a single coordinated package of standards called Advanced Clean Cars'. The regulation came into force in the same year. Further modifications to the ZEV regulation were made in 2013, 2014 and 2015, with changes in 2015 focused on ensuring that additional ZEV credits for fast refuelling ZEVs were awarded on a vehicle by vehicle basis. In 2022, the most recent regulations, Advanced Clean Cars II, was adopted, which according to CARB imposed 'the next level of low-emission and zero-emission vehicle standards for model years 2026–2035 that contribute to meeting federal ambient air quality ozone standards and California's carbon neutrality targets' [36].

Norway's incentivisation approach began in 1990 with the removal by the Norwegian Government of purchase and import tax for EVs. In 1994, an EV produced by Think, a Norwegian EV manufacturer established in 1991, was advertised at the Lillehammer Olympic Winter Games to incentivise Norwegian drivers to adopt EVs. In 1996, road tax was removed from EVs (until 2021), and tolls removed in 1997 (until 2017). In 2001, EVs were exempted from 25% VAT on purchases, in 2005 they were allowed to drive in bus lanes and in 2009, they were exempt from charges on ferries. In 2015, VAT exemptions were introduced for EV leases. 2017 was a key year in Norway as a more regulatory approach was adopted. The Norwegian parliament set a national target that all new cars sold would be zero emissions by 2025 and that residents in apartment buildings would have a right to charge. This meant apartment complexes would be required to provide some form of charging arrangements. There were further tweaks to incentives in 2018 as tolls and ferries were no longer free and instead EVs would be charged 50% of total ticket/toll prices. In 2022, it was mandated that for all public procurements, cars would need to be 100% EV [37]. In 2018, a report produced by Ecofys for the Federal Republic of Germany and the European Climate Initiative EUKI drew together evidence analysing the drivers behind EV adoption and the roles incentives played. The study noted that in Norway, 41% of EV buyers primarily adopted EVs to save money [38]. The alignment between EV adoption and saving money/financial incentives is clear in Norway, pushing adoption rates upwards.

Although it took time for incentives to take effect in Norway, the data shows that between 2012 and 2017, market share of new EVs registered (including plug-in hybrids) increased from around 3% in 2012 to just under 40% in 2017. This compares to adoption rates in Germany using the same metric of near 0% in 2012 and about 2% in 2017. The Ecofys report suggests that this growth is not just down to incentives. It may also be down to an increase in more affordable models, but when compared with Germany where there are no incentives, this policy approach clearly had some impact. The report also suggests that adoption rates are due to a 'very high degree of public and political acceptance of the importance of promoting low-emission transport'. The Norwegian utility providers also played an important role, investing in charging infrastructure early in Norway's journey. Due to its incentives and overarching approach, EVs held just over 79% of total market share in 2022. Norway seems on track to reach its zero emissions goal by 2025 [37]. The impact of California's regulatory approach is somewhat different. The EV adoption rate in California is lower than Norway. Using the first half of data for 2023, 25% of EVs sold in California were EVs. This is slightly less than three times the percentage of EVs sold across the United States, which sits at 9%. A popular website, InsideEVs, reporting on BloombergNEF forecasts, states that EV sales are projected to make up 23% of total United States car sales in 2025 and 51% by 2030 [39]. Introduced in 2022, the Inflation Reduction Act (IRA) and the associated subsidies for consumers, car manufacturers and battery producers likely impacted positively on EV adoption rates across the United States; however, with the election of Donald Trump in November 2024 and the subsequent scrapping of the IRA and other associated laws, it is not clear what the overall impact on EV adoption will be. It is likely that the difference between Norway and California is that for the latter even though there are incentives to increase adoption, an EV is still more expensive than an ICE vehicle in the United States. Charging is still a challenge, although the IRA has had an impact on this. Behaviour change, though, is proving much more difficult to shift. Those who currently own an ICE vehicle are unlikely to switch, especially if they own two or more ICE vehicles in a single household [40].

Moving our focus to the United Kingdom, the CCC stated in 2008 that for the United Kingdom to achieve its net zero targets required a switch to EVs. Incentivisation was the first approach to increase adoption alongside deploying infrastructure. This began with the plug-in car grant, launched in 2011 by the UK Government, which provided a subsidy to EV buyers for eligible plug-in vehicles. This helped to bridge the upfront cost difference between an EV and ICE equivalent. Throughout the duration of the scheme that ran until 2022, over 400,000 claims were made, with a total value of grants issued being £1.5 billion [41]. In 2014, the Electric Vehicle Homecharge Scheme was introduced to encourage the installation of charge points at home through covering 75% of costs. The scheme closed in 2022 and evolved into the EV charge point grant, targeting those who own flats or who are in rented accommodation. Between 2015 and 2020, EV policies began to rapidly change. In 2016, the Workplace Charging Scheme was introduced to provide financial support to workplaces ranging from businesses to charities. This enabled them to install EV charge points on their premises. In 2017, EVs became exempt from Vehicle Excise Duty, commonly known as road tax. This ended in April 2025. In 2018, the Road to Zero Strategy was introduced that set out to continue the current grant regime alongside a range of other targets. This included setting the target that 25% of the central government fleet would be ultra-low emission by 2022, that all new car purchases would be ultra-low emissions by default and that 100% of the central government car fleet would be ultra-low emission by 2030. A core part of this strategy was incentivising company car drivers to adopt EVs through favourable benefit in kind tax that reduced the tax burden when receiving a company car perk through work. Furthermore, the strategy launched the £400 million Charging Infrastructure Investment Fund to increase charge point deployment and, through the Automated and Electric Vehicles Bill encouraged charge points in locations like motorway service stations. The strategy also looked at various aspects of encouraging charge point deployment including using building regulations and also investing £4.5 million in the On-street Residential Charge point Scheme [42].

In 2021, the UK Government introduced a ban on the sale of new petrol and diesel vehicles which would come into force from 2030, with hybrids allowed until 2035. This formed a key part of the decarbonising transport strategy. Alongside this, the strategy brought forward the deadline of government car and fleet being 100% zero emission from 2030 to 2027. The bulk of the remainder of the strategy emphasised support for certain elements of the grant regime, driving further charge point deployment and introducing regulations to improve the consumer charge point experience [43]. 2022 saw the introduction of Taking Charge: The EV Infrastructure Strategy. As part of this, the strategy noted that the pace of change was too slow and that infrastructure needs to improve. The strategy which is still in place and is awaiting a refresh from the new Labour Government (as of August 2024) established a £950 million Rapid Charging Fund to increase the rollout of charge points at motorway service stations and the introduction of the Low Emissions Vehicles Infrastructure fund to increase charge point deployment by local authorities. The strategy also made it a requirement to access funding that local authorities develop an EV strategy. To achieve EV adoption by 2030, it stated that 300,000 public charge points are required [44] (though some estimates suggest this is the lower end of the scale and that 720,000 are needed). In 2023, the UK Government introduced the Zero Emissions Vehicles Mandate that came into force in 2024 and set out the percentage of new zero emission cars manufacturers will be required to produce each year up to 2030. This requires 80% of new cars and 70% of new vans sold in Great Britain to be zero emission by 2030, increasing to 100% by 2035. The Government also pushed back the ban of sales of ICE vehicles to 2035 [45].

The impact of these various interventions is that EV adoption and charge point deployment have increased though not at the pace compared with Norway or California. The RAC, analysing data from the Department for Transport (DfT) and Society of Motor Manufacturer and Traders, identified that between 2010 and 2024, the number of EVs on UK roads increased from just over 8,000 to 1.1 million, with rapid growth taking place between 2020 and 2024 when around 900,000 were added to UK roads. The RAC, using data from Zapmap and DfT, noted that in January 2016, there were around 3,500 chargers, whereas in March 2024 there were just under 60,000 chargers. Over this period, the proportion that were level 3 DC rapid and ultra-rapid increased [46]. The National Infrastructure Commission believes that the United Kingdom is on target to meet its 300,000 charge points required by 2030, provided that growth continues at the current trajectory. The Commission does however flag that in later years (and specifically 2029), this trajectory will be difficult to maintain [47]. EV adoption is, however, not where it needs to be, even though 19% of all new car registrations in June 2024 were EV [48]. A significant increase in EV sales as a percentage of total car sales is required to meet the 2035 target. The ZEV mandate may have an impact, combined with more aggressive EV pricing, but time will tell.

Shifting our focus to the European Union, this is complicated by the fact that Member States set their own transport policy; however, alignment in recent years on EV strategies focused on adoption and charge point targets enables us to look broadly at the history of targets across the Union and progress to date. Incentivisation in terms of encouraging EV adoption through tools like reducing VAT on EVs or subsidising EV ownership is primarily driven at a Member State level. The European Commission focuses its efforts on Europe wide legislation to set targets for EV adoption and charge point rollout. In 2020, the European Commission launched the European Green Deal which aimed for climate neutrality. A key part of the deal was the promotion of electric mobility through the setting of the target that all new cars and vans registered in Europe will be zero emission by 2035. The deal also set the target that 3.5 million public charge points would be deployed across the European Union by 2030, with a key milestone being

1 million charge points deployed by 2025 [49]. In 2021, the Commission launched Fit for 55, a package of legislation which aimed at reducing net greenhouse gas emissions by at least 55% by 2030, compared with 1990 levels. As part of this package the Alternative Fuels Infrastructure Regulation (AFIR) set targets which included recharging stations for cars and vans being installed every 60 km and that users of EVs are able to pay easily to charge [50]. The Council of the European Union adopted AFIR legislation in 2023, providing further detail on targets including that 'fast recharging stations of at least 150 kW for cars and vans need to be installed every 60 km along the European Union's main transport corridors, the so-called 'Trans-European transport (TEN-T) network' [51]. In 2021, the European Commission also adopted the Mobility Strategy which aims for at least 30 million zero emissions cars on European roads by 2030 and that by 2050 nearly all cars, vans and buses will be zero emission [52]. Furthermore, innovation in EV adoption and charge point rollout is promoted through Horizon Europe that funds research and innovative projects in these areas.

In terms of meeting EV adoption targets, EVs passed the tipping point to mass adoption in a majority of EU countries in 2024. This tipping point, Bloomberg argues, is adoption rates over 5% of EV car sales in any given country. Most European countries had reached that tipping point by 2020, with a number in Eastern Europe reaching that target by 2023. The Scandinavian countries saw the highest adoption rates, using quarter four in 2023 as a snapshot with Norway, Iceland, Denmark, Sweden and Finland all seeing adoption rates over 35%. German and French adoption rates sat at around 20%, whereas most Eastern European countries saw rates below 10% [53]. Recent data would suggest that the European Union is making progress towards its EV adoption targets. In 2023, sales of EVs amounted to 2.4 million. Available second-hand EVs are also increasing. In both 2021 and 2022, the market size grew by 70% year-on-year, reaching almost 120,000 electric cars in 2022 [54]. The European Alternative Fuels Observatory (EAFO), which conducts analysis focused on a range of areas including EVs, noted that the data around high EV adoption rates in countries across Europe correlates with where there is a comprehensive charging network and financial incentives to encourage uptake. The EAFO calls out the Netherlands and Norway as useful models for the European Union as their 'extensive infrastructure and progressive policies' are driving up EV adoption [55].

In terms of meeting charge point deployment targets, the European Union is on target. According to a briefing from Transport & Environment, most countries met their 2024 charging target by the end of 2023. Since 2020, the European Union has witnessed rapid growth in charge point deployments, seeing a 348% increase of level 3 DC charge points and a 251% increase of level 2 AC charge points between 2020 and 2023. By the end of 2023, there were over 630,000 charge points installed across Europe, a major increase compared with the 190,000 which existed in 2020. The European Union looks set to achieve its 2025 target of 1 million charge points by early 2025, though, like EV adoption rates, charge point deployment is not consistent across the European Union with Bulgaria, for example, overachieving its AFIR 2030 obligation by more than 12 times in 2023 due to fast paced charge point growth. Most other countries have overachieved their projected 2024 AFIR targets in 2023 but will need to continue investing in infrastructure to ensure they continue to meet future AFIR projected targets and, ultimately, the 3.5 million charge points required by 2030. The analysis conducted by Transport & Environment shows that 'except for Bulgaria, charging infrastructure deployment has generally aligned with EV uptake across the EU. Thus, the often-cited East-West and North-South divide in fulfilling AFIR targets is not supported by the data as the targets account for the different EV uptakes' [56].

2.10 Deploying charge points effectively: critical to the future EV transition

What should hopefully be clear from the journey to achieve decarbonisation in transport through increased EV adoption rates and charge point rollout is that this has not occurred without the intervention of international bodies, national and regional governments. Ever since the actions taken by Norway and California in the 1990s, the state has played a key role in promoting EVs and charge point deployment through regulation and incentives. What is also clear is that in promoting charge point rollout, governments have recognised it is not enough to just set an arbitrary target and that the right mix of Level 2 and 3 charge points is required to nudge EV adoption rates higher. Furthermore, the efforts of the European Union to deploy charge points along the TEN-T demonstrate that the location of charge points is equally as important as the number deployed. This is further emphasised by the UK Government's focus on various schemes to deploy charge points in residential settings. What has not been analysed in detail is the key stakeholders involved in that journey and the process required to deliver charge point rollout at a localised level. This is the focus of the following chapters. It should also be noted that there is one key lesson to take away from the history of EVs and charge points. This is that the false start of the early 1900s was not just a result of the technology not being advanced enough when compared with ICE vehicles; it was also down to the lack of infrastructure, and somewhat fiddly nature of using it. Let this be a lesson to all of us in the EV industry. We need to get this transition right to help save the planet. There is no going back!

References

[1] Gomstyn A. *The history of climate change [online]*. 2024. Available from https://www.ibm.com/blog/climate-change-history/ [Accessed 13 July 2024].

[2] UK Research and Innovation. *A brief history of climate change discoveries [online]*. 2025. Available from https://www.discover.ukri.org/a-brief-history-of-climate-change-discoveries/index.html#article [Accessed 13 July 2024].

[3] Manabe S., and Wetherald R. T. 'Thermal Equilibrium of the Atmosphere with a Given Distribution of Relative Humidity'. *Journal of the atmospheric sciences*. 1967, vol. 24 (3), pp. 241–259.

[4] Lang J. *Net Zero: A short history* [online]. 2021. Available from https://eciu.net/analysis/infographics/net-zero-history [Accessed 16 July 2024].

[5] Jackson P. 'From Stockholm to Kyoto: A Brief History of Climate Change'. *Green Our World*. 2007, vol. 44 (2), [online]. Available from https://www.un.org/en/chronicle/article/stockholm-kyoto-brief-history-climate-change [Accessed 17 July 2024].

[6] United Nations Climate Change. *What is the Kyoto Protocol? [online]*. 2024. Available from https://unfccc.int/kyoto_protocol#:~:text=In%20short%2C%20the%20Kyoto%20Protocol,accordance%20with%20agreed%20individual%20targets. [Accessed 17 July 2024].

[7] BBC. *A brief history of climate change [online]*. 2013. Available from https://www.bbc.co.uk/news/science-environment-15874560 [Accessed 22 July 2024].

[8] Climate Change Committee. *UK action on climate change [online]*. 2024. Available from https://www.theccc.org.uk/uk-action-on-climate-change/#:~:text=The%20Climate%20Change%20Act%20(2008,framework%20to%20cut%20carbon%20emissions [Accessed 22 July 2024].

[9] Greenpeace. *What is the UK doing about climate change? [online]*. 2024. Available from https://www.greenpeace.org.uk/challenges/climate-change/what-is-the-uk-doing-about-climate-change/#:~:text=The%20UK%20was%20the%20first,by%202050%2C%20from%201990%20levels. [Accessed 22 July 2024].

[10] United Nations Climate Change. *Conference of the Parties (COP) [online]*. 2024. Available from https://unfccc.int/process/bodies/supreme-bodies/conference-of-the-parties-cop [Accessed 22 July 2024].

[11] Climate Change Committee. *The benefits of the Climate Change Act [online]*. 2024. Available from https://www.theccc.org.uk/about/our-expertise/the-benefits-of-the-climate-change-act/#:~:text=The%20Act%20has%20cemented%20the,have%20been%20approved%20by%20Parliament. [Accessed 22 July 2024].

[12] European Council. *Climate change: what the EU is doing [online]*. 2024. Available from https://www.consilium.europa.eu/en/policies/climate-change/ [Accessed 22 July 2024].

[13] Australian Office of Financial Management. *Australian Government Climate Change Commitments, Policies and Programs [online]*. 2024. Available from https://www.aofm.gov.au/sites/default/files/2024-02-02/Climate%20change%20slides%20updated%20February%202024.pdf [Accessed 22 July 2024].

[14] Government of Canada. *Canada's climate plans and targets [online*]. 2024. Available from https://www.canada.ca/en/services/environment/weather/climatechange/climate-plan/climate-plan-overview.html [Accessed 22 July 2024].

[15] The London School of Economics and Political Science. *What is the 2008 Climate Change Act? [online]*.2020. Available from https://www.lse.ac.uk/granthaminstitute/explainers/what-is-the-2008-climate-change-act/ [Accessed 23 July 2024].

[16] Committee on Climate Change. *Building a low-carbon economy –The UK's contribution to tackling climate change [online]*. 2008. Available from https://www.theccc.org.uk/wp-content/uploads/2008/12/Building-a-low-carbon-economy-Committtee-on-Climate-Change-2008.pdf [Accessed 23 July 2024].

[17] The London School of Economics and Political Science. *How much do renewables contribute to the UK's energy mix and what policies support their expansion? [online]*. 2023. Available from https://www.lse.ac.uk/granthaminstitute/explainers/how-much-do-renewables-contribute-to-the-uks-energy-mix-and-what-policies-support-their-expansion/ [Accessed 23 July 2024].

[18] Ruud R. 'Sustainable Development Discourse in Norway'. *L'Europe en Formation*, 2009, vol. 2 (352), pp 143–155.

[19] Fominova S. *Top countries in the race to net zero emissions [online]*. 2022. Available from https://net0.com/blog/net-zero-countries [Accessed 23 July 2024].

[20] Department for Energy Security and Net Zero. *UK first major economy to halve emissions [online]*. 2024. Available from https://www.gov.uk/government/news/uk-first-major-economy-to-halve-emissions [Accessed 23 July 2024].

[21] European Commission. *National long-term strategies [online]*. 2024. Available from https://commission.europa.eu/energy-climate-change-environment/implementation-eu-countries/energy-and-climate-governance-and-reporting/national-long-term-strategies_en [Accessed 24 July 2024].

[22] Australian Government Department of Finance. *APS net zero emissions by 2030 [online]*. 2024. Available from https://www.finance.gov.au/government/climate-action-government-operations/aps-net-zero-emissions-2030#:~:text=The%20Net%20Zero%20in%20Government,launched%20on%2028%20November%202023 [Accessed 24 July 2024].

[23] HM Government. *Net zero strategy: Build back greener [online]*. 2021. Available from https://assets.publishing.service.gov.uk/media/6194dfa4d3bf7f0555071b1b/net-zero-strategy-beis.pdf [Accessed 24 July 2024].

[24] Evans S. *Analysis: Which countries are historically responsible for climate change? [online]*. 2021. Available from https://www.carbonbrief.org/analysis-which-countries-are-historically-responsible-for-climate-change/ [Accessed 29 July 2024].

[25] Ritchie H., Rosado P. and Roser M. *CO_2 and greenhouse gas emissions, 2023 [online]*. 2023. Available from https://ourworldindata.org/grapher/co2-emissions-transport [Accessed 29 July 2024].

[26] Ritchie H. *Cars, planes, trains: Where do CO_2 emissions from transport come from? [online]*.2020. Available from https://ourworldindata.org/co2-emissions-from-transport#:~:text=The%20data%20is%20sourced%20from,comes%20from%20trucks%20carrying%20freight. [Accessed 29 July 2024].

[27] IEA. *Global CO2 emissions in transport by mode in the sustainable development scenario, 2000-2070 [online]*. 2020. Available from https://www.iea.org/data-and-statistics/charts/global-co2-emissions-in-transport-by-mode-in-the-sustainable-development-scenario-2000-2070 [Accessed 29 July 2024].

[28] Ritchie, H. *Where in the world do people have the highest CO2 emissions from flying? [online]*. 2020. Available from https://ourworldindata.org/carbon-footprint-flying [Accessed 29 July 2024].

[29] World Population Review. *Rail usage by country 2024 [online]*. 2024. Available from https://worldpopulationreview.com/country-rankings/rail-usage-by-country [Accessed 29 July 2024].

[30] IEA. Transport [online]. 2024. Available from https://www.iea.org/energy-system/transport [Accessed 30 July 2024].

[31] Energy Saving Trust. *A brief history of the electric car [online]*. 2021. Available from https://energysavingtrust.org.uk/a-brief-history-of-the-electric-car/ [Accessed 31 July 2024].

[32] Shahan, S. *Tesla Model Y surpasses 2 million sales! [online]*. 2023. Available from https://cleantechnica.com/2023/10/10/tesla-model-y-cumulative-sales-surpass-2-million/ [Accessed 31 July 2024].

[33] Reddy A. *History of electric vehicle charging: A journey through evolution of electric vehicle charging [online]*. 2024. Available from https://evchargingexplained.com/history-of-electric-vehicle-charging/ [Accessed 02 August 2024].

[34] Hall of Electrical History Foundation. *A woman uses a hand-cranked battery charger to charge her electric Columbia Mark 68 Victoria automobile [online]*. 1906. Available from https://rarehistoricalphotos.com/electric-cars-history-pictures-1880-1920/ [Accessed 02 August 2024].

[35] U.S. Department of Transportation. *Charger types and speeds [online]*. 2024. Available from https://www.transportation.gov/rural/ev/toolkit/ev-basics/charging-speeds [Accessed 02 August 2024].

[36] California Air Resources Board. *Zero-emission vehicle program [online]*. 2024. Available from https://ww2.arb.ca.gov/our-work/programs/zero-emission-vehicle-program/about#:~:text=The%20California%20Air%20Resources%20Board,reflect%20the%20state%20of%20technology [Accessed 3 August 2024].

[37] Norsk elbilforening. *Norwegian EV policy [online]*. 2024. Available from https://elbil.no/english/norwegian-ev-policy/ [Accessed 3 August 2024].

[38] Steinbacher K., Goes M., and Jörling K. *Incentives for electric vehicles in Norway [online]*. 2018. Available from https://www.euki.de/wp-content/uploads/2018/09/fact-sheet-incentives-for-electric-vehicles-no.pdf [Accessed 3 August 2024].

[39] Mihalascu D. *California tops US EV adoption: 25% EV share of total sales in H1 2023 [online]*. 2023. Available from https://insideevs.com/news/688779/california-tops-us-ev-adoption-25-percent-share-total-sales-h1-2023/ [Accessed 3 August 2024].

[40] Morgan K. *Three big reasons Americans haven't rapidly adopted EVs [online]*. 2023. Available from https://www.bbc.com/worklife/article/20231108-three-big-reasons-americans-havent-rapidly-adopted-evs [Accessed 3 August 2024].

[41] Office for Zero Emission Vehicles. *Plug-in car grant vehicle application form and guidance notes* [online]. 2021. Available from https://assets.publishing.service.gov.uk/media/62a82a87d3bf7f037097bf05/plug-in-car-grant-guidance-vehicle-application-form-and-guidance.pdf [Accessed 3 August 2024].

[42] Department for Transport. *The road to zero – next steps towards cleaner road transport and delivering our industrial strategy [online]*. 2018. Available from https://assets.publishing.service.gov.uk/media/5b968e3ee5274a13859-deed2/road-to-zero.pdf [Accessed 3 August 2024].

[43] Department for Transport. *Decarbonising transport – a better, greener Britain [online]*. 2021. Available from https://assets.publishing.service.gov.uk/media/610d63ffe90e0706d92fa282/decarbonising-transport-a-better-greener-britain.pdf [Accessed 3 August 2024].

[44] HM Government. *Taking charge: the electric vehicle infrastructure strategy [online]*. 2023. Available from https://assets.publishing.service.gov.uk/media/6245ba40e90e075f15381cf0/taking-charge-the-electric-vehicle-infrastructure-strategy.pdf [Accessed 3 August 2024].

[45] Department for Transport. *Government sets out path to zero emission vehicles by 2035 [online]*. 2023. Available from https://www.gov.uk/government/news/government-sets-out-path-to-zero-emission-vehicles-by-2035 [Accessed 3 August 2024].

[46] The RAC. *The road to electric – in charts and data [online]*. 2024. Available from https://www.rac.co.uk/drive/electric-cars/choosing/road-to-electric/ [Accessed 3 August 2024].

[47] National Infrastructure Commission. *Infrastructure progress review 2024 [online]*. 2024. Available from https://nic.org.uk/studies-reports/infrastructure-progress-review-2024/#tab-transport [Accessed 3 August 2024].

[48] Zapmap. *EV market stats 2024 [online]*. 2024. Available from https://www.zap-map.com/ev-stats/ev-market [Accessed 3 August 2024].

[49] European Commission. *Delivering the European green deal [online]*. 2023. Available from https://commission.europa.eu/strategy-and-policy/priorities-2019-2024/european-green-deal/delivering-european-green-deal_en#making-transport-sustainable-for-all [Accessed 4 August 2024].

[50] European Council. *Fit for 55 [online]*. 2024. Available from https://www.consilium.europa.eu/en/policies/green-deal/fit-for-55/ [Accessed 4 August 2024].

[51] European Council, *Alternative fuels infrastructure: Council adopts new law for more recharging and refuelling stations across Europe [online]*. 2023. Available from https://www.consilium.europa.eu/en/press/press-releases/2023/07/25/alternative-fuels-infrastructure-council-adopts-new-law-for-more-recharging-and-refuelling-stations-across-europe/ [Accessed 4 August 2024].

[52] European Commission. *Mobility strategy [online]*. 2021. Available from https://transport.ec.europa.eu/transport-themes/mobility-strategy_en [Accessed 4 August 2024].

[53] Randall T. *Electric cars pass the tipping point to mass adoption in 31 countries [online]*. 2024. Available from https://www.bloomberg.com/news/articles/2024-03-28/electric-cars-pass-adoption-tipping-point-in-31-countries [Accessed 4 August 2024].

[54] International Energy Agency. *Global EV Outlook 2024 [online]*. 2024. Available from https://ourworldindata.org/energy [Accessed 4 August 2024].

[55] European Alternative Fuels Observatory. *Regional insights into electric vehicle adoption across Europe: New data from Eurostat [online]*. 2024. Available from https://alternative-fuels-observatory.ec.europa.eu/general-information/news/regional-insights-electric-vehicle-adoption-across-europe-new-data [Accessed 4 August 2024].

[56] Transport & Environment. *Public charging in Europe: Where are we at? [online]*. 2024. Available from https://www.transportenvironment.org/uploads/files/2024_04_AFIR-Implementation.pdf [Accessed 4 August 2024].

Chapter 3

Deploying EV infrastructure: the overall process and key stakeholders

David Trousdale[1]

Now that we have looked in detail at how electric vehicles (EVs) and charge point infrastructure have become a key element of road based decarbonisation of cars and light goods vehicles, it is important to start looking at the end-to-end process of installing charge points and the associated stakeholders involved in this process. This then enables us to look in further detail at the key steps and stakeholders as part of future chapters. It should be noted that my professional and personal experience of charge point planning and deployment has heavily shaped this chapter, supplemented by conversations with charge point operators (CPOs), local authorities in a United Kingdom context and consultancies. It should also be noted that although this has been written based primarily on United Kingdom experiences, I have tried, through research on case studies beyond the United Kingdom, to make sure the process is applicable in other Western countries. This is not an exhaustive analysis and more detailed guidance is available from other sources, though it is a useful starting point to understand the process.

3.1 The overall process

When looking at deploying EV charge point infrastructure, there are eight key steps which should be followed; however, prior to this, it is important to consider the charging use case or cases a charge point deployment supports. These heavily influence all steps and will impact the charge point technologies deployed. These use cases are en route charging, destination charging and residential/work-based charging. En route charging focuses on drivers who are travelling longer distances and require a quick (15–30 minutes) charge to enable them to continue on with their journey. Charge points are level 3 DC chargers with the ability to charge at 50 kW or more per hour. Destination charging focuses on drivers who are travelling to a location and will be stationary for several hours (from 1 hour 30 minutes to 4 hours) while they do other things. They will not rely on the charger as the only means to

[1]Principal Advisor - Climate Change and Sustainability, Newcastle City Council, United Kingdom

charge their vehicle so expect slower charging rates. Charge points are level 2 AC with the ability to charge at between 7 and 22 kW per hour. Residential/work-based charging focuses on drivers who will leave their vehicle potentially for 7 or more hours where they live or work. Drivers will rely on these charge points primarily to charge their vehicle but may supplement this with destination and en route charging. Charge points can be level 1 or 2 AC, typically charging at rates of between 1.5 and 7 kW per hour. All three charging use cases follow the eight steps outlined later in this section, with two caveats. The first is that work-based charging tends to focus on a single site with the company making most if not all of the decisions. As a result, the process followed is much simpler and primarily relates to whether the company owns the land where the charge points will be deployed, whether there is space on the land for charge points and whether the site has power capacity to handle the additional load. The second is that residential charging can be broken down into two user sub-groups. One group is those who have off-street private parking meaning the decision to install a charge point is primarily theirs. The other group is those who rely on on-street public charge points or those in public car parks in residential areas. Deploying charge points for this group follows a multi-step process. Work-based and private off-street residential charging will not be the focus of this chapter.

When looking at the process that applies to the three charging use cases (excluding work-based and private off-street residential charging), the initial stage is the planning stage which is made of multiple steps, the first of which is the decision to deploy charge point infrastructure. This involves determining the 'why' behind deployment and explores what the justification is for charge point deployment (e.g., filling a gap in the market, just transition, land grab), which charging use case is the target for deployment and the high-level timeline for delivery. Following this, the next step determines the 'where' in terms of the geographic target area. Key questions include whether the deployment is part of a broader plan to expand into a new geography, build on an existing network, target a new user group or rationalise/refresh a charge point network. The following step then determines the 'what' behind deployment via completion of a business case to further determine the rationale and secure funding for deployment. Deployment then moves into the 'how', setting out project plans for deployment at a local geographic level, though still at a high level of detail, and the governance and approvals required to pass through the various steps to complete installation. Completion of all of these steps enables deployment to move into the build stage.

The first and only step in the build stage is the deployment of charge points. The step is lengthy and can take several months or even years to complete. Detailed planning of the build at a site level (as opposed to a local area level during the planning stage) takes up most of the time, with the actual build being relatively quick in comparison. Following the build stage is the operations stage that contains two steps. The first is operations, maintenance and monitoring and covers all the activities related to managing the charge point network. The second is upgrading and decommissioning. This covers activities related to ensuring charge points do not become outdated or defective and that if they do, a plan exists to repair, replace or decommission them. Now that we understand the stages and steps at a high level, we will deep dive into each in more detail.

3.1.1 The initial decision to deploy the infrastructure

Having seen multiple charge point deployments in my career, it is often the first step of the planning stage, the initial decision to deploy charge point infrastructure, which determines the success or failure of a charge point deployment and, in some cases, network. Success or failure is primarily driven by demand and utilisation of charge points. Demand is the proportion or number of EV drivers in an area or geography who want to charge. Utilisation is the number or proportion in the same area/geography who actually charge. Demand and utilisation are intrinsically linked with the 'why' behind charging. Failure of deployed charge points is usually because a CPO did not understand the 'why' in detail behind deployment. On the success side of the scale are the CPOs who know their target use case inside out and ruthlessly pursue deployment. They know when charge points need to be deployed in the future to enable the growth of their network in alignment with increased EV adoption and growth in charge point competition. A good example, in my view, of a CPO in the United Kingdom doing this is Osprey. Their business model is clearly focused on en route charging, with deployments targeting high-demand areas. They have recently been awarded the accolade of 'Best United Kingdom charging network' [1], which reflects their success in charge point deployments. At the failure side of the scale are those CPOs that try to target all use cases and deploy charge points haphazardly with little direction or plan. An example of this is Charge your car, also based in the United Kingdom. They have been ranked either the bottom or near the bottom of Zapmap EV charging network ranking 2 years in a row (2021 and 2022) [2]. Although this is down to a myriad of reason ranging from charge point reliability to ease of use, it must also be acknowledged that the network is a hotchpotch of charge points owned by bp pulse, another United Kingdom-based CPO. The network developed with little strategy or plan, largely driven by local authorities wanting to install charge points without understanding the reason for deployment or the long-term strategy around demand and utilisation. These charge points were then inherited by bp pulse.

Understanding the 'why' is primarily driven by understanding the charge point use cases in detail as this determines everything else. It is critical to do this at the start of a deployment journey as changing direction part way through can have major consequences. And even if a CPO understands the 'why' in detail, it must constantly assess itself against an ever-changing context of EV adoption that determines demand, utilisation and charge point competition. Starting with en route charging, understanding the 'why' behind this determines the likely locations an en route charge point should be deployed in. It also determines the connection socket type and charging rate and the number required. Furthermore, the 'why' determines the infrastructure (including power network upgrade) and cost, the rollout strategy for other charge points across a broader network and, ultimately, the CPO's profitability. Taking a motorway or freeway network as an example, a CPO focused on en route charging would want to understand why en route charging is required along a route. This would be based on the nature of the route (is it a strategic route like Trans-European Transport Network (TEN-T) or a more provincial route which connects to the strategic route), traffic flow along the route and proportion of vehicles that are EV now and in the future, competition along the route and available sites to deploy. Understanding this would then help shape the number and type of charge points (several level 3 100 kW plus charge points as a minimum if on a strategic route, slower and fewer in number if on a provincial route), where they can be deployed based on available land (existing service stations or developments nearby the route) and the investment required (£1 million plus in this example). But even analysing all of this is not enough. The 'why' needs to delve deeper to understand what pricing model should be adopted at site, the impact on utilisation and the growth in future competition. All of these elements feed into later steps in the process but understanding this in detail at the start before the investment wheels have started turning should minimise the risk of failure further along in the process.

Understanding the 'why' behind destination charging prompts similar questions for the CPO including what is the purpose of the destination and how long drivers tend to stay there, the location of the destination in relation to a strategic or provincial road network and existing charge points deployed at or around the destination. Using a shopping centre or mall as an example, users are likely to remain there for several hours, but not overnight and are unlikely to expect their car to be fully charged while they are there; however, if the location is near to a motorway (e.g., Lakeside in the United Kingdom), demand may result from en route drivers stopping for lunch or shopping while completing a longer journey. Combining all of these elements would point towards charging infrastructure that is level 2 with maximum charging speeds of 22 kW. The number of charge points is likely to be more than ten, though the specifics of the site would further refine this. As users are likely to be repeat customers, an app with the option of a reward-based system or flexible pricing would positively impact on utilisation. Utilisation at the location would, however, need to be frequently monitored to determine whether more charge points are required or whether units need to be decommissioned. Innovative business models (like EV valet parking) might also work effectively in these locations. Understanding these points early on should ensure success.

Understanding the 'why' for residential on-street public charging is much more complicated than en route or destination charging. This is because there are various reasons why residents want to charge publicly and these lend themselves to multiple different charging solutions. Some using residential charge points may solely rely on these meaning they always need them available on their street or nearby. Others may have access to charge points at work or destination chargers where they regularly shop or engage in recreational activities. This level of reliance will impact the numbers required, the type and rate of charge. Furthermore, the cost of charging in a residential setting plays a more important role as those without off-street private chargers are well aware that domestic electricity tariffs catering for EVs are the cheapest way to charge (in the United Kingdom at least) due to being exempt from higher value-added tax (VAT) and also being able to offer peak and off-peak rates, with the latter being cheaper. On-street charge point users want similar solutions in their area, which further impacts on the solutions that can be offered.

In summary, understanding the 'why' at the start of the charge point installation journey is key to minimising the risk that what is installed is not in demand and is poorly utilised. Although there is no one answer to the 'why' behind installation, the key is understanding the charging use case in detail. Logical reasons for installing charge points in each use case are as follows:

- **En route charge points**: there are gaps along a strategic or provincial transport route either because there are no existing charge points or because the current numbers are insufficient to meet demand/are over utilised. Deploying charge points in this setting would likely see good utilisation.
- **Destination charge points**: there are locations that are in high demand from a recreational activities perspective. Examples might include cinemas, shopping centres/malls, hotels, gyms, sports centres and bowling alleys and supermarkets/grocery stores where EV drivers spend several hours whilst doing other things. CPOs should always consider whether the location is close to a strategic or provincial road network (this might result in a hybrid en route/ destination charge point arrangement). They should also consider the charge point competition in an area that may impact on demand and utilisation.
- **Residential charge points**: there are areas in an urban conurbation where a significant number of residents do not have access to off-street private parking and there are gaps in the existing charge point networks. The socio-demographic makeup of an area also lends itself to high demand and good utilisation, covered in more detail in Section 3.1.2.

Understanding the 'why' in detail then naturally lends itself to the next step in the planning stage, understanding the 'where' at a local area level.

3.1.2 Determining the geographic area to deploy

It is not enough to understand the justification behind deployment as deciding to deploy charge points based on target use cases will lend itself to types of location but not the specifics of the geography where charge points will be installed. This is all about understanding the actual location on a street, by a strategic/provincial road network or in a car park. Analysing the geographic area where charge points should be deployed has developed over the past decade as a result of a greater understanding of spatial mapping tools, their availability and the use of machine learning and artificial intelligence to ingest data and draw out key conclusions. This overall process can be described as data-driven spatial analysis and ultimately predicts the potential demand and utilisation across a specific geography. This is then translated into streets, car parks or plots of land on maps that show the greatest potential for charge point deployment. The ability to complete this analysis can be achieved using a myriad of tools either in-house by CPOs or local/state governments or via consultancies. A commonly used free geographic information system, QGIS, has made it possible for anyone with a basic understand of spatial data to analyse this to generate demand maps using various data sources. Other paid for tools like ArcGIS are just as powerful for completing this task. The outputs generated can be displayed in many ways, however. Figure 3.1 shows one example of this type of modelling and represents EV charge points that are within a 5-minute walking distance for EV users. This is important as it helps understand how far residents are willing to walk to use on-street charge points.

Various consultancies also specialise in completing this analysis on behalf of a client who might be a CPO or local authority. Many, like Urban Foresight, use available GIS software and combine this with their knowledge of an area or market to generate bespoke detailed maps which then enable the client to make decisions on the potential deployment locations for charge points [3]. These are then produced as finalised reports showing maps, lists of areas with the highest potential for deployment and specific deployment locations in those areas based on various factors. It is then the responsibility of the decision-makers in the CPO to analyse these lists to determine where to deploy, reflecting on the 'why' determined in Section 3.1.1.

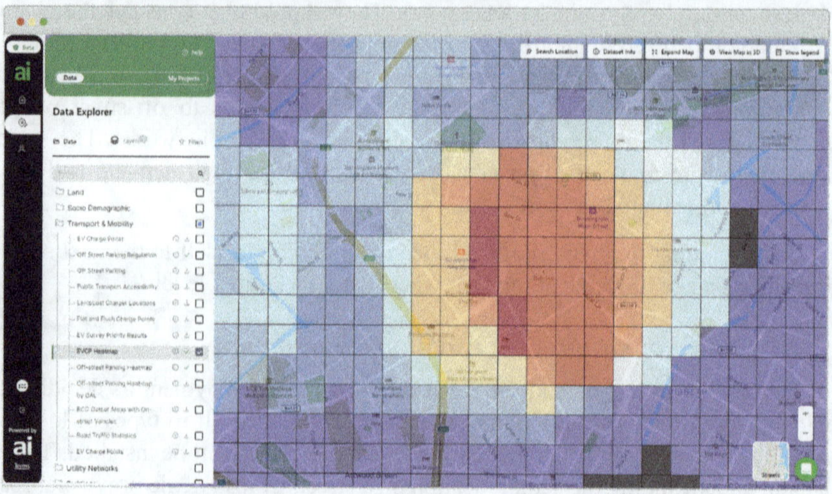

Figure 3.1 Walking Distance from EV charge points dataset as visualised in the LAEP+ Platform. The colour scale represents EV charge points that are within a 5-minute walking distance for EV users. Red squares represent 1 or more EV charge points and blue squares represent zero EV charge points within 5-minute walking distance. Empowering Birmingham City Council's EV Charging Strategy with LAEP+ and Granular Data [online]. 2024. Available from https://www.advanced-infrastructure.co.uk/case-study/empowering-birmingham-city-councils-ev-charging-strategy-with-laep-and-granular-data [Accessed 25 February 2025].

Although the data analysed is varied, this can be broken down into the following six broad groupings. Data is often available, some of which is free, across these areas:

- **Socio-demographic data**: this covers information related to age, income levels, access to off-street parking, current EV ownership via vehicle registration data, propensity to own an EV and consumer trend data (derived from credit agencies like Experian). Analysing this data spatially gives an indication of who might utilise charge points in an area.
- **Existing charge point data**: this covers what competition exists in an area both in terms of existing and planned sites. The data can be broken down by type and charging rate as this is a good indication of charging use cases targeted in a specific area. Analysing this data helps determine the saturation of charge points in a specific area.
- **Existing amenities, recreational features and transport networks**: this covers features like existing car parks, shopping centres, retail parks, gyms, bowling alleys, cinemas, hotels and other locations where people spend their time. Transport networks include major and minor roads, train and other transport stations, park and ride facilities and interchange points. Understanding these features helps determine what is driving demand which will ultimately determine charging type and rate.
- **People movement data**: this is not always available but covers the number of vehicles travelling in an area or along a route and the proportion of which are EV.
- **Power infrastructure**: this covers what power is available in an area, whether upgrades are planned and if there is capacity, whether it has been allocated. This helps in determining what headroom is available in specific locations and whether power upgrades are required to satisfy demand from newly installed charge points.
- **Land ownership**: this covers who owns land in an area and gives an indication of either the permissions needed or the ownership options required to deploy charge points.

Analysis of this data is required to produce demand maps which can then be used to create site long lists of where charge points should be deployed based on the data. Completing this activity requires prioritisation of different data sources depending on the targeted use case. For example, if en route charging is the focus the existing transport network and charge points, movement data, power infrastructure and land ownership would be prioritised. Transport network, movement data and land ownership would carry the highest weighting. This is because placement near to the strategic road network where there is a high anticipated EVs is key to en route charge point utilisation. The availability of land to build charge points is also key. All other factors, though important in this example, might influence a refined list but are not core to the initial location selection.

The completion of spatial analysis enables the translation of this data into site long lists of potential locations where charge points could be deployed depending on the use case. These long lists are best produced in table format with key site features noted as columns and each location entered as lines. This enables CPOs to quickly scan through and identify potential sites. More detailed shortlisting and finalisation of sites is done during the 'planning deployment at a local level' step. In completing the 'why' and 'where' of charge point deployment, the next step of the planning stage is the 'what', completed through a business case.

3.1.3 The business case including seeking funding

The purpose of the business case is to provide justification for deployment of charge points. This step draws on analysis from previous steps in the planning stage and forms the firm foundation to progress into more detailed planning and then installation. There are various examples of business case structures, all of which, according to the Association for Project Management, should evaluate the benefit, cost and risk of alternative options whilst providing a rationale for the preferred solution [4]. A commonly used business case methodology is the Five Case Model, an approach established by HM Treasury which is used both in the United Kingdom and beyond. Its structure ensures that a proposal has a good strategic fit, has a strong economic and commercial case, is affordable and fundable over time and can be successfully delivered by an organisation and its partners. These elements align well with the process of planning for charge point deployment. A point that should be noted is that it tends to focus on public sector projects; however the broad structure and guidance is applicable in multiple contexts, especially those where there is significant change as is the case where charge points are being installed. The underlying methodology is broken down into five key parts:

- **Strategic case**: this sets out the case for change and demonstrates how a proposal provides strategic fit. This requires a clear understanding of the rationale, drivers and objectives of a proposal and should be SMART. This stands for Specific, Measurable, Achievable, Relevant and Time constrained. Core to this is making a compelling case for an intervention through demonstrating an understanding of the business as usual (BAU), business needs in terms of related problems and opportunities, potential scope of the project, the likely benefits, risks, constraints and dependencies associated with the proposal. This strongly links with the 'why' in Section 3.1.1.
- **Economic case**: this sets out how the proposal will deliver value for money versus other proposals, the value to society, including wider social and environmental effects. A key part of this is producing a short list of options which should include the BAU, a realistic and achievable 'do minimum' that aligns with the strategic case requirements, the preferred option and any other options which should be considered. Each option is analysed using a cost–benefit analysis to ensure the preferred proposal offers value for money versus other proposals.
- **Commercial case**: this establishes that the preferred option is commercially deliverable in terms of procurement and supply chain. This is reliant on a good

understanding of the marketplace and what is achievable in terms of goods and services to be provided by the supply chain.

- **Financial case**: this demonstrates the affordability and funding of the preferred option, including the support of stakeholders and customers, as required. Core to this is ensuring there is a detailed understanding of the capital, revenue and whole life costs of the proposal and its impact on the balance sheet, income and expenditure of the organisation.
- **Management case**: this determines that robust arrangements are in place for the delivery, monitoring and evaluation of the scheme, including feedback into the organisation's strategic planning cycle. Ultimately, this determines whether the proposal is viable and that assurances are in place with regard to standard elements of project management, including managing change, contract management, benefit realisation and risk management [5].

Completion of these steps should ensure that the proposal has a strong business rationale and should therefore be pursued. It should be noted that the Five Case Model is used at all three levels of business case development: strategic outline business case, outline business case and full business case. The business case step is complete when approval is granted to progress with the proposal, translating it into a project.

Using the Five Case Model in planning the deployment of charge points would likely involve the following aspects. The strategic case would focus on the strategic fit of charge point deployment in terms of delivering a CPO's desired goals. For example, if a CPO's focus is purely on en route charge points, the strategic fit would test the alignment of the proposal with this business focus. The rationale, drivers and objectives of a charge point deployment proposal would likely focus on planned charge points delivering growth in a specific area where there is an en route charging gap or the existing competition is failing to fulfil demand. Demonstrating the case for deployment would analyse the BAU in a specific area in terms of current and anticipated demand and utilisation This would also include analysis of existing charge point competition with an understanding of their demand and utilisation if known. Problems and opportunities for charge point deployment would be considered in the geography as would the potential scope for deployment, benefits of deploying in this area, risks that might occur with deployment, constraints (most likely power capacity and land availability) and dependencies (an example might include a planned power grid upgrade).

Using the same hypothetical en route scenario, the economic case would focus on the different costs of different deployments so as to ensure value for money for the business owner or shareholders. The appraisal of different options would start with the BAU which would focus on what a geographic area would look like without the CPO deploying additional charge points. The 'do minimum' would focus on the minimum number of charge points to satisfy demand and utilisation modelling at the bottom end of the range, assuming this modelling shows an upper and lower range. The preferred option would likely focus on installing charge points that align with the demand and utilisation modelling at the upper end of the range. Further permutations of the 'do minimum' and preferred option would be explored outlining different mixes of charge point speeds to meet anticipated demand and deliver target utilisation. A cost–benefit analysis of each of these options would then be completed to compare them, with the selected option delivering value for money when considering the numbers and costs of different charge points deployed and anticipated power network upgrade costs versus meeting the demand in a specific area.

The commercial case in this scenario would focus primarily on whether the CPO was able to draw on its supply chain to deploy the charge points it needs to deliver its preferred option. This was a challenge for CPOs during the coronavirus pandemic as charge point manufacturers were struggling to source materials to meet demand. The supply chain for CPOs also includes engaging with power networks to upgrade connections in anticipation of additional load derived from new charge points. In my past experience, this is a major barrier that can place a significant delay on charge point deployment planning. The financial case using this example would primarily focus on the affordability of the capital investment in charge point deployment at scale either at a specific site or across multiple sites versus the revenue generated based on anticipated utilisation. This would be used to understand the return on investment and ultimately determine the viability of the deployed charge points. The installation of a charge point is broken down into two parts, the capital expenditure which includes the charge point unit itself and the associated bay infrastructure (e.g., signage, protective barriers) and development expenditure which includes power network connection costs, electrical and civil materials and labour. The operational expenditure includes the cost of power import, maintenance, charge point management system and payment system, insurance, administration and VAT.

A level 2 charge point can cost anywhere between £8,000 and £15,000 which covers the charge point itself and the installation of two bays. The range is primarily due to variations in power connection cost [6]. A level 3 DC charge point unit alone costs around £71,000* [7] and power network upgrade costs are substantially more when an ultra-rapid level 3 DC charge point is installed. These can range anywhere between £50,000 and £300,000 plus depending on the size of the connection. Development costs are largely the same as a level 2 charge point and can add an additional £5,000 to installation based on two bays. Taking these numbers and applying to an en route charging hub comprised of 8 charge points and 16 bays, the capital and development costs can be anywhere between £1 and £2 million. Operational expenditure, when broken down per kWh can be anything between £0.40 and £0.60 with power import costs and VAT being the bulk of this cost [6]. Using the average rapid charging cost to the customer derived from Zapmap of £0.70 per kWh [8], a CPO will generate between £0.10 and £0.30 net revenue per kWh. Taking the middle of this range (£0.20) and an assumed capital and development cost of £1.5 million, a charging hub would need to sell 7,500 MW, or just under 26 MW (26,000 kW) a day just to break even after the first year.

*An Alpitronic Hypercharger HYC 300 kW DC station is €83,999 excluding VAT. Using the exchange rate of €0.85 to £1 on 17 August 2024, this results in a United Kingdom cost of £71,590.76.

If we assumed each charging session dispensed 50 kW, the charging hub would need to see 520 sessions per day or 32 sessions per bay. In this example, I've been somewhat unfair in terms of the financials as expecting a site to break even after the first year is unrealistic. Most CPOs would expect a site to break even after year 3 or 4. A charge point is expected to have a life span of 10 years. These calculations bring to life the challenging aspects of making the financial case for a charging hub stack up and demonstrate that utilisation is key to success.

The management case for this example would primarily focus on whether the CPO has the project management capability to deliver the charging hub including the ability to liaise with key stakeholders such as the planning authority and power network provider. It would also cover the CPO's capability in terms of monitoring the success of the charging hub once installed and whether steps need to be taken if utilisation falls below the anticipated level required to maintain site profitability. This might include discounting charging to push up utilisation. The management case should also ensure the CPO has a robust operational management team covering customer support, contract and risk management. Looking at all five cases together, a proposed charge point installation would require a strong justification for deployment including a strong strategic case in terms of the en route charging hub filling a gap in the market that aligns with the CPO's organisational goals. Deployment would need a robust economic case in terms of the site delivering value for money based on anticipated demand and utilisation and a strong commercial case in terms of the CPO being able to draw on its supply chain to deploy charge points. It would also need a detailed financial case in terms of the charging hub being able to demonstrate that the site will break even and become profitable after a number of years and a strong management case in terms of the CPO having the capability and capacity to install, operate and maintain the hub.

Once the CPO has approval to progress beyond the business case, the next step in the planning stage is the detailed planning of charge point deployment at a local level.

3.1.4 Planning deployment at a local level

Planning deployment at a local level builds on all previous steps and starts with the CPO finalising the site locations at a granular level, having developed a long list during the geographic areas analysis in Section 3.1.2. This involves analysing and ranking the long list based on various factors such as anticipated demand and utilisation, site proximity to power connection (and upgrade cost if known), number and type of charge points which can be installed at site, land availability and anticipated ease of install. Each CPO will have a different way to weight these factors when prioritising sites. The result will be a shortlist of sites. The CPO then looks at each shortlisted site in detail, producing an information pack outlining key features of the site. This includes proposed site designs showing bays and street furniture (e.g., bollards, posts) supported by site images and bird's-eye views, number and types of charge point to deploy, power diagrams showing connection points, nearby feeder pillar containing metering and cable layout and other notable features significant to understanding the site in detail. Figure 3.2, provided

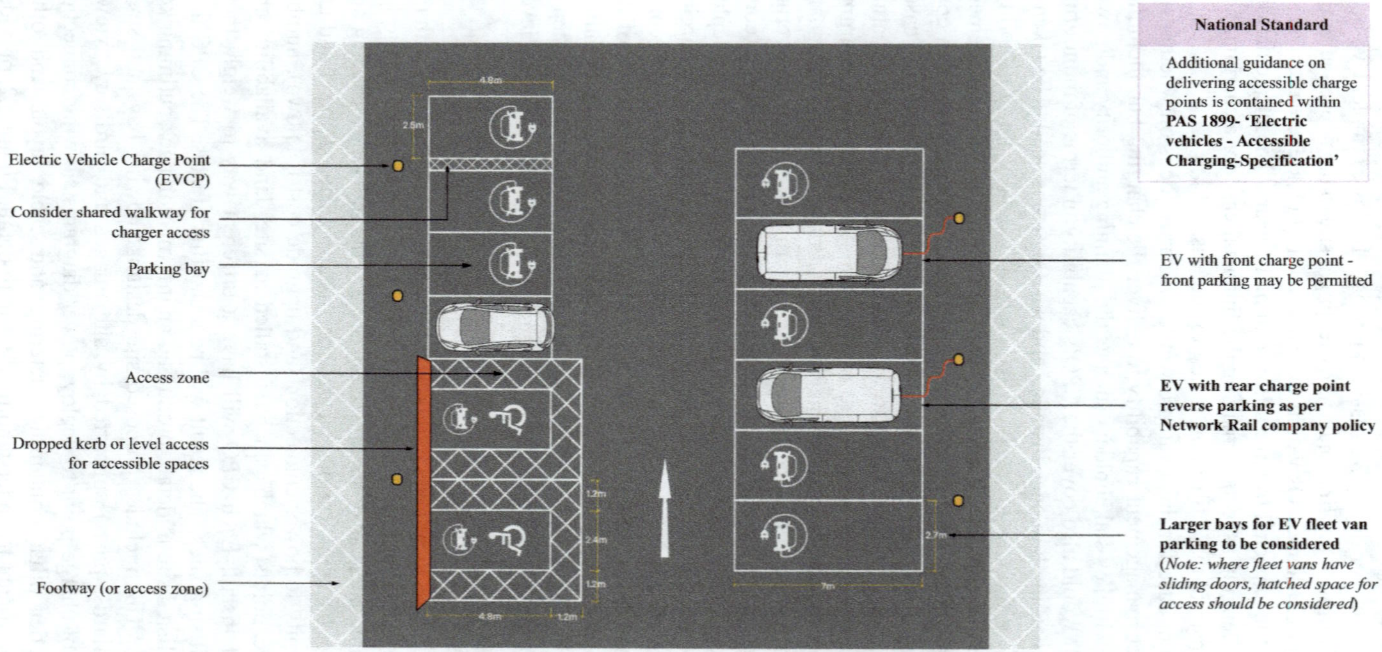

Figure 3.2 A proposed site layout for EV charge points at Network Rail train stations, offices and depots [Source: Network Rail. Electric Vehicle Charging Points and Associated Infrastructure [online]. 2023. Available from https://www.networkrail.co.uk/wp-content/uploads/2023/11/NR-GN-CIV-200-13-Electric-Vehicle-Charging-Points-and-Associated-Infrastructure.pdf [Accessed 20 August 2024]].

by Network Rail, the United Kingdom's rail network maintainer, shows how a site layout might look. It should be noted during this step that detailed planning, including final site designs and construction phase plans, does not take place and is completed during the charge point build stage. The CPO will then reach out to the local power provider (distribution network operator (DNO) in the United Kingdom) to seek a connection quote. They may have engaged earlier in the process; however, as the sites are known in better detail, the power provider will be able to issue a more detailed quote and timeline for site upgrades. The CPO may decide to conduct site visits or carry out site assessments remotely using photos and digital representations of the site.

The CPO also engages with relevant stakeholders during this step to ensure they are bought into the process. Where the CPO does not do the work in-house, they will engage with contractors to ensure they have the capacity and capability to carry out the installation work. The CPO will seek quotes for this work across the sites. The CPO will engage with its supply chain to ensure that the charge point units and associated materials are available when site deployment starts. The CPO will engage with the landowner to establish an agreement regarding installation. This might include discussing purchasing, leasing or renting land for charge points and bays. The CPO will engage with the local planning authority (usually the local authority or state government) to understand what planning is required. They may also engage with the local governing body to understand whether traffic regulations or rules need to be introduced to enforce compliance with EVs parking in designated EV bays. Where the CPO has been contracted to deliver charge points on behalf of a customer the customer may have already engaged with stakeholders on the CPO's behalf or may have internal stakeholders the CPO needs to engage with. For example, a local authority will propose the CPO works with the internal transport department to cover off traffic regulations.

Once these elements are complete, the CPO is then able to develop a detailed project plan for all sites and determine the final financial cost of all elements to deliver the sites, based on capital and development costs using quotes and agreed prices. These elements plus the business case outlined in Section 3.1.3 are the key gateway documents, which, once approved, and all associated stakeholders are happy, enable the CPO to move to the deployment stage. It should be noted that these documents may, and often do, evolve during the build stage.

3.1.5 Governance and approvals

The final step of the planning stage is securing approvals and determining the governance arrangements for the next stages. In most cases, the process will involve a mix of internal and external approvals. The key internal approval is securing the finances to progress the project to the build stage and the associated project management plan to deliver the deployments. In Section 3.1.4, the CPO developed detailed costs that require capital to progress and a detailed project timeline governing delivery. The CPO may source capital through investors or via payment from a customer to build the network on their behalf. If the latter is the case, the customer may need to sign off the build plans to release funds or only pay for work completed once they are satisfied with it (during the build stage). This governance would form part of the external approvals. The CPO may also need to gain internal approval for its risk and change management plans governing delivery of the project.

The external approvals a CPO requires in addition to the aforementioned ones noted earlier in this section include finalised site location approval from the landowner, planning approval (if applicable) and traffic regulation approval (if required). Site location approval is sought from the landowner and will involve finalising the land ownership or leasing/renting agreement. Planning approval where required is gained from the local planning authority and will depend on local planning rules. In the United Kingdom, for example, planning permission is required where a charge point is installed on-street, in a conservation area or on/next to a listed building, where permitted development rights have been removed or if more than one charge point is provided for a parking bay or if the charge point unit is taller than 1.6 m. Traffic regulation approval, if applicable, is required in on-street charge point deployments where bays need to be designated specifically for EV charging. This is sought from the local transport authority. Once these approvals have been gained the CPO can move into the build phase.

At this point in the process, the CPO sets out the governance process for the remainder of the project. This establishes a number of gateways through which the CPO must pass through. Some of these are required within the CPO and are therefore internal, whereas others need approval from external organisations to progress through.

- **Approval of final site detailed designs (internal to the CPO)**: This either comes at the end of the planning stage or at the start of the build stage.
- **Approval to start the build (internal and external to the CPO)**: Once gained the CPO can start the install.
- **Approval of build alignment with designs (internal and external to the CPO)**: The CPO will sign off that they are happy the design aligns with the built layout.
- **Safety approvals (external to the CPO)**: The CPO will seek approvals for electrical safety of the charge point site.
- **Charge point system and commissioning approvals (internal and external to the CPO)**: The CPO will seek approvals that the charge point operates as expected and has been properly commissioned to go live (e.g., not being left in standby mode). These approvals will also include charge point go live.
- **Release of payment approval if applicable (external to the CPO)**: If the CPO is delivering for a customer, they may have to complete all of the outlined approval steps to receive funds.
- **Approval to change pricing if applicable (external to the CPO)**: A CPO, if delivering and operating charge points on behalf of a customer, will likely need approval if changes are made to agreed pricing.
- **Approval to upgrade charge points (internal and external to the CPO)**: A CPO will need approval if charge points become obsolete and require physical

or software upgrades. If this is the case, the CPO may need to gain approvals again as upgrades progress.

- **Approval to decommission a charge point (internal and external to the CPO)**: The CPO requires this where charge points are being taken out of action.

A CPO establishes these gateways in their project management plan.

3.1.6 Deploying charge points

The only step in the build stage is the deployment of charge points. Providing the designs have been finalised as part of the planning stage, the CPO will appoint a contractor to carry out the work. Some CPOs may have this capability in-house. The build stage should be planned to coincide with the power network upgrade; otherwise, there is the risk that the charge points sit dormant while the connection is being commissioned or the opportunity for revenue is lost if charge points are installed a significant time after the connection. An important part of the build stage is the civils work. The contractor/CPO will develop a construction phase plan which includes this and sets out how the build will take place safely. A core part is the electrical cabling for the site which involves digging feeder channels from the main feeder pillar which, in turn, is connected to the power network. They may choose to install more channels than are required so as to prepare the site for future expansion. The channels might include wired Ethernet connections between charge points. Once the feeder channels are in place, the contractor will install the charge points where they need to be, mark out the bays and install the surrounding street furniture including signage, barriers and parking blocks. The contractor may need to put in place traffic management if working on-street or impeding traffic flow off-street. In the United Kingdom, the entire process is completed in adherence to the Construction (Design and Management) Regulations 2015, which ensures health and safety measures are put in place throughout the build.

The CPO will then check the build aligns with the original designs and conduct tests of charge points before accepting that the build is complete. Tests will include that the charge points draw power at the correct rate from the power network, that this rate can be remotely changed and that the charge point connects to the charge point management system. Where this connection is made further tests include that it can be identified, that the charge point can be remotely started and stopped and that it can be safely operated both on site and remotely. Safe control covers triggering emergency functions like cutting power to the unit and resetting it. Electrical tests will result in electrical certification. If the unit has the ability to pay for charging the payment terminal will be tested to ensure it works. The ability to change pricing remotely will also be tested. Completion of all of these tests will enable the CPO to formally commission the charge point, making it live for public use. At this point, the CPO will provide various documents to the landowner or customer, including user manuals, manufacturer contact information, warranty information, safety guidelines, electrical certificates, functional demonstration materials showing how the charge point works and a checklist of all the checks carried out during installation and testing [9]. The CPO may choose then to conduct one more test once the unit is live to ensure it is working as expected. If the CPO is delivering charge points on behalf of a customer, they might need to seek approval for successful completion of both the build and commissioning.

3.1.7 Operations, maintenance and monitoring

The first step of the operations stage is operations, maintenance and monitoring of charge points and involves several activities that CPOs must carry out to ensure the units work as expected and customers receive a satisfactory service. Customer management is the activity that takes up most of the CPO's time and involves the CPO setting and changing pricing, processing charging session payments including charging for overstaying or blocking of charge point bays, managing customer queries and complaints, issuing refunds where customers are over charged and remotely helping customers where charging cables become locked, charge point resets and where the charge point does not operate as expected. CPOs are expected to provide a helpline which in most cases is accessible by web, app or phone 24 hours a day, 7 days a week to help with queries. The second key activity is maintaining charge points. Each manufacturer will provide different guidance on how to maintain their charge point unit; however, as a broad rule, it is recommended every six months that the cables and sockets/connection points are inspected for dirt or damage. It is also recommended that the charge point units themselves are cleaned, especially around air intake areas for level 3 DC charge points to remove dust and build-up of organic material (a filter might need to be changed). Furthermore, the unit itself must be checked to see if there are signs of damage (including from the weather) and/or water ingress and that the bay and surrounding street furniture is as expected (e.g., barriers are not bent or damaged from being struck). Units should also be checked to make sure key functions work, including being able to start and end a charge and being able to stop a charge in an emergency. The maintainer may open the charge point unit to inspect internal wiring to ensure everything is expected. Monitoring of charge points is also an important activity and may give an indication of maintenance needs based on data transmitted by the charge points. Monitoring is on-going and will ensure that:

- Charge points are functioning correctly and that faults are quickly identified. This includes whether connectivity to the charge point is constant and there are no outages.
- Metrics are available (some of which may be specified by national regulations) and as a bare minimum should show reliability and uptime of a charge point, charging events by time of day and length of charging session, energy consumed per charging event and price per kWh.
- Capacity and utilisation at the site are based on number of actual sessions and not projections.
- Faults are repaired and units brought back online within expected and agreed time periods and in line with manufacturer guidance or service level agreements with a client. The CPO should check before repairing whether the fault falls within the warranty period for the unit and is therefore covered [9].

In addition, it is good practice to develop monthly reports summarising this data and combining it with customer data to give a detailed overview of operations. If charge points are delivered on behalf of a customer, it is likely that this reporting is specified in the contract. Providing the CPO operates, maintains and monitors charge points effectively and efficiently, they should expect to see good demand and utilisation.

3.1.8 Upgrading and decommissioning

The final step of the operation stage is upgrading or decommissioning existing charge points. Upgrades may be physical or virtual. Regarding the former, key reasons include swapping the socket to a new standard (e.g., North American Charging Standard) or if the payment terminal requires upgrading to accept a new payment method (e.g., contactless payment). Furthermore, the communication module may require upgrading to 5G if 3G/4G is switched off locally. Regarding the latter, upgrades normally take place remotely, unless a physical install is required in some instances if software on the unit becomes corrupted or an entirely new operating system needs to be installed. Upgrading remotely primarily involves installing a new version of software. Remote upgrades require little planning as the software can be pushed out over the air over night to minimise disruption. Physical upgrades require more planning and will normally involve localised signage being placed near or on the charge point stating when the work will be carried out and that the charge point will be out of action. The upgrade should be able to take place at one site over a single day, minimising utilisation disruption. The CPO might choose to contract the work out or complete it themselves.

Decommissioning usually takes place when a charge point has reached its end of life, has become damaged beyond repair or a site is no longer viable meaning the charge points must be removed. In a minor number of cases, this may be done because utilisation is not as expected and the cost to operate is greater than the income generated. Once this has been identified, the CPO will switch power off to the unit physically or remotely and display a sign on the unit stating that it is out of order. The CPO may choose to leave the unit as is until a site upgrade is planned or they may schedule a date when the charge point will be removed. This may correspond with the installation of a new charge point. To remove the charge point, a CPO will likely contract an electrical engineer to disconnect the charge point and make safe the connection points to the underground power cables. If ducting was installed, the duct will be closed. If no ducting exists, the connection point may be tarmacked over. If a new charge point is being installed in the same place, the electrical engineer may temporarily seal off the connection point in preparation for the new charge point to be installed at a later date.

3.2 Key stakeholders

Throughout the planning, build and operation stages, there are multiple stakeholders who play an important role in charge point deployment and operation.

Chapter 10 covers each stakeholder in detail; however, this section provides a high-level overview of their role in the process.

3.2.1 The charge point operator

The CPO plays a key role in every part of the planning, build and operation stage, either as a contractor to another entity (e.g., a local authority or state government) or as the primary operator with agreement from the landowner to lease/rent land where the charge points are installed. The CPO may choose to own the land the charge point is installed on. The primary drivers of the CPO are to increase demand and utilisation, thereby generating a profit for the business and their shareholders. In some instances, a CPO may enter into an agreement to install charge points in a location as a 'land grab' to secure a power network connection for a future expansion or to prevent a competitor from installing charge points. During the planning stage, the focus of the CPO is to identify the sites with the best anticipated demand and utilisation. During the build stage, the CPO's efforts focus on achieving cost efficiencies and value for money by installing charge points as efficiently as is quickly and practically possible so as to start generating revenue. During the operation stage, the role of the CPO is to operate and maintain the charge points, minimising maintenance costs and downtime whilst pushing up utilisation by making charging as attractive as possible (this might include making changes to the price). The CPO will only decommission or upgrade a unit when it is absolutely necessary, preferring to operate the charge points until their end of life, which is usually 10 years.

3.2.2 The landowner

The landowner's primary role is to provide access to the land where the charge points will be deployed, either by selling the land to the CPO or leasing/renting it out. Throughout the installation process, the landowner is heavily involved in the planning stage. The CPO will identify them when looking for potential sites to deploy charge points and may have initial conversations during long-listing. More detailed discussions take place during shortlisting when the site lists are finalised. The CPO and landowner will enter into negotiations to sell or lease/rent the land, with the latter being done based on an annual fee, though in some cases the landowner may insist on a revenue share where an income is earned based on every kW dispensed at the site. A hybrid payment agreement mixing a fixed lease/rental fee with revenue share when utilisation achieves a certain level might also be agreed. If the land is sold, the landowner's role stops there. If a leasing/rental agreement is put in place, their role diminishes during the build and operation stages. They fulfil a high-level monitoring role, commercially managing the agreement to ensure it is functioning as expected. Disputes between both parties may occur if rent and/or revenue share ceases or if the land is not used as agreed.

3.2.3 The local authority, council or state government

The local authority, council or state government can fulfil multiple roles. In a limited number of cases, they act as a CPO and assume the role outlined in

Section 3.2.1. In some cases, they act as the client to a CPO, appointing them to deliver a charge point network on their behalf. In most geographies, the role they fulfil is as the planning authority during the planning stage. This role changes depending on local planning laws and may not be required in some instances. Looking first at the role of the local authority, council or state government as a client appointing a CPO, they play an active role throughout all three stages. During the planning stage, they will determine the long list, put this out to tender and then work with the appointed CPO following contract award and signature to finalise the short list. When they finalise the tender, they determine the rental or revenue share agreement (or both). They then work with the CPO to plan for installation, approving timelines, budgets and designs, supporting in securing power connections and preparing traffic regulation orders. They also play a key role as the client approving planning applications if required for install. During the build stage, their role is reduced to one where they monitor the progress of the installs to ensure they are delivered against the delivery programme. During the operation stage, they fulfil a similar role, monitoring the demand, utilisation and associated rental/revenue share payment for the charge points to ensure they are as is expected. Disputes may occur between the CPO and authority, council or state government if payments do not reflect utilisation or if the CPO is failing to maintain the service level agreements (e.g., charge point uptime) as agreed in the contract.

Looking second at the local authority, council or state government's role as the planning authority alone, their involvement in all three stages is diminished. During the planning stage, provided this is required under local law, they approve or reject the planning application for charge points. During the build and operation stages, they monitor the charge point deployments to ensure they align with the original planning application and enforce the original plans and designs if the CPO deviates from what was agreed.

3.2.4 The power provider or DNO

The power provider/DNO's role is to provide the CPO or local authority with cost estimates relating to connections and network upgrades during the planning stage, and then complete those connections or upgrades during the build stage. They are usually engaged at a high level during the planning stage when determining the geographic area to deploy as they can give an indication of the power capacity and headroom in an area. This helps determine the most suitable locations for installation. In some instances, this information can be provided by a spatial power planning platform. They are also further engaged by providing upgrade cost estimates during development of the business case so as to develop the detailed costings underpinning the deployment. The power provider/DNO will then be engaged during the planning deployment at a local level step to provide finalised costs for the shortlisted sites. During the deployment of charge points step as part of the build stage, the DNO completes the power connections and network upgrades for each installation. This can take a number of months, which means that the CPO should build their deployment schedule around the power provider/DNO's connection schedule. They may also be involved during the operations stage if a charge point or site is upgraded and additional power is required. Like during the planning stage, costings are provided which are used to base the investment decision on.

3.2.5 The regional or national transport body/department

The regional or national transport body's role can depend on how hands on they are in the planning, build and operation stages. In some cases, they work with the local authority, council or state government to procure a CPO on a regional or multi-regional scale. The logic in doing this is that the coordinating role achieves scale and with scale comes the likelihood of a better commercial deal that benefits multiple authorities. Rarely, though, will they act as the procurer, leaving this to the authority, council or state government. When they do fulfil this role, this tends to be if local government does not have capacity or where a national transport agency needs to deploy charge points along the strategic road network, working with service station owners to procure a CPO. In other cases, they assume a more distant role, setting charge point policy direction based on national/multi-national policy and strategy guidance and determine deployment targets regionally or nationally. This direction plays a key role during the planning stage as the CPO or local authority will align with policy and establish a site long list that supports meeting the broader strategic targets. During the build and operation stages, they monitor what was installed and align this with progress towards overall targets. In a limited number of cases, they fulfil both the procurement coordinator and charge point monitor role.

3.2.6 The national government including multi-national organisations

The core role of national government and multi-national governing organisations like the European Union is to set policy and strategy direction. As a result, their role in the process is small but influences all stages. CPOs, local government and regional transport bodies will refer to the policy guidance set by them. Furthermore, national government will set building and safety regulations and standards guiding the CPOs during the build stage and consumer regulations and standards that determine how the charge points function during the operation stage. They might have a more influential role beyond setting out policy and strategy where charge points are being deployed along a strategic road network; however, they defer to the transport department or transport authority that engage directly with the CPO through all three stages.

3.2.7 Consultancies

Consultancies by their very nature can be involved throughout all three stages, but only ever as advisors. During the planning stage, consultancies can advise on the long-list exercise utilising their modelling and analytic experience to determine demand and propose suitable locations with anticipated good utilisation. They may consult on developing the business case for the CPO. They also provide advice on the design aspects of planning, including detailed site layouts, electrical schematics and civils planning. During the build phase, consultancies can project and manage the installation on behalf of the CPO, commissioning contractors to complete the work if the CPO does not have that capability. During the operation stage, consultancies might advise on evaluating the utilisation of charge points and proposing options to increase this. They might advise on upgrade or expansion plans. They may also in certain cases advise the CPO on securing investment to grow including acquiring another CPO.

3.2.8 Residents/citizen groups

Resident or citizen groups can play an influential role in the planning, build and operation stages. During the planning stage, they have the ability to feed into the governance approval process for deployment, either supporting or rejecting proposed sites. Local authorities may choose as part of the planning stage to consult with the local community, presenting them with the opportunity to make their opinions known. Community groups and residents often have strong views when on-street charge points are being installed, objecting in some cases where there is limited parking and the provision of dedicated EV bays will restrict this further. If the community is not at least consulted in some way, this might result in poor utilisation when the proposed sites become operational, especially in destination and residential charging use cases. Resident or citizen groups must be kept abreast of developments during the build stage so as to keep them on side. They have the ability to be disruptive if there is no local support for charge point sites. During the operation stage, residents or citizen groups should be engaged via a communications campaign to demonstrate, specifically in the case of destination and residential charge points, the benefits of charging locally. This can ensure there is good demand for and utilisation of the charge points installed.

Now that we have developed an understanding of the end-to-end process of planning, building and operating charge points and the key stakeholders throughout this process, we can now turn our attention to the policy and strategy environment in detail. This plays an important part in the planning stage, shaping the business justification for deployment.

References

[1] Electric & Hybrid Vehicle Technology International. *Osprey Charging celebrate doubling its number of rapid EV chargers in the UK* [online]. 2023. Available at https://www.electrichybridvehicletechnology.com/news/osprey-charging-celebrate-doubling-its-number-of-rapid-ev-chargers-in-the-uk.html [Accessed 12 August 2024].

[2] Zapmap. *Zap-Map reveals the best and worst rated EV charging networks in the UK* [online]. 2022. Available from https://www.zap-map.com/news/best-worst-rated-ev-charging-networks-uk [Accessed 12 August 2024].

[3] Urban Foresight. *Make good things happen* [online]. 2024. Available from https://urbanforesight.org/ [Accessed 13 August 2024].
[4] Association for Project Management. *What is a business case* [online]. 2024. Available from https://www.apm.org.uk/resources/what-is-project-management/what-is-a-business-case/ [Accessed 15 August 2024].
[5] HM Treasury. *Guide to developing the project business case* [online]. 2018. Available from https://assets.publishing.service.gov.uk/media/66449468ae748c43d3793bb8/Project_Business_Case_2018.pdf [Accessed 15 August 2024].
[6] Trousdale D. 'EV charge point utilisation modelling: Its critical importance in the future of the transition to EV'; *ITS European Conference*, Lisbon, Portugal, June 2023.
[7] Charging Shop. *Alpitronic Hypercharger HYC 300 kW DC station* [online]. 2024. Available from https://chargingshop.eu/product/alpitronic-hypercharger-hyc-300-kw-dc-station/ [Accessed 17 August 2024].
[8] Zapmap. *What's the price of charging an electric car on the top 10 rapid charging networks?* [online]. 2024. Available from https://www.zap-map.com/ev-stats/rapid-charging-prices [Accessed 17 August 2024].
[9] Network Rail. *Electric Vehicle Charging Points and Associated Infrastructure* [online]. 2023. Available from https://www.networkrail.co.uk/wp-content/uploads/2023/11/NR-GN-CIV-200-13-Electric-Vehicle-Charging-Points-and-Associated-Infrastructure.pdf [Accessed 20 August 2024].

Chapter 4
The strategy, policy and funding environment

David Trousdale[1]

Chapter 3 set out the end-to-end process and the key stakeholders involved in the deployment of charge points; however, what underpins the entire process is the policy, strategy and funding environment that sets the direction of travel in different geographies. For most countries, this ultimately derives from the Paris Agreement, signed in 2015, which states that the signatories must reach net zero by 2050. A key mechanism to do this is through reducing road-based transport emission by transitioning cars and vans to electric vehicles (EV), and through rolling out the infrastructure to enable this transition. This chapter outlines the current (as of 2024) strategies, policies and funding in place from a charge point infrastructure perspective and demonstrates the difference in approaches between countries. The focus is not, however, on the policies, strategies and funding aimed at increasing EV adoption, though in many cases these are closely linked with the need to roll out charge point infrastructure. This chapter also covers EV infrastructure targets and outcomes to be achieved, technology requirements and regulations, funding available, progress to date, challenges experienced and potential changes in policy and strategy on the horizon.

4.1 The United Kingdom

The United Kingdom was the first major economy to set legally binding carbon budgets through the Climate Change Act in 2008. By law, the United Kingdom's emissions must be net zero by 2050. Since 2008, several policy and strategy documents have been produced that set the direction of travel, alongside a funding regime to enable deployment. It is expected that since the Labour victory in July 2024, the policy, strategy and funding environment will change; however, the broad direction of travel is anticipated to stay the same. It should be noted that in this section the focus is primarily on the United Kingdom Government and not the devolved governments.

[1]Principal Advisor - Climate Change and Sustainability, Newcastle City Council, United Kingdom

4.1.1 Powering Up Britain: The Net Zero Growth Plan [1]

Powering Up Britain: The Net Zero Growth Plan emphasises the progress to date made by the United Kingdom Government through successive strategies outlined in this chapter. It notes that the Government has supported the rollout of public charge points, seeing this triple in 4 years from 10,300 charge points in January 2019 to over 37,600 charge points in February 2023, with more than 7,500 of these being level 3 DC rapid charge points. The Government also launched the Local Electric Vehicle Infrastructure Fund Pilot (LEVI) which allocated £57 million of public and private funds to 25 local authorities across England. This set out to deliver 4,400 charge points targeted specifically at drivers with no off-street parking. Progress is ongoing regarding delivery of the pilot. In addition, it set out to improve the experience of customers accessing charge points of 8 kW and above and existing charge points of 50 kW and above by making it possible to pay by contactless card. Furthermore, the strategy mandated that payment roaming would become commonplace and that pricing at charge points would be more clearly marked, enabling drivers to compare costs more easily. The Government also mandated 99% reliability for level 3 DC rapid charge points and a 24/7 free helpline for drivers needing charge point support.

4.1.2 Net Zero Strategy: Build Back Greener [2] and Decarbonising Transport: A Better, Greener Britain [3]

The Net Zero Strategy: Build Back Greener, published in 2021, set out a funding target regarding charge points. This involved committing an additional £620 million to support the rollout of on-street charging infrastructure, building on £1.3 billion already announced by the Government in November 2020 [4]. Part of this funding was deployed through LEVI and the On-Street Residential Chargepoint Scheme (ORCS); however, it is not clear whether the full £1.9 billion was ever allocated and spent. The Decarbonising Transport: A Better, Greener Britain Strategy, published in 2021, made further reference to the £1.3 billion of funding available to install charge points and introduced the £950 million Rapid Charging Fund (RCF), its aim being to rollout 6,000 charge points across the strategic road network by 2035. ORCS was also highlighted alongside LEVI as a policy lever to rollout on-street charge points. The Strategy also stated that the Department for Transport would publish an EV infrastructure guide for local authorities to help them on their journey installing charge points. The Strategy noted that building regulations would be introduced (see Section 4.1.4) which state that all new-build residential and non-residential buildings with associated parking spaces would have a charge point and that regulations would be introduced to improve the consumer experience of public charging (see Section 4.1.7) and ensure the charge points are smart. Furthermore, the Government would work with Motability and the British Standards Institution to develop accessibility standards for charge points (see Section 4.1.8). It should also be noted that the Government would work with the Office of Gas and Electricity

Markets (Ofgem) to improve the process by which charge points are connected to the power network.

4.1.3 Taking Charge: The Electric Vehicle Infrastructure Strategy [5] and Transitioning to Zero Emission Cars and Vans: 2035 Delivery Plan [6]

Taking Charge: The Electric Vehicle Infrastructure Strategy, published in 2021, referenced numerous targets already noted in Section 4.1.2. The overarching aims of the Strategy were to accelerate the rollout of Level 3 DC rapid charge points, to support the charge point sector to thrive, address barriers to private sector rollout and give people confidence in the public network. New targets included that large fuel retailers would be mandated to provide a minimum number of charge points at specific sites to satisfy current and anticipated future demand, that local authorities would be obligated to produce local charging strategies outlining how they would deliver charge point growth in their area and that a dedicated support programme would be provided to local authorities to ensure they have the expertise and support they need to rollout infrastructure. An additional fund of £500 million derived from the Levelling Up Programme would also be allocated to enable rollout. The Strategy also set out to enable innovations in charge point business models and technology including vehicle to grid and wireless charging. Further guidance was provided on how the Government would work with Ofgem to ensure charge points integrate with the energy system including ensuring that the vast majority of charging is 'smart' and done at off-peak times so as not to overload the grid and that connection costs are reasonable so as not to deter deployment. These aims would be partly achieved through the use of Ofgem's £300 million Green Recovery Scheme. The Government would work with Ofgem to accelerate vehicle to everything charging. It would also mandate that all charge points sold in the United Kingdom meet minimum device level requirements and that they would be smart. Other aspects of the Strategy specifically related to planning included considering amendments to the Transport Planning Practice Guidance regarding charge points. The aims of these changes were to streamline and ensure consistency in planning and delivery and consult on measures to make Traffic Regulation Orders (TROs) more straightforward. Critically, the Strategy introduced the target of between 300,000 and 700,000 charge points needed by 2030 to meet transport decarbonisation ambitions. The lower target is reliant on more work-based charge points, efficient driving and lower mileage, whereas the higher target is based on the inverse of all of these factors.

The Transitioning to Zero Emission Cars and Vans: 2035 Delivery Plan, published in 2021, reiterates several strategy focus areas stated in the Taking Charge Strategy with the addition of the Charging Infrastructure Investment Fund. This was launched in 2018 by HM Treasury [7] and comprised a £400 million fund to leverage private sector investment to drive charge point rollout. The fund was managed by Zouk Capital on a commercial basis, with the first allocation of £70 million to rollout 3,000 rapid charge points in 2019.

4.1.4 Building Regulations: Approved Document S [8]

Approved Document S made amendments to the Building Regulations (2010) by setting requirements for the installation of EV charge points in new buildings, major renovations and certain existing structures. This included that for every new home, including those created from a change of use, and providing there is associated parking within the development, an EV charge point must be provided. For residential buildings with more than ten parking spaces and which are undergoing a major renovation, they must have at least one charge point per dwelling with associated parking. Passive charge point infrastructure should also be installed in all spaces to enable installation of charge points in the future. For all new non-residential buildings with more than ten parking spaces, a minimum of one charge point and passive infrastructure covering one in five of the total number of parking spaces must be installed. For all non-residential buildings undergoing a major renovation that will have more than ten parking spaces, the same regulation applies. It should be noted however that developers installing covered car parks are not required to install charge points, only that cable routes are provided for potential future installs [9].

4.1.5 Rapid Charging Fund [10]

The RCF aims to expand ultra-rapid level 3 DC charge points across the motorway network. The £70 million RCF pilot focused on identifying barriers to achieving this objective, particularly by funding upgrades to grid connections at motorway service areas in England, where high costs hinder EV charge point expansion. The £950 million fund is open to motorway service operators to upgrade electricity connections along motorways and A-roads so as to enable private sector investment in new charge points. The Office for Zero Emission Vehicles also supports the private sector through funding to achieve a minimum of 6 level 3 DC high-powered charge points at every motorway service area in England, originally by 2023 but this deadline has been pushed out into the future [11].

4.1.6 Local Electric Vehicle Infrastructure Fund [12] and On-Street Residential Chargepoint Scheme [13]

The LEVI Fund and ORCS focus on providing funds to local authorities to deploy on-street charge points. LEVI, which is currently in the process of delivery, provides funding in the form of capital to contribute to the cost of delivering charge points. Funding is also provided in the form of a capability fund for local authorities to recruit individuals to plan and deliver charge points. LEVI's aims are to deliver a step change in how on-street charge point infrastructure is deployed across England and to accelerate the 'commercialisation of, and investment in, the local charging infrastructure sector' [12]. The overall fund, worth £400 million [14], included a pilot phase that continues to be delivered in nine local authorities [15]. The current phase of funding (as of September 2024) is being allocated to 81 local authorities [16] of which 44 have received payment to begin the rollout [17]. ORCS, which provided a total of £52 million [18], remained open until March 2024 and funded grants of up to £200,000, covering 50% of project costs. The local authority needed to match the additional 50% and confirm that they would be able to complete the project by March 2025. Due to smaller capped funding per project, ORCS operated on a smaller deployment scale when compared with LEVI.

4.1.7 Public Charge Point Regulations [19]

The Public Charge Point Regulations 2023 focuses on four key areas related to the consumer experiences. These are to ensure that consumers can easily locate a charge point, that they can easily pay to charge across multiple public charge points, that they have confidence that when they reach the charge point it is in good working order and that they are able to compare prices across various charge point operator (CPO) networks. A summary of the regulations is captured in Table 4.1.

Table 4.1 Summary of public charge point regulations

Policy area	**Requirements**	**Lead time from regulations coming into force**
Contactless	New public charge points of 8 kW and above and existing charge points of 50 kW and above must offer contactless to consumers. Proprietary networks that open their charge points for public use will have 1 year from the date that the charge point becomes public to offer contactless.	1 year
Roaming	CPOs must enable consumers to pay through at least one roaming provider at their charge points.	2 years
99% reliability	Rapid charge points must be 99% reliable, measured as an average across each charge point operator's level 3 DC rapid network. Information on reliability compliance must be published on the CPO's website.	1 year
Helpline	A free to use 24/7 staffed telephone helpline must be available and advertised at all charge points.	1 year
Open data	All data must be accurate and charge point operators must use the open charge point interface (OCPI) to hold and share their data. Reference and availability data must be made publicly available and in a machine-readable format. Government bodies, distribution network operators (DNOs), transmission owners and electricity system operators must have access to all data.	1 year
Pricing metric	The maximum price of a charging session must be displayed clearly in pence per kWh. The price can be displayed either on the charge point or through a separate device.	Immediately

Source: Department for Transport, Office for Zero Emission Vehicles. *Public Charge Point Regulations 2023 guidance* [online]. 2023. Available from https://www.gov.uk/government/publications/the-public-charge-point-regulations-2023-guidance/public-charge-point-regulations-2023-guidance. This information has been licensed under the Open Government Licence. To view this licence, visit https://www.nationalarchives.gov.uk/doc/open-government-licence/

4.1.8 British Standards Institute (BSI) PAS 1899:2022 [20]

The British Standards Institute Electric Vehicles Accessible Charging Specification – PAS 1899:2002 sets out the specifications that support the building of inclusive charge point infrastructure in the United Kingdom. PAS 1899 aims to ensure inclusive design principles that address accessibility for all, including people with a range of disabilities. PAS 1899 outlines minimum accessibility requirements for all public charge points and provides best practice guidance for charge points near accessible parking bays. The roles and responsibilities for public charge point accessibility involve multiple stakeholders, including designers, local authorities, landowners, and operators. The primary responsibility for meeting PAS 1899 requirements falls on the buyer of the charge point, who may be an operator, service provider or local authority. Buyers must ensure that charge points are designed and installed in line with accessibility standards and that surrounding environments are appropriately modified. PAS 1899 also affects other stakeholders, like designers and manufacturers, who need to adhere to design and maintenance requirements. Although voluntary, PAS 1899 serves as a benchmark for accessible charge point provision in the United Kingdom and potentially abroad. PAS 1899 looks at the following:

- The physical aspects of the surrounding environment to the fixed charge point.
- The location, placement and spacing of charge points within the streetscape/public realm and relative to other infrastructure or objects.
- Factors to be taken account of in the design and specification of accessible charge points and their more immediate surrounding areas.
- Information, signals and indicators provided.

PAS 1899 covers tethered and non-tethered charge points, and those that can be attached to existing street furniture like charge point bollards/columns and wall mounted/lamp column-mounted chargers.

4.1.9 Progress to date and challenges ahead

The various strategies and policies have had success in the United Kingdom, resulting in an increase in the number of level 2 and 3 DC charge points. The RAC, using data from Zapmap and DfT, noted that in January 2016 there were around 3,500 chargers, whereas in March 2024 there were just under 60,000 charge points. Over this period, the proportion that were rapid and ultra-rapid increased [21]. The National Infrastructure Commission believes that the United Kingdom is on target to meet its 300,000 charge points required by 2030 provided that growth continues at the current trajectory. The Commission does however flag that in later years (and specifically 2029), this trajectory will be difficult to maintain [22]. A detailed report produced by the Environment and Climate Change Committee of the House of Lords raised a number of challenges ahead with charge point growth [4]. These included that the geographic spread of charge points means there are areas where there are significant gaps. Figure 4.1 highlights the difference between regions and also challenges the National Infrastructure Commission's narrative by showing that charge point deployment is not keeping up with EV adoption rates. The report also

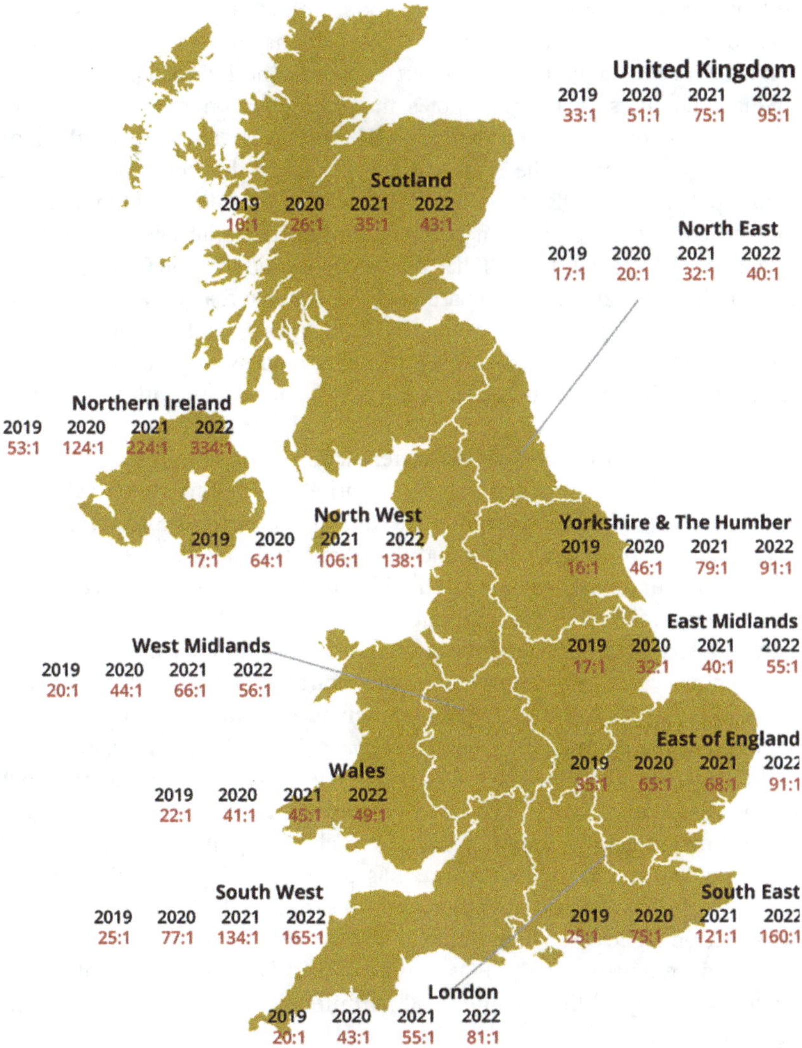

Figure 4.1 *Change in ratio of cars to public charge points across the United Kingdom 2019–2022 [Source: Environment and Climate Change Committee, House of Lords. EV strategy: rapid recharge needed [online]. 2024. Available from https://publications.parliament.uk/pa/ld5804/ldselect/ldenvcl/51/5102.htm [Accessed 06 September 2024]. This information has been licensed under the Open Government Licence. To view this licence, visit https://www.nationalarchives.gov.uk/doc/open-government-licence/].*

noted that a lack of confidence in charge point infrastructure, alongside high EV upfront costs, is impacting on adoption rates which may have an impact on demand and utilisation in the long run without a significant increase in charge point deployment numbers. The report notes that the focus on rollout should be on 'lower-powered, on-street AC charging' that involves 'cheaper installation for operators, less demand on the grid, and cheaper prices for consumers'. Concerns were raised that once ORCS and LEVI funding ends, local authorities would struggle to continue deploying at pace. Regarding en route charging, the report noted that the delivery of the RCF has been delayed and that the Government must ensure that funds are allocated so that the whole strategic road network benefits, not just specific motorway service stations. With regard to planning, the report highlighted that 'simplified and standardised planning processes are needed' to prevent confusion for CPOs when rules are interpreted by local authorities differently in different geographies [4].

A major concern raised was the difference in the cost of charging between private and public charging and that this is primarily due to higher value-added tax (VAT) rates on public charging, the ability for domestic charging to take advantage of cheap off-peak rates and that domestic rates can be locked in and fixed more favourably when compared with fluctuating prices charged by CPOs. Smart charging and vehicle to grid require further focus from the Government to give the industry confidence and prevent the infrastructure needing to be replaced in the coming years. An additional major concern is both the ability of the transmission network to be upgraded in anticipation of increased demand and the ability of DNOs to upgrade the local grid at pace. Their role is critical in ensuring that CPOs are able to draw on the power they need locally to support public charging. DNOs also focus on prioritising different grid upgrades and need to focus on fast tracking those related to charge point deployments. The Government, according to the report, has a role to play here in committing resources to anticipated grid upgrades so as to support DNOs in this process. The report's section on DNO challenges concludes that 'DNOs and other stakeholders such as local authorities and CPOs could benefit from improved data sharing, online tools to support the planning and permitting process and more pro-active engagement between DNOs and customers' [4].

4.1.10 Anticipated changes in policy and strategy

The various challenges highlighted in Section 4.1.9 demonstrate that if the Government is to meet its charge point deployment targets by 2030, changes in policy and strategy will need to take place. The list outlined below, though not exhaustive, gives a view on the potential changes that may take place:

- **Further funding for local authorities to roll out additional level 2 on-street infrastructure**: this recognises that local authorities play a critical role in localised deployment and for them to continue at pace, specifically in deploying on-street charge points, requires additional funding in the form of something akin to LEVI phase 2.

- **Further investment in rapid level 3 DC charge points where there are gaps in the network**: ensuring level 3 DC rapid charging is available to all is key to en route charging. Investment in infrastructure where there are gaps in the network is critical to enabling increased adoption, especially in strategic routes running through rural areas with little existing infrastructure.
- **Equity of public and private charging costs**: this would ensure that charging is fairer for all and could be achieved by removing VAT on public charging; however, this is highly unlikely due to lost government revenue. An alternative would be to increase VAT on home EV charging tariffs though there would be considerable resistance to this from consumers.
- **Major changes to how DNOs plan for and deliver upgrades**: a key barrier to future progress is DNOs not being able to move at pace in terms of upgrading local grid connections to handle the additional load from charge points. It is likely that changes will be made to how DNOs prioritise grid upgrades with regional bodies like combined authorities having greater say in planning and upgrade prioritisation. Further public investment in grid upgrades is also expected to ensure the pace of deployment increases.
- **Further promotion of and investment in smart charging and vehicle to grid**: smart charging is key to managing the load on the grid, whereas vehicle to grid should in theory help smooth the peaks and troughs assuming enough vehicles are plugged in and allow for remote control of their batteries. Their role is important as part of the future transition of the grid is to renewables. This can only be achieved if smart charging and vehicle to grid becomes more common through greater investment in its capabilities and greater adoption by charge point manufacturers.

Though not technically related to charge point infrastructure, per mile road-based charging for EVs may be introduced in the future to recoup lost revenue from fuel excise duty as a result of greater adoption of EVs versus internal combustion engine (ICE) vehicles. The impact this might have on charge points, however, is that it may result in slowing adoption rates which, in turn, will push demand and utilisation down, thereby reducing the pace at which charge point rollout needs to take place. The obvious negative of this would be that road-based transport decarbonisation would slow with the potential outcome being the Government missing its overarching net zero targets.

4.2 The European Union

The European Union has an important role in setting the strategy and policy direction for charge point infrastructure for the 27 Member States. It is each State's responsibility to interpret the policy and strategy direction set by the European Union and apply this in their specific geographic area. Sections 4.3 and 4.4 covers how two Member States, France and Germany, have aligned their policies, strategies and funding with that set by the European Union. This section covers the policies and strategies specifically set by the Union, as developed by the European Commission, which is its legislative body.

4.2.1 European Green Deal [23]

Introduced in 2021, the European Green Deal set out a path to Europe becoming climate neutral by 2050. There are several strands to the Green Deal, though the focus on transport is to make it sustainable for all. Regarding charge point infrastructure, the Strategy states that targets regarding the number of charge points to be deployed will be mandatory and that sufficient public charging capacity will be deployed to meet the demand for the rise of EVs as adoption rates increase across the 27 Members States over the next 5 years. Delivery of the charge point infrastructure aspects of the European Green Deal is enabled by two key pieces of legislation, the Alternative Fuels Infrastructure Regulation (AFIR) and Fit for 55.

4.2.2 Fit for 55 [24] and Alternative Fuels Infrastructure Regulation (AFIR) [25]

Fit for 55 makes the target of reducing European Union emissions by at least 55% by 2030 a legal obligation and also states that the European Union over the coming years will work on new legislation to make it net zero by 2050. There are many focus areas that form part of the Fit for 55 package; however, looking specifically at charge point infrastructure, the targets set include the requirement for installation of level 3 DC rapid charge points along every 60 km of the Trans-European Transport Network (TEN-T), that users of charge points are able to pay easily and that infrastructure across the European Union is easy to use. Looking in further detail at the ease of use of charge points, the legislation states that different payment and price display options must be taken into account. The legislation also offers some flexibility in terms of the spread of charge points along the TEN-T, allowing some leeway to adapt the total power of charge point hubs for sections with low traffic volumes and that the distance between charge point hubs can be extended along sections with very low traffic. Broadly speaking, Fit for 55 sets the direction of travel. AFIR provides greater legislative detail.

The original AFIR legislation was enacted in 2014 as Directive 2014/94/EU and set out the European Union's approach to public charge point deployment. The updated AFIR legislation repealed Directive 2014/94/EU and established new regulations for public charge points. Core elements are that deployment of well signposted charge point hubs along the TEN-T should be located on or within 3 km driving distance of the nearest exit of a TEN-T road. The charge point hubs should be complemented with publicly accessible fast charge points in urban nodes and that charge points send and receive data in real time, enabling the flow of information between key stakeholders. Further elements include that the charge points should be interoperable, using smart metering to optimise charging and support bidirectional charging and that price transparency is key as drivers should be provided with accurate price information. Furthermore, roaming capability is mandatory and that Member States should monitor the development of the charge point market across their geography. Users of charge points should be able to pay easily and conveniently to charge their vehicle. Charge points should be designed and built with accessibility in mind (in terms of individuals with reduced mobility or disabilities). Member States should upgrade the power grid to support the additional power load from charge points and ensure that technical specifications for charge points are standardised across the European Union. Specific details include that:

- Charge point hubs along the TEN-T core road network are deployed in each direction of travel with a maximum distance of 60 km between them:
  - By December 2025, each hub will offer charging rates of at least 400 kW and include at least one charge point with an individual power output of 150 kW.
  - By December 2027, each hub will offer charging rates of at least 600 kW and include at least one charge point with an individual power output of 150 kW.
- Along the TEN-T comprehensive road network, the same deployment distance applies with differing targets:
  - By December 2027, along at least 50% of the length of the TEN-T, each charge point hub offers a power output of at least 300 kW and includes at least one charge point with an individual power output of at least 150 kW.
  - By 31 December 2030, each charge point hub offers a power output of at least 300 kW and includes at least one charge point with an individual power output of at least 150 kW.
  - By 31 December 2035, each charge point hub offers a power output of at least 600 kW and includes at least two charge points with an individual power output of at least 150 kW.
- Price information shared with the user includes price per kWh, price per minute, price per session and any other price component that applies.
- Every 3 years from 2024 onwards, Member States shall assess how the deployment and operation of charge points could enable EVs to further contribute to the flexibility of the energy system, including their participation in the balancing market, and to the further absorption of renewable electricity.
- CPOs must make available for their charge points:
  - Static data including location, number of connectors, number of parking spaces for people with disabilities, contact information for the CPO and opening hours.
  - Charge point unit data including charge point identification codes, types of connector, type of current, maximum power output and vehicle type compatibility.
  - Dynamic data including operational status, availability, ad hoc price and how renewable the electricity supplied is.
- All charge points are equipped at least with socket outlets or vehicle connectors of Type 2, Mennekes (for level 2 charging) and connectors of the

combined charging system, CCS/Combo 2 (for level 3 DC high power charge points). AFIR does not prohibit other connector types (e.g., CHAdeMO) [26].

Each Member State must report on their progress annually including details on the total aggregated charge point power output and the number of publicly accessible charge points deployed and the number of EVs and plug in hybrid vehicles registered in their territory on 31 December of the previous year.

4.2.3 *Sustainable & Smart Mobility Strategy [27] and Recharge and Refuel [28]*

Published in 2021, the Sustainable & Smart Mobility Strategy provided numerical targets for the number of charge points to be installed across the Member States. This stated that 1 million public charge points are required by 2025 and 3 million by 2030. This figure was increased to 3.5 million in 2022 [29]. These targets are being delivered by Recharge and Refuel, a European flagship programme under the Recovery and Resilience Facility.

4.2.4 *Progress to date and challenges ahead*

The European Automobile Manufacturers' Association (ACEA) produced a report analysing the progress made to date regarding charge point deployment. As of April 2024, just over 632,000 charge points have been installed by the end of 2023, serving 3 million EVs and 410,000 are needed annually to reach the 3.5 million target by 2030. Distribution across Member States is uneven, with 61% of all charge points found in the Netherlands, France and Germany. Only 13.5% offer level 3 DC charging capabilities and this needs to expand as a percentage of the overall total to meet the target. The report also notes the importance of the swift implementation of the AFIR to make sure that infrastructure deployment keeps pace with demand [30]. Transport and Environment, an influential not for profit organisation that specialises in analysing and shaping government policy, disagrees with ACAE. They note that in most countries across the European Union, the annual AFIR targets required to meet the overarching 2030 target have been met, with Bulgaria having delivered over 677% of its required charge points in 2024. The 27 Member States as a whole are slightly above their 2024 charge point target and slightly below their 2025 target. The report shares the same point as the ACAE, that investment needs to continue and increase to reach the 2030 target but differs in that Transport and Environment are confident that the European Union can meet this challenge. The only caveat raised is that growth is inconsistent across geographies, with Greece and Ireland needing to invest heavily in infrastructure to reach their 2024 and future target [31].

It is not only charge point deployment which is the main challenge holding back transport decarbonisation but rather EV adoption rates themselves. The European Union, like the United Kingdom and other Western nations, is experiencing a blip in EV uptake. EV sales as a share of total vehicles sold across all

Member States have dropped from an average of 15% in 2023 to around 12% at the end of quarter one of 2024 [32]. This, it is believed, is the result of high EV costs when compared with ICE equivalents and the fact that drivers are deciding to own their cars for longer, thus putting off a purchase. It is likely that part of this challenge will be overcome when more affordable EVs are launched from 2025 onwards; however, there is a risk that a more permanent slow down sets in as vehicle manufacturers switch production to ICE vehicles to maintain profits and counteract the fall in EV sales. A further challenge may come in the form of pressure from car manufacturers being placed on the European Union to water down its EV mandate or push it back beyond 2035. This pressure might force adoption rates downwards, reducing demand for charge points, thereby slowing down deployment rates. European car manufacturers have already had a degree of success in lobbying Member States and the European Union to put tariffs on the import of affordable Chinese manufactured EVs that were managing to make inroads in car sales. That being said, it is unlikely that the influence of car manufacturers will succeed in driving a major policy shift away from EVs and charge point infrastructure.

4.2.5 Anticipated changes in policy and strategy

It is unlikely that the European Union will change policy and strategy direction. What will likely happen is tweaks and changes to the existing direction of travel. This might include:

- **Relief for European car makers struggling to make the transition to EV**: this might take the form of subsidies or partial exemptions from mandated EV sales targets, though it is unlikely that this will reduce demand for charge points or the need to continue charge point growth as per AFIR and the Mobility Strategy.
- **A targeted funding programme for those countries struggling to meet their charge point targets**: this could take the form of something similar to LEVI or the RCF in the United Kingdom.
- **Extension of the AFIR and Mobility Strategy driven charge point deployments beyond 2030**: this may occur to bring this in line with the sales ban of ICE vehicles which is scheduled to take effect in 2035.

4.3 France

France has a supportive policy and strategy environment which enables the installation of public charge points. Its aims and objectives are articulated through various policies, strategies, funding streams, laws and plans.

4.3.1 Law on Mobility (LOM) [33]

The Law on Mobility, enacted in 2019, established three main pillars with the overarching aim to promote more accessible, more affordable and cleaner transport modes. The third pillar, initiating the transition towards cleaner mobility, encompasses the rollout of charge points and sets the goal of road transport carbon neutrality by 2050, enshrined in French law. To increase the installation of charge points, the law sets out the target of increasing the number of public charge points by a multiple of five by 2022, making charge points mandatory in some car parks, establishing the right to access a charge point and aiming to reduce the cost of installation by half. Specific details of LOM include:

- Setting the target of 100,000 charge points installed by 2022.
- Making it a requirement that all public buildings including those with large car parks must allocate a number of parking bays to charge points.
- Increasing the responsibility of municipalities and local authorities in relation to planning and deploying charge point infrastructure. This includes on-street and off-street locations.
- Encouraging the installation of destination charge points at shopping centres and transit hubs.
- Making it a legal requirement that charge point hubs are installed at every 80–100 km along French highways and major roads.
- Ensuring interoperability of charge point networks through a common payment system and user interface.

The LOM also introduced various financial incentives to businesses and local authorities installing charge points including the ADVENIR Programme (see Section 4.3.4). Local authorities are also able to apply for grants to fund charge point deployment, with funding partly targeted towards low demand areas so as to ensure charging equitability and prevent the neglect of rural less populated areas. An additional part of LOM includes simplifying the installation process including reducing the complexity regarding obtaining permits and approvals. CPOs are encouraged through LOM to form public–private partnerships to deploy charge points.

4.3.2 France Recovery Plan (France Relance) [34]

Launched as a result of the Financial Act in 2021, the French Recovery Plan (FRP) comprised a budget of €100 billion, sought to enable the economy to return to its pre-pandemic level by the summer of 2022 and to improve the competitiveness of French businesses, while preparing industry for the long-term challenges ahead. Charge point deployment formed part of the ecology theme of the FRP under the strand of supporting the transition to clean vehicles. The FRP set the target to reach 100,000 charge points installed in public service areas on motorways and the national road network by 2022. To achieve this, financial assistance was provided in the form of €100 million of funding. This subsidy was designed to encourage CPOs and local authorities to deploy charge points by covering part of the installation cost.

4.3.3 EV Infrastructure Charging Programme (ADVENIR) [35]

ADVENIR, originally launched in 2016, focused on charge point installation in private car parks; however, the renewal of the Programme in 2018 set out to install

13,700 new charge points, including 3,000 on public roads. ADVENIR was renewed again in 2020. The renewed Programme involved establishment of a €320 million budget to install 175,000 charge points by 2025 which can be accessed by local authorities and CPOs. The Programme is based on the Energy Savings Certificate mechanism and is financed by companies operating in the energy sector [36]. The grant available for charge points is capped at 30% of the supply and installation cost and can be no greater than €1,000 excluding VAT for slower level 2 charge points (3.7–11 kW) increasing to €9,000 excluding VAT for ultra-rapid level 3 charge points (greater than 140 kW). An additional grant subsidy of €300 can be accessed for in demand charge points [37].

4.3.4 Building Control Regulations [38]

Any industrial, residential or commercial building with car parking spaces where a building permit has been applied for since the start of 2017 requires that wiring and ducting is installed to enable the deployment of charge points. This was strengthened further by the LOM which extended the wiring requirement in 2021 to buildings under renovation. In addition to this, Right to Charge (*Droit à la prise*) was introduced in 2021. This gives any tenant or owner the right to install a charge point in an open or enclosed car park where they have the right to park [39].

4.3.5 Progress to date and challenges ahead

France is making major progress regarding charge point deployment and has met its AFIR growth target for 2024 and 2025. For 2026, France has met 90% of its growth target [31]. 400,000 are needed by 2030 to meet the overall AFIR target [40]. In mid-2023, France reached the milestone of 100,000 charge points and will need to install roughly 50,000 a year to reach the 2030 target. Data for 2023 indicated that by the end of that year there were 118,000 charge points [41]. As of April 2024, France has installed just over 129,000 charge points and is on track to meet its target of 400,000 charge points by 2030 if current installation rates continue [42]. This is supported by analysis by Transport and Environment, which is confident that the European Union, and France, whose targets feed into those across all Member States, will meet its overall target of 3.5 million charge points by 2030 [46]. In terms of level 3 DC rapid charge points with charging speeds greater than 50 kW, France is also broadly on track to deliver its target of 25,000 charge points by 2027, having installed 19,000 by April 2024 [48]. Assuming the funding and support regime provided by the French Government continues, charge points deployment should continue at pace. Furthermore, France, unlike other European nations, can rely on affordable energy as a result of its combination of renewable energy and nuclear power. This means it is not exposed to fluctuations in global energy prices, ensuring consistent and affordable per kWh charging costs. This should mean that this factor does not have an impact on EV adoption and, ultimately, will help support demand and utilisation. However, France, like other Member States, continues to see adoption influenced by a lack of affordable EVs. The rise of the sub-€25,000 EV in 2025 should go some way to ensure that the adoption of EVs continues, pushing up demand [43]. However, a recent news story about European car manufacturers from September 2024 [44] puts this at risk as they threaten to stop the development of more affordable EVs to focus on ICE vehicles. Time will tell whether this occurs and adoption rates fall, impacting demand for charge points.

A key area of challenge for France is the lack of an accessible charge point standard akin to PAS 1899:2022 in United Kingdom. AVERE, The European Association for Electromobility, recommends that this is tackled by proposing that the European Union establishes a standard across all Member States. This would ensure 'harmonisation of European accessibility requirements for the charging infrastructure' and would allow manufacturers of both automobile and charging equipment infrastructure to design and produce charge points in compliance with the standards that are in force in all Member States [45].

4.3.6 Anticipated changes in policy and strategy

Provided financial incentives and government support continues, it is anticipated that there are only minimal changes to French charge point policy and strategy. The main likely change is the adoption of an accessibility standard for charge points. The policy and strategy changes relevant to the European Union highlighted in Section 4.2.5 also apply in France with the exception of targeted funding as this is not needed due to the financial incentives currently available. A less likely, though potential risk is the rise of far-right political parties in France who oppose green policies including the promotion of EVs over ICE vehicles. Their rise to power, which came close to occurring in the Summer of 2024 following President Emanuel Macron's snap election, would have had a major impact on EV adoption and charge point deployment [46]. This did not come to pass but remains a risk in France and other European countries.

4.4 Germany

Like France, Germany has a number of policies, strategies and funding schemes that enable charge point deployment. In relation to funding, a number of programmes are no longer available.

4.4.1 Charging Infrastructure Masterplan [47]

Germany developed a charge point masterplan, its aim being to integrate charge points with the electricity system. As part of this, the German Ministry of Transport and Digital Infrastructure is investing €6.3 billion through various schemes and programmes to establish a publicly accessible charge point network which is intended to foster private investment in further infrastructure. Progress in terms of installation is tracked via the National Control Centre for Charging Infrastructure, established in 2019.

4.4.2 Deutschlandnetz Initiative [47]

Under the Fast Charging Act, this commits the German Ministry of Transport to installing 1,100 charging hubs as part of the initiative. These hubs need to be capable of charging a vehicle at rates of at least 300 kW. The Initiative targets two separate en route charging scenario. The first is in specific regional locations, whereas the second targets charge point deployment on the autobahn network. Installers who are successfully awarded the work through a procurement process are then given financial incentives to install and operate the infrastructure. In September 2023, ten bidders were successful and are installing 8,000 charge points across 900 locations. It should be noted that in Germany, charge points installed next to motorways are subject to federal law, whereas those next to state and provincial roads fall under local state law.

4.4.3 Installing on motorways [47]

Motorway service stations are included in the Government's level 3 DC charge point rollout plans. Ninety per cent are run by private operators. The right to build charge points at service stations requires a public concession and formal consent must be granted.

4.4.4 Installing alongside state and provincial roads [47]

Where publicly accessible, charge points are being installed in locations alongside state or provincial roads installation must comply with zoning laws, ease of use and traffic safety and number of parking spaces to be converted to charge point bays. They must also comply with urban design principles, charge point accessibility, utilisation and environmental protection measures. A special use permit is required for charge point operation. During installation, The Civil Engineering Office might be required to approve charge point installation in underground car parks and the Monument Protection Authority if installation is located near a listed historic building.

4.4.5 Privately owned locations with public charge points [47]

Federal state construction law allows for the construction of charge points on privately owned land, though local construction laws must be followed. A special use permit is not required on private land.

4.4.6 Installing charge points in flats and apartments [47]

In 2021, the Condominium Property Modernisation Act was introduced, which enables tenants to install charge points in underground car parks without their landlord's consent. Flat owners tend to have no right to build a charge point without approval from the apartment block owners' association.

4.4.7 German charging station regulations [47]

The charging station regulations that came into force in 2023 state that all charge points must ensure payment can be made via bank card, either via a physical terminal or, where possible, via contactless. Charge points must also adhere to the requirements set in the Payment Services Supervision Act in relation to customer authentication, primarily granting customers using payment authentication services the same rights as they'd expect from their bank. Furthermore, the regulations cover minimal technical and accessibility standards, interoperability and the ability for charging without membership to a CPO (ad hoc charging). The regulations cover similar requirements to AFIR around pricing.

4.4.8 Subsidies and incentives [47]

The E-charging Infrastructure for Electric Vehicles Programme in Germany (Öffentlich zugängliche Ladeinfrastruktur für Elektrofahrzeuge in Deutschland) was introduced in 2021 and establishes a €500 million fund that runs until 2025 with the aim being to install at least 50,000 charge points, 20,000 of which are level 3 DC charge points [47]. At the time of writing, no further funding rounds are planned. A further €400 million was made available in autumn 2023; however, the German national budgetary crisis meant that only €12.3 million of funding was allocated [48]. Due to its federal structure, there are other options to seek funding at state level. For example, the Funding Programme for E-Charging Infrastructure for Electric Vehicles in Bavaria ("Förderrichtlinie Ladeinfrastruktur für Elektrofahrzeuge im Bayern") exists with the aim being to expand publicly accessible charging infrastructure. Refreshed in 2021, the Programme establishes a €20 million fund that runs until the end of 2024. The aim of the fund is to finance as many charge points as possible under a regime where a maximum funding amount is granted based on the type of charge point, with level 2 charge points receiving less than level 3 DC charge points. Funding is also available for grid connection. To receive the grant, applicants must cover 40% of the cost [49]. Funding prerequisites include that the charge point must be publicly accessible and powered using renewable energy, that it must operate for 6 years and that it must operate to the technical specifications outlined in the funding guidelines.

Further funding for charge point deployment is available via the European Investment Bank and German state-owned bank Kreditanstalt für Wiederaufbau (KfW) with the latter offering subsidies up to the value of €50 million per project and the former grants of €30 million for environmental protection projects. Charge point infrastructure falls into this category [47].

4.4.9 Progress to date and challenges ahead

Similar to France, Germany is subject to the AFIR growth targets and has met those for 2024 but is slightly behind its target for 2025 and further behind for 2026 [31]. 1 million is needed by 2030 to meet the overall AFIR target [50]. As of September 2023, over 97,000 level 2 charge points and 18,000 level 3 DC charge points had been installed [47]. As of June 2024, there are over 130,000 publicly accessible charge points of which 19% are level 3 DC charge points. Bavaria has the highest number of charge points followed by Nordrhein-Westfalen. This is an increase of roughly 11,000 charge points when compared with 2023 [51]. Bearing in mind that the overall target according to AFIR is 1 million charge points, it is clear that Germany is not on track and is likely to significantly miss this target. Around 145,000 charge points (roughly the same total number of charge points installed between now and when records started in 2015) would need to be installed annually to reach the 2030 target. Funding is also constrained as mentioned in Section 4.4.8. This would indicate that the pace of growth is unlikely to increase. Furthermore, Germany, like other European countries, is experiencing faltering EV adoption. In 2023, 700,200 EVs were registered, 16% less than the previous year and below the goal for 2023 of 1 million EVs registered. This, it is believed, is due to the ceasing of incentives to encourage the adoption of EVs [52]. The German economy is also experiencing near flat growth, with current estimates showing projections of 0.2% for 2024 [53]. The challenge ahead is how to increase the charge point deployment rate with limited funding/subsidies, in a stagnant economic climate where EV adoption is faltering. Adding further to this is a political environment where a surge in right-wing anti-green parties (specifically Alternative for Germany) may make this even more difficult [54].

4.4.10 Anticipated changes in policy and strategy

The current challenges Germany is facing as outlined in Section 4.4.9 look to continue for the next couple of years unless a major change in fortune occurs. What is proven by other countries in Europe is that a generous funding regime is required to foster charge point growth, accompanied with financial incentives that encourage EV adoption. Therefore, Germany must change its policy and strategy approach and identify further funding to subsidise charge point growth and encourage EV adoption. This will obviously be difficult in a challenging economic climate; however, it is the only option. The German Government will need to find ways to raise revenue to fund these additional incentives. One avenue might be additional taxation of its domestic car manufacturing market. This is unlikely because this industry has a powerful voice in Germany and is also experiencing its own challenges, with the recent example of Volkswagen threatening to stop production of cars in Germany altogether [55].

4.5 The United States

The growth of charge points in the United States has, up until fairly recently, been driven by individual states championing charging infrastructure, with California being a good example of a state promoting rollout. However, in the past 3 years, the Federal Government began to play a major role in financing infrastructure deployment. This had an impact, with the Infrastructure Investment and Jobs Act (IIJA) having kicked off a period of increased support, investment and growth. That being said, the election of President Trump in November 2024 resulted in this federal support coming to a halt.

4.5.1 Infrastructure Investment and Jobs Act (IIJA) [56]

In 2021 the $1.2 trillion IIJA was launched with the aim of fostering growth and creating jobs. As part of this Act, $7.5 billion of funding was set aside to support the rollout of a national public charge point network. The fund aimed to deploy

500,000 charge points. $5 billion was allocated to states as part of the National Electric Vehicle Infrastructure Formula Programme (NEVI). Funding was at state level and was intended to support the deployment of charge points at strategically important points along alternative fuel corridors (mainly interstate highways). Installed charge points would form part of a broader interconnected network. Remaining funding as part of NEVI was made available for deployments along public roads and in other public locations. The $2.5 billion of funding was distributed by a competitive grant programme aimed at deployment of charge points along designated alternative fuel corridors and publicly accessible locations in communities. The funding was split in half for each of these different use cases with the latter targeted specifically at rural, underserved and overburdened communities. Eighty percent of funding was provided by the Federal Government, with the remaining 20% provided by private and/or state funds [57]. The policy requirements of NEVI included that where charge points are installed along an interstate highway, this should be done every 50 mi and that where these are deployed, there should be a minimum of 4 level 3 DC charge points with charging rates of 150 kW. Once these are installed along a route or area, any remaining funding can be used to deploy charge points along public roads and in other publicly accessible locations, including parking lots, public buildings and rest areas.

4.5.2 Inflation Reduction Act (IRA)[58]

The IRA came into force in 2022 with the primary aim being to encourage the transition to EV. A bulk of the Act related to incentivising EV uptake through tax credits for the transition of private vehicles and commercial fleets and the development of clean heavy goods vehicles. The parts of the Act relating to charge points included investment tax credits for charge point installation. This provided 30% tax credits up to $100,000 and 20% over $100,000 to cover the upfront cost of installing charge points for businesses and property owners. Residential and commercial installations could take advantage of these tax credits. Tax credits were accessed when installing public charge points provided the location was in an eligible census tract. These were generally, low-income communities for purposes of the New Market Tax Credit, or a location that was not in an urban area.

4.5.3 State initiatives

A number of states have charge point strategies that complement the federal programmes outlined in Sections 4.5.1 and 4.5.2. Two examples of states include California and New York. California's charge point strategy, entitled California's Deployment Plan for the National Electric Vehicle Infrastructure Programme, has the goals to install, using NEVI funding:

- 250,000 charge points, including 10,000 level 3 DC charge points by 2025.
- 1.2 million charge points, including 37,500 level 3 DC charge points for light goods vehicles and 157,000 charge points for medium and heavy good vehicles by 2030.
- Charge points at pace whilst ensuring equitable outcomes [59].

Grants, partly funded through NEVI and other funding sources, are provided via California Energy Commission's Clean Transportation Programme (CTP) which is published annually as part of an investment plan. During the year 2021–2022, \$842.2 million of funding was made available. For the years 2022–2023 and 2023–2024, \$372.8 million was made available. To deploy NEVI funding, California divided the alternative fuel corridors based on the interstate highways network into multiple segments based on an analysis of gaps in the current network, future charge point needs and geography whilst also reflecting the need to equitably deploy charge points. It is not clear whether the Plan will continue now that NEVI has been stopped.

New York State's EV strategy, Charge NY, aims to expand the charging network by installing 10,000 public charge points by 2025 and 50,000 by 2030 with at least 35% being deployed in disadvantaged communities. The Strategy seeks to build a closer relationship between public and private entities to share cost and increase the pace of deployment [60]. The scheme initially invested \$750 million in charge point deployment in 2020 [61]. New York State took advantage of the NEVI Programme and is expanding the network of level 3 DC charge points along key travel routes, particularly north of Interstate 84, extending to the Canadian border and Buffalo. The budget sought from NEVI covering the period 2022–2024 was \$100.7 million and supports a total of 2,034 mi of charging corridors across the state. In addition to this, Charge Ready NY 2.0 has been launched which used NEVI funding to encourage deployment by business, organisations and municipalities. \$4,000 per charge point is available to cover installation in public locations. Furthermore, the EV Make-Ready Programme offers rebates up to 90% of electrical infrastructure and installation costs for commercial level 2 and 3 DC charging stations at workplaces, public locations, multifamily residences and locations used by fleets. The National Grid is also supporting rollout via a \$200 million fund for level 2 and 3 DC charge point deployment projects. Other smaller funding schemes are available. Those mentioned in this section are the largest and most subscribed [62]. It is not clear if these programmes will continue now that federal funding has stopped.

4.5.4 Other federal programmes

There are various other federal funding programmes that aim to support the rollout of charge points. The below list is not exhaustive but gives an overview of a number of these. It is not clear what impact the Trump administration will have on these schemes:

- **Charging and Fuelling Infrastructure Discretionary Grant Programme administered by the Federal Highway Administration** [63]: this fund aims to deploy publicly accessible charge points in strategic locations. There are two strands to the Programme. The first is targeting community charge point deployment on public roads, schools, parks and in publicly accessible parking facilities. The second is deploying charge points along alternative fuel corridors. The Programme opened in 2024 and was funded through the Bipartisan

Infrastructure Law. This fund provided $2.5 billion over 5 years for this Programme to strategically deploy EV charge points in urban and rural communities in publicly accessible location. This includes downtown areas and local neighbourhoods, particularly in underserved and disadvantaged communities. 80% of the funding was provided by the Federal Government. The remaining 20% came from the applicant.

- **Port Infrastructure Development Programme (PIDP) administered by Maritime Administration** [64]: established by the Maritime Administration, the PIDP is intended to improve charge point infrastructure in port areas. Eligible projects are those located in port boundaries or directly associated with port operations and multi-modal connections. The grant total for 2022 to be spent in the fiscal year 2023 was $662 million with projects ranging in size from $1 million to $165 million. The funding is primarily targeted at light, medium and heavy goods vehicles.
- **Clean Cities and Communities Programme** [65]: Clean Cities and Communities (CC&C) is a United States Department of Energy (DOE) partnership to advance clean transportation nationwide. CC&C was established in 1993 to reduce greenhouse gases. Since then, emissions have been reduced by 72 million tons. Various projects across a range of clean transportation areas have been funded with examples including the 'Solutions for curbside-charging electric vehicles for planned urban growth' project run by the University of North Carolina at Charlotte and 'Upper Midwest Inter-Tribal Electric Vehicle Charging Community Network' project run by Native Sun Community Power Development. Since its inception, $570 million worth of funding has been awarded. In 2023, 16 projects were selected with a total funding value of $32.5 million [66].

4.5.5 President Biden's 500,000 charge point goal [67]

As part of the Bipartisan Infrastructure Law's $2.5 billion Charging and Fueling Infrastructure (CFI) Discretionary Grant Programme, the Biden-Harris Administration announced at the start of 2024 $623 million in grants to help build out a charge point network across the United States. The overarching aim of funding was to deploy a national charge point network comprising at least 500,000 publicly available charge points by 2030. CFI complemented NEVI in funding charge point deployment. To provide a consistent charging experience for users, charge points deployed using CFI funds had to adhere to the same minimum standards established for NEVI-funded charge points – including requirements that CFI-funded chargers were made in America as well as installed and maintained in accordance with recognised workforce standards.

4.5.6 Progress to date and challenges ahead

As of August 2024, there were over 192,000 publicly available charge points across the United States, with approximately 1,000 charge points deployed per week [68]. Based on that deployment rate, roughly 52,000 are being deployed annually, meaning that provided the rate is maintained, the target number of 500,000 will be reached by 2029. 64% of Americans now live within 2 mi of a public charge point and those who live closest have a positive view of EVs [69]. One of the major challenges that may hold back demand and utilisation of charge points is the fact that the American public are sceptical that the United States will be able to deploy the necessary infrastructure to support the wide scale adoption of EVs. A further challenge is the political divide between Democrats and Republicans. According to the Pew Research Center, Democrats are much more likely than Republicans to consider buying an EV. 26% of Republicans who live within 1 mi of a charge point state that they are very or somewhat likely to consider an EV for their next vehicle purchase. This falls to 15% for Republicans who live more than 2 mi away from a charge point. 60% of Democrats living in close proximity to a charge point are likely to consider an EV as their next vehicle purchase, though this drops to 50% when their nearest charge point is over 2 mi away. This reflects a broader challenge in the United States, the fact that EV adoption and charge point deployment are partisan, more so than in any other Western country. Following the election of Donald Trump in November 2024, the various funding programmes put in place by the previous administration have been significantly changed, wound back or cancelled.

At a deployment level, the key challenge is the range of costs related to charge point deployment. These can be anywhere between $122,000 and $440,000 per level 3 DC charging hub and this range is primarily related to connecting locations to the power network. Rural locations cost more to connect on average than urban ones. The total cost to deploy 500,000 level 3 DC charge points could be as much as $74 billion [70]. The cost to connect charge points needs to be reduced to make deployment more affordable and achievable. A further challenge is ensuring charge points are installed where there is demand. Walmart, for example, has a plan to develop its own charge point network at US stores by 2030, recognising that there is demand (90% of Americans live within 10 mi of a Walmart or Sam's Club store). However, further work needs to be done across the entire country to ensure charge points are deployed where they are needed. IIJA and IRA funding went some way to address this but more granular data analysis at charge point utilisation level is required to ensure effective deployment [71].

4.5.7 Anticipated changes in policy and strategy

The election of Donald Trump in November 2024 resulted in various funding programmes being significantly changed, wound back or cancelled. It is anticipated that charge point deployment will slow, though support from state governments is anticipated to continue.

4.6 Canada

The Canadian Government and its provincial governments are supportive of charge point deployments and offers a range of incentives and grant funded projects to encourage rollout. The programmes noted below are those with the greatest impact; however, other smaller programmes may exist which have not been included.

4.6.1 Canada's Electric Vehicle Availability Standard [72]

Announced in 2022 the Standard applies to light-duty vehicles (passenger cars, sport utility vehicles (SUV) and light trucks). Under the Standard, manufacturers and importers must meet annual regulated sales targets. The Standard recognises the importance of charge point deployment in achieving the transition to EVs. The EV targets should prompt private sector investment in charge point infrastructure; however, there is a need for public grant funding and incentives to increase the pace of deployment. Funding schemes include the Zero Emission Vehicle Infrastructure Programme (ZEVIP), Canada Infrastructure Bank's Charging and Hydrogen Refueling Infrastructure Initiative and the Electric Vehicle and Alternative Fuel Infrastructure Deployment Initiative. Regarding standardisation, the Standard notes that most manufacturers and CPOs are moving to the North American Charging Standard (NACS) port, enabling most drivers to use the same port. Standards being set by Measurement Canada will also enable EV drivers to pay for the kW they use as opposed to the length of time they are connected to a charge point.

The Standard established credits for investment by manufacturers in level 3 DC charge points based on a model where a credit is earned for each CAN $20,000 invested. Earning credits is achieved as long as all stations are rated at least 150 kW, charge points are operational for 5 years after installation with a compatible charging port and that charge points must go live between 1 January 2024 and 31 December 2027.

4.6.2 Electric Vehicle Charging Infrastructure for Canada Strategy [73]

The Electric Vehicle Charging Infrastructure for Canada Strategy, launched in 2024, states that if the number of EVs on Canadian roads grows as projected, there will be 5 million by 2030 and 21 million by 2040. This will require 679,000 level 2 and level 3 DC charge points across the country by 2040, with 40,000 required to be installed annually to meet this target. Significant capital investment is required to achieve this and public funding has a role to play. The Strategy notes that charge point deployment is primarily focused on urban areas and that to ensure equitable distribution, deployment should also focus on rural and remote communities. Rollout across Canada will require significant grid upgrades to meet demand. Estimates range from CAN $26 billion to CAN $294 billion over the 2025–2040 period and this covers upgrades to generation, transmission and distribution. Interestingly, the Strategy notes that if Canadians on average drive 25% fewer kilometres, the number of charge points required is 58,000 lower in 2030 and 168,000 lower in 2040.

4.6.3 Zero-Emission Vehicle Infrastructure Programme (ZEVIP) [74]

ZEVIP funds the deployment of charge points across Canada, specifically targeting installation in public places, on-street, in multi-unit residential buildings (MURBs), at workplaces and at depots for vehicle fleets. Projects must be 50% funded and can be no greater than CAN $5 million in total, including the 50% grant. If the funding is being sought by indigenous businesses and communities, this group can seek up to 75% of project costs. Utility companies and CPOs can take advantage of this funding. Deployment is aimed at three charge point use cases. The first is en route charging. The requirements include the deployment of charge points within 1.6 km of Transport Canada's National Highway System Designation which includes highways, core routes, feeder routes and/or northern/remote routes. A minimum of 2 level 3 DC charge points with charging rates of 100 kW or more or a minimum of 1 level 3 DC charge point of 100 kW or more, co-located alongside existing charge point infrastructure with a power output of at least 150 kW, must be installed to qualify for funding. Charge points must be available 24/7. The second use case is destination or residential charge points focused on deployments in communities where there are charge point gaps. Installation must include a minimum of 2 50 kW level 3 DC charge points or a minimum of 20 level 2 or 3 DC charge points if installing less than 2 level 3 DC 50 kW charge points. Like en route charge points, these must be available 24/7. The third use case is private charging and involves deploying charge points in private buildings which can be MURBs, workplaces or locations where fleet vehicles are serviced.

4.6.4 Electric Vehicle and Alternative Fuel Infrastructure Deployment Initiative (EVAFIDI) [75]

Launched in 2016, EVAFIDI set out to establish a coast-to-coast network of EV level 3 DC charge points along core routes and highways. Fifty-eight organisations received funding valued at CAN $260 million in total. The funding catalysed growth in charge point investment and enabled organisations to develop the business case to justify investment in infrastructure. 1,052 charge points funded by the scheme are currently operational out of a total of 1,090 planned installations.

4.6.5 Canadian Infrastructure Bank (CIB) funding for EV charge points [76]

The CIB's Charging and Hydrogen Refuelling Infrastructure (CHRI) Initiative was created to invest CAN $500 million in large-scale EV charging and hydrogen refuelling infrastructure across Canada. CIB's investment is intended to complement private capital investment in charge point deployment. Repayment of CIB investment is based on a shared utilisation risk model where repayment increases in line with increased utilisation. For sites where utilisation exceeds expectations, interest rates increase. As utilisation increases, the CIB's return will increase proportionally with interest rate step-ups at predetermined target utilisation levels. CIB investment can be sought alongside ZEVIP to maximise the effectiveness of the Federal Government's contribution. A minimum investment of CAN $10 million is required to take advantage of the Initiative.

4.6.6 The Canadian Electrical Code [77] and National Building Code

The Canadian Electrical Code is applicable across all provinces and sets out a number of provisions of a technical nature, noting that standards must be adhered to to ensure compliance. Those which relate to the installation location state that it is permitted to install charge points indoors in detached residential garages, enclosed or underground parking structures, repair and non-repair commercial garages, agricultural buildings, and similar rooms or other locations where an EV can charge. It is also permitted to install charge points at residential car ports and driveways, on-street curbside, open parking structures, parking lots, commercial charging facilities and similar locations.

Regarding Building Standards, there are currently no requirements in the National Building Code to include EV ready requirements; however, it is likely that this will change in the future. British Columbia is the only province to implement EV ready building code requirements for new MURBs [78]. These differ depending on the city, though in most cases they set the requirement that for new constructions, 100% of parking bays must be EV ready. By this, the code means that necessary cabling and infrastructure is installed but not the charge point [79].

4.6.7 Provincial incentives and other mandates

Across the six provinces, there are different levels of strategic direction beyond that set by the Federal Government. Five out of the six have an EV mandate and have developed and implemented multiple policies relating to installation or operation of public charge points or electrification targets and at least two relevant policies on installation or operation of EV public charge points [77]. The most significant funding programmes to support charge point deployment include Quebec which has established CAN $514 million of funding to deploy 116,000 charge points, British Columbia which has set up a fund of CAN $26 million and Ontario, a similar fund of CAN $91 million. In addition to this, seven global automotive manufacturers have agreed to install 30,000 charge points across North America from 2025 onwards [72]. Looking in more detail at Quebec as an example of a province with a detailed strategy, it predicts that by 2030 there will be 2 million EVs on provincial roads and to meet this increase requires an expansion in the number of deployed charge points. The Québec Electric Vehicle Charging Strategy 2023–2030 sets out a scalable action plan aimed at deploying accessible, reliable and high performing charging infrastructure to meet the needs of EV drivers by 2030 and beyond. The two key targets set by the Strategy include that by 2030:

- 35% of parking spaces in multiple dwelling buildings will be adapted to allow the installation of charge points, constituting 600,000 parking spaces.
- Québec will have a total of 6,700 public level 3 DC charge points and 110,000 level 2 public charging stations [80].

Regarding the CAN $514 million of investment, CAN $265 million will be made available over the next 5 years to CPOs to install level 2 and 3 DC charge points. CAN $108 million will be set aside for upgrading the electrical infrastructure in multiple dwelling buildings. The remaining funding is targeted at heavy goods vehicle charging infrastructure, fleet charge point deployment, charge points for tourists and a pilot project focused on autonomous EV charging [80].

4.6.8 Progress to date and challenges ahead

According to Electric Autonomy Canada, the number of charge points installed in Canada at the start of 2024 was over 27,000 [81]. Progress is being made towards the targets set as part of the Electric Vehicle Charging Infrastructure for Canada Strategy. This notes that by 2040, 679,000 level 2 and level 3 DC charge points will be required. This means that between now and the future date, around 43,000 charge points need to be installed annually. Using this metric, growth needs to ramp up massively. Year-on-year growth is increasing with the installation of level 2 charge points increasing by 34.2% and level 3 DC charge points by 26.6% when compared with 2023, where over 16,500 level 2 and just under 4,000 level 3 DC charge points existed in total. However, to reach the 2040 target will require year-on-year growth in excess of 100% for the next couple of years at least. This ramping up is a major challenge which will likely require additional funding. Like the United States, Canada is also experiencing the challenge of connecting more remote locations to the grid, especially on highways distant from large urban areas. These locations are proving costly to upgrade the required power connection infrastructure. This is also challenges for indigenous communities who often live in rural areas. To ensure these communities do not get left behind, further funding will likely need to be provided to support their transition to EVs. Urban areas also face challenges in terms of deployment due to high population density and limited space. This is a challenge also experienced in most European cities. To overcome this, CPOs need to develop strong business cases that justify the investment, demonstrate charging solutions are appropriate for urban on-street use cases and that there is a viable return based on good predicted utilisation [82]. Charge point reliability is also flagged as a challenge in Canada, partly due to connectivity and the range of extreme weather conditions that charge points experience [83,84]. Reliability (and the perception of charge points being usable) plays a role in influencing EV adoption, so must be addressed either through better maintenance, charge point redundancy or mitigation of extreme conditions (e.g., canopies over charge points). A final point to note is that the variation in charge point pricing across Canada is flagged as a challenge that may influence EV adoption, especially if EV drivers feel charging is much more expensive (and much more variable in price) compared with ICE vehicles. For example, analysis done by the Canadian Broadcasting Corporation noted that less than 25% of public charging stations across the country are currently (as of 2024) billing by kilowatt hour and that pricing is inconsistent across the country. This creates a myriad of issues as different cars draw kilowatts at different speeds depending on a range of factors meaning that one vehicle could draw 40 kW in a session of 1 hour, whereas another car might draw 50 kW over the same period. Under a pay by time model, the former would be paying more per kilowatt than the latter. A driver might also be charged for being plugged in but not charging at the end of a session. One challenge Canada seems not yet to be facing is a slowdown in EV adoption rates. Zero-emission vehicles are gaining momentum in Canada, with close to 185,000 new registrations in 2023. This represents a 49% increase from 2022, and accounts for 11% of all new motor vehicles registered that year. [85]. Provided this continues, this should ensure demand for charge points continues, as will projected utilisation.

4.6.9 Anticipated changes in policy and strategy

Canada held elections in May 2025, with the result being that the Liberals returned to power under Prime Minister Mark Carney, albeit still as a minority government. Although at the time of writing the new government is still taking shape, it seems unlikely the direction of travel will change. An action they should consider is further investment (either in terms of subsidies or private funding) to support rollout. Furthermore and as stated in Section 4.6.8, changes need to be made to how charge point pricing is structured across Canada so that a uniform pricing rule is introduced which charges by kilowatts used.

4.7 Australia

EV adoption in Australia has been slow when compared with other Western nations, as has charge point deployment. EVs accounted for 7.2% of all vehicle sales in 2023, increasing from 3.1% in 2022 [86]; however, this is significantly less than the United Kingdom which saw 16.5% of all car sales being EV in 2023 and 16.6% in 2022 [87]. Australian drivers as a whole prefer ICE sports utility vehicles or light commercial vehicles, with pickups being particularly popular. In terms of charge points, there are over 2,000 charging sites in Australia; however, it is not clear from the data how many charge points each site hosts [88]. The Australian Government, state governments and local governments recognise the importance of EVs and associated infrastructure in reducing emissions, and have developed a number of strategies and policies to encourage EV adoption and charge point deployment.

4.7.1 National Electric Vehicle Strategy (NEVS) [89]

The overarching vision of the NEVS is to increase the uptake of EVs to reduce Australia's emissions and improve the well-being of its citizens. A key objective relates to establishing the resources, systems and infrastructure to enable rapid EV uptake. To support this, the Australian Government is expanding the rollout of charge point infrastructure through the AUS $500 million Driving the Nation Fund and the National EV Charging Network, both of which establish a national network of charge points installed at 117 sites on major highways at an average interval of 150 km. Regarding the former, there are four elements to the Programme. The first is the aforementioned National EV charging network where AUS $39.3 million will support rollout across the 117 sites. The second is hydrogen highways where up to AUS $80 million will be invested in hydrogen infrastructure for heavy goods vehicles. The third is the dealership and repairer initiative for vehicle electrification nationally (DRIVEN) Programme where AUS $60 million will support the installation of charge points at car dealerships and workshops. The fourth is the Driving the Nation Programme where AUS $70.7 million will co-fund initiatives that will reduce Australia's road transport emissions. This investment of AUS $250 million brings the total investment, including that invested in 2022 of AUS $250 million to AUS $500 million [90]. Looking specifically at the elements that support the rollout of a public charge point network, the National EV Charging Network focuses on deploying Level 3 DC charge points with minimum charging speeds of 75 kW when four vehicles are charging simultaneously. Deployment complements existing and planned infrastructure, prioritising known blackspots, often in regional and remote communities. The Driving the Nation Programme will fund grants focused on delivering innovative charging solutions for Australians living in communal flats and in remote and regional communities.

Other enablers to deliver the vision in NEVS include investment in, development and deployment of a national EV charge point mapping tool to enable easy access to charge points. Although states and territories are developing these locally, the purpose of this tool will be to facilitate infrastructure, energy and telecommunication planning around EV charge points at a national level, enabling cross-state certainty of charging for EV drivers and also through providing information for future investment. Furthermore, the tool will ensure that EV drivers have the correct information about each charge point, enabling them to easily plug and pay regardless of where they are. The map will also identify gaps that will help inform charge point investment to improve the network. NEVS sets out proposed updates to the National Construction Code agreed by Commonwealth, State and Territory Building Ministers in October 2022 which includes that new buildings are designed, constructed and fitted out to enable the installation of charge points. For those who live in multi-residential buildings, the Government will conduct a study that delivers publicly accessible guidance and assessment tools aimed at supporting governments, building owners, owners' corporations and occupants. This will help them to understand charge point energy demand and management, safety requirements, costs and optimal charging solutions based on factors such as building age, location, size and availability of physical and technical infrastructure.

NEVS also flags that work is being done in implementing pathways for EV integration including ensuring that Australia's electricity system is ready for wide-scale EV adoption. Areas of focus for integration include ensuring national standards and communication protocols for EV supply equipment, cybersecurity and smart functionality relevant to Australia. Other areas of focus include a common mechanism for charge point data sharing, nationally aligned service and installation rules and streamlined network connection processes for consumer energy resources, including charge points. Commonwealth, state and territory governments will work together to align reporting and sharing of charge point data and collaborate to develop standards. One further point is that NEVS notes the importance of charge point accessibility for those with a disability and states that advice will be developed by the Australian Building Codes Board focused on making EV charging easier and safer in the built environment, including charge point requirements for accessible parking spaces.

4.7.2 State and territory policies

Each of the eight states and territories are at different stages in developing their EV charge point infrastructure strategies and policies. Focusing on two states, both New South Wales and Queensland have EV strategies.

Looking first at New South Wales, they launched their Electric Vehicle Strategy in 2021 to support and accelerate the uptake of light EVs through an AUS $633 million fund which also includes measures focused on heavy goods vehicle EVs and local EV manufacturing [89]. The Strategy aims to increase sales of EVs to more than 50% of new car sales by 2030–2031. To enable this, the New South Wales Government is investing:

- AUS $149 million in level 3 DC charge points at 100 km intervals across all major state highways, and every 5 km in urban areas.
- AUS $20 million in EV destination charge point grants focused on installation at tourist locations, including motels, restaurants and attractions.
- AUS $10 million in EV ready building grants to assist eligible New South Wales apartment buildings by co-funding the assessment and installation of EV infrastructure upgrades.
- AUS $10 million in EV kerbside charging grants to install charge points in areas of New South Wales with the least access to private off-street parking [91].

Deployment of charge points across all of these funding schemes should be based on the state master plan which shows where future charge points would bring the most benefit, ideal zones for public charge points and current and indicative future networks of charge points across the state, including level 2 and level 3 DC charge points [92]. The Strategy does not set targets for the number of charge points to be installed

Turning to Queensland, the state launched a Zero Emission Vehicle Strategy in 2022 with a time period of 2022–2032. The Strategy aims to accelerate Queensland towards a cleaner greener transport future whilst ensuring the state energy network supports the transition to achieve net zero emissions by 2050 [103]. AUS $10 million is being set aside as part of the Strategy to co-fund the deployment of public charge points to sit alongside the Queensland Electric Super Highway [93]. The Super Highway Programme aims to support 53 level 3 DC charge points along the highway route by the end of 2024. The deployment of public charge points away from the highway is focused on publicly accessible sites. Only level 3 DC charge points are eligible for funding. Each location must have a minimum of four charging bays which can all charge at a minimum of 75 kW per bay. Data requirements state that each charging session should include time and duration, kWs used and the number of sessions at each site [94]. Again, like New South Wales, no charge point targets have been set.

4.7.3 Installing and operating charge points

The legal framework for the installation and operation of charge points is determined by state or territory regulations. There is yet to be harmonisation or standardisation across the country. Key aspects that should be considered include environmental and planning regulations that relate to the location and scale of installation, building codes and standards that primarily focus on electrical safety regulations and communication standards, electricity regulations relating to power supply, compliance and grid capacity, and data privacy [95]. To understand the detail requires a deep dive into the minutiae of planning, building standards and electricity supply laws and regulations at state level. This goes beyond the focus of this chapter though further detail can be found on each territory or state's planning and building codes website.

4.7.4 Progress to date and challenges ahead

This is difficult to assess for Australia as there are no charge point targets set. We know from the introduction to Section 4.7 that EV adoption is increasing in Australia. We also know from state-by-state analysis that the number of charge points is increasing [88] but because there is no yardstick against which to compare this to (in terms of numbers required), it is impossible to assess progress. This in itself is a major challenge that must be overcome as a link between EV adoption and charge points must be established so that when adoption grows the number of charge points grows in tandem. If this is not the case, then adoption may stall because EV drivers do not have enough places to charge their car. Conversely, if charge point installation outpaces EV adoption, this will result in lower utilisation and less profitable sites meaning less investment. These targets should be set at federal government level, in liaison with state and territory governments. In addition to this is the lack of consistency in charge point placement along highways with each state or territory opting for different intervals. The intervals set by NEVI of 150 km may be too great, differing greatly from other Western countries. The danger is that these longer intervals result in greater charging demand which might result in lengthy waits for charge points if not enough are provided. This may further impact on EV adoption. The national charge point mapping tool is helpful in determining deployment locations; however, without estimates of the number of charge points required in an area, its value is limited. From the analysis of existing policies and strategies, it seems there is not a great enough focus on destination or residential charging at slower speeds using level 2 charge points and that most investment is in level 3 DC en route charge points. This is a challenge as it misses an opportunity to charge EVs slowly but more cheaply when they are stationary for prolonged periods of time. Slower residential and destination level 2 charge points do not require as great a grid upgrade as level 3 DC charge points and can also play a role in vehicle to grid smart energy systems.

A further challenge is the lack of a consistent charge point technical specification or standard across Australia. NEVI recognises this and notes that a consistent standard is required relating to installation, communication protocols for EV supply equipment and cybersecurity and smart functionality. Furthermore, consistency is required in terms of a common mechanism for charge point data sharing, nationally aligned service and installation rules and streamlined network connection processes for consumer energy resources [89]. The fact that these do not yet exist is a challenge as it is much harder to ensure standardisation after charge points have been installed as those that do not comply may need physical upgrades to equipment and software updates. It is also not clear in the Strategy whether enough focus has been placed on funding grid connections in remote locations as this will be key to charge point installation.

Although EV adoption is growing, Australian drivers prefer ICE SUVs or light commercial vehicles like pickups [86]. Until those who prefer these types of vehicle can be converted to an EV equivalent, there is a risk that adoption growth may stall, undermining the case for further infrastructure. Australia, due to its size and disparate nature of urban conurbations, faces the additional challenge of EV vehicle ranges being more of an issue for Australians who drive long distance journeys. To address this en route, charge points along highways must be well positioned, have adequate charging speeds and offer adequate connection points to satisfy demand.

All in all, it feels the Australian Government must do more to plan for the growing number of EVs across the country.

4.7.5 Anticipated changes in policy and strategy

Similar to Canada, Australia entered into a federal election year in 2025, resulting in the re-election of the Labor Government under Prime Minister Anthony Albanese. As the new administration takes shape, it is not clear whether the focus will continue as is or if it will change. That being said, it is clear from the challenges noted in Section 4.7.4 that the Australian Government needs to set charge point targets in numbers, establish standardisation and harmonisation in charge point technical specifications and requirements, continue to install en route level 3 DC charge points and focus efforts on residential and destination charging infrastructure.

4.8 New Zealand

EV adoption in New Zealand is strong with figures showing that in August 2024, 11.8% of all new car sales were EVs. In 2023, EVs made up 27.2% of all new car sales [96]. There are various policies and strategies that promote both EV uptake and charge point deployment.

4.8.1 Charging Our Future: National Electric Vehicle Charging Strategy For Aotearoa New Zealand 2023 – 2035 [97]

Launched in 2023 the EV Charging Our Future Strategy sets out policy interventions that support an equitable transition to a low emissions transport system in which accessible, affordable, secure and reliable charging infrastructure is available to everyone who needs it. Specifically, the Strategy considers charge point infrastructure in both public and private locations, taking into account different charging use cases. It also primarily focuses on light EVs while identifying initial areas for action to support heavy goods vehicle EV charging. It supports an equitable transition by considering the needs and experiences of current and future EV users from different backgrounds whilst also considering the needs and experiences of non-EV users in the transport system. The Strategy sets out five outcomes that are needed to equitably decarbonise road-based transport:

- **Outcome 1**: charge points are integrated into energy supply and infrastructure systems ensuring that it remains affordable, reliable, secure and safe to charge. This includes actions to minimise the stress on the electricity network through supporting greater availability of vehicle and electricity use data. This data can be used to deliver more efficient network investments and increasing transparency of electricity network capacity to encourage efficient investment in public and private charge point infrastructure. It also includes continuing to co-fund support for charge point projects that demonstrate or incorporate innovative solutions to manage electricity capacity constraints. Furthermore, co-funding includes promoting the benefits and supporting the uptake of smart charge points.
- **Outcome 2**: EV users from diverse backgrounds can use accessible, affordable, secure and reliable charge points when and where they need them. This includes actions to improve the equity of and access to home charge points through commissioning research into aspects of equity of access to charging. It is also important to explore solutions that increase the provision of public charge points in locations with limited off-street parking, leveraging research into charging needs across demographics and geographies. In addition, this includes identifying Māori partners with a specific interest in charging and/or the most appropriate entities to engage with and conduct further targeted engagement. Further actions relate to accommodating for regional variation in charging needs and energy supply by monitoring the expansion of the public charge point network in line with EV adoption forecast levels. Exploring the introduction of regionally specific targets for charge point infrastructure is also key, based on the findings of the research programme and providing additional government support to assist in the planning and installation of public charge point infrastructure that meets the needs of rural communities. Additional actions include exploring the role of existing vehicle service suppliers in improving regional and rural charge point provision alongside other options for charge point locations and supporting innovative technologies that increase network resilience in rural locations.
- **Outcome 3**: New Zealand's charge point network is underpinned by integrated and streamlined cross-sectoral planning and standards. This includes actions to improve standardisation and interoperability of charge points including promoting national consistency and reliability of service. It also includes

promoting a customer-centred approach to charging, exploring policy options to ensure chargers are efficient and safe and supporting and enabling data sharing where appropriate to ensure interoperability. Further actions include supporting local authorities to deploy charge points and developing systems and support networks to share best practice between local authorities, industry and central government. Actions relating to optimising data capture and use include exploring the viability of mandatory data provision for those charge points co-funded by the Government and reviewing the current data capture configuration of the EVRoam live map-based database, and considering what other information would be beneficial to the user of EV charge points. EVRoam collects real-time information about monitored charge points and shares this via apps and websites. EVRoam also enables a clear picture of how and where New Zealand's EV charge point network is developing. Actions relating to housing and urban development planning include exploring the costs and benefits of introducing charge point requirements for new developments, investigating potential changes to planning strategies and providing guidance material for local councils, landowners and developers.

- **Outcome 4**: New Zealand's charge point network functions effectively, can adapt and evolve over time, and is attractive to users, operators and investors. This includes actions to accelerate commercial investment through working with investors, CPOs and other key stakeholders to support investment in public charge points. In addition, this includes enabling data access and sharing where appropriate to accelerate commercial investment, and ensuring public funds are targeted at areas where commercial investment is unsuitable. Working with DNOs and CPOs is critical to make the network connection process (including pricing) efficient and supportive of rollout. The action relating to enabling innovations in new technology and business models focuses on continuing to co-fund the demonstration of innovative charge point technologies and working with industry to address barriers to uptake where they exist.
- **Outcome 5**: the national charge point network supports the transition to, and use of low emissions transport modes across the wider transport system. Actions relate to progressing work on heavy goods vehicle charging, working with stakeholders to identify where heavy goods vehicle charging infrastructure is needed and researching into what a charge point network would look like for heavy goods vehicles. Other actions relate to decarbonising other modes across the system and ensuring a coordinated investment approach focused on researching the present and future charging needs for heavy goods vehicles, aeroplanes, trains and ships.

The Strategy sets two key targets, that there will be a charging hub every 150–200 km on main highways by 2028 and that it will support the installation of at least 600 EV charge points in rural locations by 2028. The charging hubs are intended to complement the existing charge points installed at 75 km intervals. Deployment is funded via the Low Emission Transport Fund which in

2023 included a NZ $105 million package to support charge point deployment at hubs and in rural and remote communities. Launched alongside the EV Charging Our Future Strategy, the Charging Our Future: Government EV Charging Work Programme sets out the details behind the actions in the final Strategy, work underway and steps to be undertaken in the next 1–3 years [98].

4.8.2 Delivering a nationwide network [99]

In 2017, the New Zealand Transport Agency set the vision to build a nationwide charge point network with level 3 DC charge points deployed at 75 km intervals. The charge points installed and those that will be deployed as part of Charging Our Future must meet three minimum requirements. They must be safe, be designed and intended for use by the general public and, if deployed on-street, they must be endorsed by the road controlling authority. They must be reliable, meaning they are monitored in real time for reliability by the CPO and, for level 3 DC charge points, be accessible 24/7. They must also be interoperable meaning that level 3 DC charge points must use Chademo and CCS Type 2 connectors. Level 2 charge points must use Type 2 sockets. Both level 2 and 3 DC charge points must use a universal payment system. All public charge points in New Zealand should connect to EVRoam. This supports CPOs to complete a gap analysis and determine where new charge points should be installed [100].

4.8.3 Supercharging EV Infrastructure Work Programme [101]

In 2024, the New Zealand Government launched the Supercharging EV Infrastructure Work Programme that set the goal to install 10,000 charge points across the country by 2030. This is underpinned by NZ $257 million to fund deployment. Key elements of this Programme include development of a co-investment model for charge point deployment, removal of red tape and regulation including the requirement to seek consent for public charge points and introduction of standards to encourage charging during off-peak periods. Furthermore, the Programme states that Government will work with the Electricity Authority on addressing barriers such as connection costs and ensuring consistent approaches to EV charging connections across all 29 electricity distributors in New Zealand. It will also establish a Cross Agency Taskforce to drive the work programme and engage with industry. Various funds have been put in place to support infrastructure rollout including the Low Emissions Transport Fund Public EV Charging Urban Infill Programme which established a NZ $15 million fund to encourage innovation and increase the accessibility, convenience, and availability of infrastructure in urban areas in New Zealand to help meet and get ahead of demand. This Programme and other funding programmes are supported by the Community EV Charger Planner tool to help CPOs and government agencies identify where the best place to deploy charge points should be [102]. In addition to this tool, New Zealand provides in-depth guidance for CPOs and charge point installers regarding public, residential and commercial charge points [103].

4.8.4 Regional EV charging plans

Different regions of New Zealand are at different levels of progress regarding their own EV plans. Wellington City Council has a developed plan to install charge points called the Charged Up Capital Programme which is part of the Council's First to Zero initiative. Charge points are being deployed in phases, having started in 2022 and running until 2025. Where they are deployed, they aim to be convenient and accessible for existing and new EV drivers, offer value for money for the Council and result in minimal disruption to other users of car parks both during installation and after [104]. Other regions, like Waikato, have produced guidelines for EV charging stations on council land in the Waikato region [105].

4.8.5 Progress to date and challenges ahead

According to the Energy Efficiency and Conservation Authority (EECA) as of August 2024, there were 1,248 public charge points across New Zealand, with 333 added in 2023, a peak year since records began in 2015. 2022 saw 183 charge points installed [106]. At this current rate, and even if every year involved the same number of installations as 2023, New Zealand will significantly miss its target of 10,000 charge points by 2030. This presents a supply-side challenge because EV growth in New Zealand is rapid. This should mean a higher demand for charge points. EV ownership is growing fastest in urban areas where there is less access to off-street parking, especially in the bigger cities. According to a survey conducted by the EECA, two in five respondents did not think there is enough public charge points on the routes they travel and one in five has range anxiety. These numbers will likely increase unless charge point growth continues. One positive is that the government did largely meet its target to install public chargers every 75 km on the state highway network.

Looking at the challenges in this context, it is clear that the number and rate of deployment of charge points must increase rapidly to meet the 2030 target and that alongside this, further investment beyond what is already committed must be made to enable this. Significant private sector CPO investment is required to increase deployment. A further challenge is ensuring that charge points are distributed across New Zealand to allow equitable access. This is currently not the case as certain regions have higher numbers than others. As of April 2024, Canterbury and Auckland have the highest number of charge points, followed by Wellington and Waikato. From this analysis, urban areas like Auckland and Wellington will need more than the 295 and 197 charge points they currently have. And of those they need, a greater number will need to be residential level 2 charge points that charge at lower speeds but are sufficient to meet the charging needs of users when vehicles are stationary for several hours. That being said, most of New Zealand's charging is done at home, which means there is likely less of a case for charge point deployment. This contributes to a further challenge: the commercial return on investment. The percentage of EVs as part of all vehicles sold annually is 19.1%. The total number of EVs in 2023 was 73,000 [107]. This relatively low number when compared with other countries impacts on the commercial viability of sites as a result of low demand and utilisation. This is a major issue in rural areas that have a low population density and poor or non-existent access to the grid. Charge points are currently only viable in areas of high demand (e.g., on state highways) and due to the fact that most charging is done at home, other locations may struggle to be profitable, which means that investment is not forthcoming. A further challenge is the resources required to deliver electrical network upgrades and upgrades to supply. Regarding network upgrades, the capital costs to install a level 3 DC charging hub with multiple charge points can be as great as connecting a small town to the grid. Electricity distribution businesses will need to invest to upgrade the grid and will need to charge CPOs to cover costs, making the commercial case for deploying charge points more challenging. Regarding supply, this will need to be managed to prevent the grid being overwhelmed. New Zealand has sufficient capacity to meet increased demand but smart charging solutions will be required to minimise pressure on the grid. Other solutions to smooth the demand peak include battery storage and on-site solar generation.

4.8.6 Anticipated changes in policy and strategy

The analysis in Section 4.8.5 demonstrates that if New Zealand is to meet its target of 10,000 charge points by 2030, a number of changes need to be made. This includes further co-investment in charge point deployments by the New Zealand Government and CPOs. This would continue the direction of travel already started by the Coalition Government following its victory in 2023. The Coalition Government is unlikely to change direction, having set its target in 2023, though it might push the target date back to 2035 if deployment does not increase. Further investment in upgrading the grid will also be required to enable charge point deployment and power demand management. Further government funding (in most cases without co-funding) will be required in rural and disconnected areas to deploy charge points so as to ensure equitable access. More localised strategies (and updates to those that exist) at regional level, with accompanying funding, will also be required to drive deployment at area and street level.

References

[1] Climate Change Committee. *UK action on climate change* [online]. 2008. Available from https://www.theccc.org.uk/uk-action-on-climate-change/#:~:text=The%20Climate%20Change%20Act%20(2008,framework%20to%20cut%20carbon%20emissions [Accessed 01 September 2024].

[2] HM Government. *Net zero strategy: Build back greener* [online]. 2021. Available from https://assets.publishing.service.gov.uk/media/6194dfa4d3bf7f0555071b1b/net-zero-strategy-beis.pdf [Accessed 02 September 2024].

[3] Department for Transport. *Decarbonising transport: A better greener Britain* [Online]. 2021. Available from https://assets.publishing.service.gov.uk/media/610d63ffe90e0706d92fa282/decarbonising-transport-a-better-greener-britain.pdf [Accessed 03 September 2024].

[4] Environment and Climate Change Committee, House of Lords. *EV strategy: Rapid recharge needed* [online]. 2024. Available from https://publications.parliament.uk/pa/ld5804/ldselect/ldenvcl/51/5102.htm [Accessed 03 September 2024].

[5] HM Government. *Taking charge: The electric vehicle infrastructure strategy* [online]. 2021. Available from https://assets.publishing.service.gov.uk/media/6245ba40e90e075f15381cf0/taking-charge-the-electric-vehicle-infrastructure-strategy.pdf [Accessed 04 September 2024].

[6] HM Government. *Transitioning to zero emission cars and vans: 2034 delivery plan* [online]. 2021. Available from https://assets.publishing.service.gov.uk/media/60f9a3918fa8f5042aecd384/transitioning-to-zero-emission-cars-vans-2035-delivery-plan.pdf [Accessed 04 September 2024].

[7] Infrastructure and Projects Authority, HM Treasury. *Charging Infrastructure Investment Fund* [online]. 2018. Available from https://www.gov.uk/government/publications/charging-infrastructure-investment-fund [Accessed 04 September 2024].

[8] Ministry of Housing, Communities and Local Government, Department for Levelling Up, Housing and Communities. *The Building Regulations 2010: Infrastructure for the charging of electric vehicles: Approved Document S* [online]. 2021. Available from https://www.gov.uk/government/publications/infrastructure-for-charging-electric-vehicles-approved-document-s [Accessed 04 September 2024].

[9] Podpoint. *Electric vehicle (EV) charging legislation for developments in the United Kingdom* [online]. 2024. Available from https://pod-point.com/guides/business/ev-charging-legislation-new-build-uk?srsltid=AfmBOo-peWdCzvmi4dGeusPep5AojRYPtKYyn51zFCuyU-NzfubOZp0KO [Accessed 04 September 2024].

[10] Office for Zero Emission Vehicles, Department for Transport. Rapid charging fund [online]. 2021. Available from https://www.gov.uk/guidance/rapid-charging-fund [Accessed 02 September 2024].

[11] Ames C. *Rapid charging cash a little slow to arrive* [online]. 2023. Available from https://www.highwaysmagazine.co.uk/Rapid-charging-cash-a-little-slow-to-arrive-/13640 [Accessed 02 September 2024].

[12] Energy Savings Trust. *Local electric vehicle infrastructure fund* [online]. 2024. Available from https://energysavingtrust.org.uk/grants-and-loans/local-electric-vehicle-infrastructure-scheme/ [Accessed 02 September 2024].

[13] Office for Zero Emission Vehicles. *On-Street Residential Chargepoint Scheme guidance for local authorities* [online]. 2024. Available from https://www.gov.uk/government/publications/grants-for-local-authorities-to-provide-residential-on-street-chargepoints/grants-to-provide-residential-on-street-chargepoints-for-plug-in-electric-vehicles-guidance-for-local-authorities [Accessed 02 September 2024].

[14] Cenex. *Local Electric Vehicle Infrastructure (LEVI) Fund* [online]. 2024. Available from https://www.cenex.co.uk/projects-case-studies/local-electric-vehicle-infrastructure-levi-fund/ [Accessed 02 September 2024].

[15] Energy Savings Trust. *Local Electric Vehicle Infrastructure pilot scheme expanded* [online]. 2023. Available from https://energysavingtrust.org.uk/local-electric-vehicle-infrastructure-pilot-scheme-expanded/ [Accessed 02 September 2024].

[16] Cenex. *Local authorities receive allocation of chargepoint funding in next stage of local electric vehicle infrastructure fund* [online]. 2023. Available from https://www.cenex.co.uk/press-releases/local-authorities-receive-allocation-of-chargepoint-funding-in-next-stage-of-local-electric-vehicle-infrastructure-fund/ [Accessed 02 September 2024].

[17] Energy Savings Trust. *£185 million of LEVI chargepoint funding awarded to local authorities across England* [online]. 2024. Available from https://energysavingtrust.org.uk/local-electric-vehicle-chargepoint-funding-awarded/ [Accessed 02 September 2024].

[18] HM Government. *Powering up Britain: The net zero growth plan* [online]. 2023. Available from https://assets.publishing.service.gov.uk/media/642556c560a35e000c0cb167/powering-up-britain-net-zero-growth-plan.pdf [Accessed 02 September 2024].

[19] Department for Transport, Office for Zero Emission Vehicles. *Public Charge Point Regulations 2023 guidance* [online]. 2023. Available from https://www.gov.uk/government/publications/the-public-charge-point-regulations-2023-guidance/public-charge-point-regulations-2023-guidance [Accessed 04 September 2024].

[20] British Standards Institute. *PAS 1899:2022 – Electric vehicles – accessible charging – specification* [online]. 2022. Available from https://www.bsigroup.com/siteassets/pdf/en/insights-and-media/insights/brochures/pas_1889_final.pdf [Accessed 05 September 2024].

[21] The RAC. *The road to electric – in charts and data* [online]. 2025. Available from https://www.rac.co.uk/drive/electric-cars/choosing/road-to-electric/ [Accessed 5 September 2024].

[22] National Infrastructure Commission. *Infrastructure Progress Review 2024* [online]. 2024. Available from https://nic.org.uk/studies-reports/infrastructure-progress-review-2024/#tab-transport [Accessed 03 August 2024].

[23] European Commission. *Delivering the European green new deal* [online]. 2021. Available from https://commission.europa.eu/strategy-and-policy/priorities-2019-2024/european-green-deal/delivering-european-green-deal_en [Accessed 06 September 2024].

[24] European Council. Fit for 55 [online]. 2024. Available from https://www.consilium.europa.eu/en/policies/green-deal/fit-for-55/ [Accessed 07 September 2024].

[25] European Parliament. *Regulation of the European Parliament and of the council on the deployment of alternative fuels infrastructure, and repealing Directive 2014/94/EU* [online]. 2023. Available from https://data.consilium.europa.eu/doc/document/PE-25-2023-INIT/en/pdf [Accessed 07 September 2024].

[26] European Alternative Fuels Observatory. *Recharging systems* [online]. 2024. Available from https://alternative-fuels-observatory.ec.europa.eu/general-information/recharging-systems [Accessed 08 September 2024].

[27] European Commission. *Sustainable & smart mobility strategy* [online]. 2021. Available from https://transport.ec.europa.eu/document/download/be22d311-4a07-4c29-8b72-d6d255846069_en?filename=2021-mobility-strategy-and-action-plan.pdf [Accessed 08 September 2024].

[28] European Commission. *Recharge and refuel* [online]. 2021. Available from https://commission.europa.eu/system/files/2021-06/b3-recharge_and_refuel.pdf [Accessed 08 September 2024].

[29] European Commission. *Launch of the Route 35 platform | Our priorities for achieving our electrification goals | Speech by Commissioner Thierry Breton* [online]. 2022. Available from https://ec.europa.eu/commission/presscorner/detail/en/SPEECH_22_7785 [Accessed 08 September 2024].

[30] European Automobile Manufacturers' Association. *Automotive insights – charging ahead: accelerating the roll-out of EU electric vehicle charging infrastructure* [online]. 2024. Available from https://www.acea.auto/files/Charging_ahead-Accelerating_the_roll-out_of_EU_electric_vehicle_charging_infrastructure.pdf [Accessed 08 September 2024].

[31] Transport and Environment. *Most EU countries on track to meet charging targets* [online]. 2024. Available from https://www.transportenvironment.org/articles/most-eu-countries-on-track-to-meet-charging-targets [Accessed 09 September 2024].

[32] Transport and Environment. *Stagnation and growth: The European EV market* [online]. 2024. Available from https://www.transportenvironment.org/uploads/files/2024_06_EV_market_briefing.pdf [Accessed 09 September 2024].

[33] International Energy Agency. *Framework law on mobility* [online]. 2024. Available from https://www.iea.org/policies/8845-framework-law-on-mobility [Accessed 09 September 2024].

[34] Jaravel X. and Montout S. *Evaluation of the national recovery plan (France Relance) – summary of the final report* [online]. 2024. Available from https://www.strategie.gouv.fr/sites/strategie.gouv.fr/files/atoms/files/note_de_synthese_-_france_relance_-_en.pdf [Accessed 10 September 2024].

[35] International Energy Agency. EV infrastructure charging program (ADVENIR) [online]. 2021. Available from https://www.iea.org/policies/2716-ev-infrastructure-charging-program-advenir [Accessed 10 September 2024].

[36] Watson Farley & Williams. *The future of EV charging: Spotlight on France* [online]. 2024. Available from https://www.wfw.com/articles/the-future-of-ev-charging-spotlight-on-france/ [Accessed 10 September 2024].

[37] Avere France. *Charging point open to the general public on the road* [online]. 2024. Available from https://advenir.mobi/borne-voirie/ [Accessed 10 September 2024].

[38] Watson Farley & Williams. *The future of e-charging infrastructure: France* [online]. 2020. Available from https://www.wfw.com/articles/the-future-of-e-charging-infrastructure-france/#:~:text=What%20is%20legally%20required%20to,plan%20local%20d'urbanisme [Accessed 10 September 2024].

[39] Dusanter C. *Right to take joint ownership: law, procedure and limits* [online]. 2024. Available from https://opera-energie.com/droit-prise-copropriete/ [Accessed 11 September 2024].

[40] Pécout A. *Electric cars: France reaches milestone of 100,000 charge points* [online]. 2023. Available from https://www.lemonde.fr/en/economy/article/2023/05/06/electric-cars-france-reaches-milestone-of-100-000-charge-points_6025693_19.html [Accessed 09 September 2024].

[41] Carlier M. *France: Public electric vehicle supply equipment 2014-2023* [online]. 2024. Available from https://www.statista.com/statistics/1364027/france-public-electric-vehicle-charging-points/#:~:text=France%3A%20public%20electric%20vehicle%20supply%20equipment%202014%2D2023&text=In%202023%2C%20there%20were%20around,of%20electric%20vehicle%20supply%20equipment. [Accessed 09 September 2024].

[42] Colaluce L. *eMobility report: France records 37,850 EVs and already has 129,525 public charging points* [online]. 2024. Available from https://mobilityportal.eu/frances-q1-2024-emobility-insights/ [Accessed 11 September 2024].

[43] Mobility Portal. *European companies announce EVs for less than €25,000. An assured destination?* [online]. 2023. Available from https://mobilityportal.eu/european-companies-evs-25000/ [Accessed 11 September 2024].

[44] Bloomberg. *Europe pulls back from EV shift as consumers snub pricey models* [online]. 2024. Available from https://www.bloomberg.com/news/articles/2024-09-06/europe-pulls-back-from-ev-shift-as-consumers-snub-pricey-models?embedded-checkout=true [Accessed 11 September 2024].

[45] AVERE. *Position paper – Charging for All – A vision for an accessible European charging infrastructure* [online]. 2024. Available from https://www.avere.org/blogpages/policy-details/2024/02/27/Position-Paper-Charging-for-All-A-Vision-for-an-Accessible-European-Charging-Infrastructure [Accessed 11 September 2024].

[46] Adler K. *The far right came close to power in France. What about the rest of Europe?* [online]. 2024. Available from https://www.bbc.co.uk/news/articles/c4ng03lnv0vo [Accessed 11 September 2024].

[47] Watson Farley & Williams. *The future of EV charging: Spotlight on Germany* [online]. 2024. Available from https://www.wfw.com/articles/the-future-of-ev-charging-spotlight-on-germany/ [Accessed 12 September 2024].

[48] Platini I. *Deployment of charging points in Germany: Government or companies, who pays for it?* [online]. 2024. Available from https://mobilityportal.eu/charging-points-germany-government-companies/ [Accessed 14 September 2024.

[49] State Ministry for Economic Affairs, Regional Development and Energy. Publication BayMBl. 2021 No. 739 of 20.10.2021 [online]. 2021. Available

from https://www.verkuendung-bayern.de/baymbl/2021-739/ [Accessed 14 September 2024].

[50] European Alternative Fuels Observatory. *Germany target tracker* [online]. 2024. Available from https://alternative-fuels-observatory.ec.europa.eu/transport-mode/road/germany/target-tracker [Accessed 11 September 2024].

[51] National Leitstelle Ladeinfrastrktur. *ÖFFENTLICH ZUGÄNGLICHE LADEINFRASTRUKTUR – Report der Nationalen Leitstelle Ladeinfrastruktur – Mrz – Jul 2024* [online]. 2024. Available from https://nationale-leitstelle.de/wp-content/uploads/2024/07/oeLIS_Report_2024_Mrz-Jul.pdf [Accessed 14 September 2024].

[52] Platini I. *End of the German environmental bonus: How does it impact the eMobility market?* [online]. 2024. Available from https://mobilityportal.eu/environmental-bonus-german-emobility-market/ [Accessed 14 September 2024].

[53] Parker J. and McGuinness D. *German economy is in 'troubled waters' – ministry* [online]. 2024. Available from https://www.bbc.co.uk/news/world-europe-68361717 [Accessed 14 September 2024].

[54] Parker J. *Rise of far right in Germany's east isn't over yet* [online]. 2024. Available from https://www.bbc.co.uk/news/articles/cm2ng0nyj2no [Accessed 14 September 2024].

[55] Connolly K. *An 'earthquake' at Volkswagen – and a crisis for Germany?* [online]. 2024. Available from https://www.theguardian.com/business/article/2024/sep/07/an-earthquake-at-volkswagen-and-a-crisis-for-germany [Accessed 14 September 2024].

[56] PricewaterhouseCoopers. *Modernizing, decarbonizing and securing America's infrastructure* [online]. 2024. Available from https://www.pwc.com/us/en/services/consulting/cybersecurity-risk-regulatory/library/infrastructure-investment-jobs-act.html [Accessed 16 September 2024].

[57] Energy Marketers of America. *Infrastructure investment and jobs act – EV charging* [online]. 2022. Available from https://www.energymarketersofamerica.org/pdfs/issuebriefs/Infrastructure_Investment_and_Jobs_Act_EV_Charging_State_Talking_Points.pdf [Accessed 16 September 2024].

[58] Marcos R., Hawkins H., and Guerra R. *Understanding the Inflation Reduction Act (IRA)* [online]. 2023. Available from https://assets.kpmg.com/content/dam/kpmg/xx/pdf/2023/06/understanding-the-inflation-reduction-act.pdf [Accessed 17 September 2024].

[59] California Department of Transportation and California Energy Commission. *California's deployment plan for the National Electric Vehicle Infrastructure Program* [online]. 2022. Available from https://dot.ca.gov/-/media/dot-media/programs/esta/documents/transportation-electrification/nevi/2022-ca-nevi-deployment-plan-a11y.pdf [Accessed 17 September 2024].

[60] New York State. *Charge NY Program* [online]. 2024. Available from https://www.nyserda.ny.gov/All-Programs/ChargeNY [Accessed 17 September 2024].

[61] Chargelab. *New York State EV charging rebates* [online]. 2024 Available from https://chargelab.co/rebates/new-york-state#:~:text=%24750%20million%20investment&text=The%20state's%20goal%20is%20to,the%20end%20of%202021%20alone. [Accessed 17 September 2024].

[62] Kadiri I. *New York's EV charging incentives* [online]. 2024. Available from https://www.ampeco.com/fr/ev-charging-grants-incentives/unlocking-the-power-of-electric-mobility-new-yorks-ev-charging-incentives/ [Accessed 17 September 2024].

[63] U.S Department of Transportation. *Charging and Fueling Infrastructure Grant Program* [online]. 2024. Available from https://www.transportation.gov/rural/grant-toolkit/charging-and-fueling-infrastructure-grant-program [Accessed 18 September 2024].

[64] Kadiri I. *Port Infrastructure Development Program (PIDP) by Maritime Administration* [online]. 2024. Available from https://www.ampeco.com/us/ev-charging-grants-incentives/port-infrastructure-development-program-pidp-by-maritime-administration/ [Accessed 18 September 2024].

[65] Office of Energy Efficiency and Renewable Energy. *About Clean Cities and Communities* [online]. 2024. Available from https://cleancities.energy.gov/about/ [Accessed 18 September 2024].

[66] Office of Energy Efficiency and Renewable Energy. *Funded Projects* [online]. 2024. Available from https://cleancities.energy.gov/projects/ [Accessed 18 September 2024].

[67] U.S. Department of Transportation. *Biden-Harris Administration announces $623 million in grants to continue building out electric vehicle charging network* [online]. 2024. Available from https://www.transportation.gov/briefing-room/biden-harris-administration-announces-623-million-grants-continue-building-out [Accessed 18 September 2024].

[68] U.S. Department of Transportation. *Investing in America: Number of publicly available electric vehicle chargers has doubled since start of Biden-Harris Administration* [online]. 2025. Available from https://highways.dot.gov/newsroom/investing-america-number-publicly-available-electric-vehicle-chargers-has-doubled-start#:~:text=Since%20the%20start%20of%20the,chargers%20being%20added%20each%20week. [Accessed 18 September 2024].

[69] Bestvater S., and Shah S. *Electric vehicle charging infrastructure in the U.S.* [online]. 2024. Available from https://www.pewresearch.org/data-labs/2024/05/23/electric-vehicle-charging-infrastructure-in-the-u-s/ [Accessed 18 September 2024].

[70] Gamage T., Tal G., and Jenn A.T. *The costs and challenges of installing corridor DC fast chargers in California* [online]. 2023. Available from https://www.sciencedirect.com/science/article/pii/S2213624X23000238 [Accessed 18 September 2024].

[71] Cavale S., and Roy A. *Walmart plans own EV charger network at U.S. stores by 2030* [online]. 2023. Available from https://www.reuters.com/business/retail-consumer/walmart-plans-own-ev-charger-network-us-stores-by-2030-2023-04-06/ [Accessed 18 September 2024].

[72] Government of Canada. *Canada's Electric Vehicle Availability Standard (regulated targets for zero-emission vehicles)* [online]. 2024. Available from https://www.canada.ca/en/environment-climate-change/news/2023/12/canadas-electric-vehicle-availability-standard-regulated-targets-for-zero-emission-vehicles.html [Accessed 19 September 2024].

[73] Natural Resources Canada. *Electric vehicle charging infrastructure for Canada* [online]. 2024. Available from https://natural-resources.canada.ca/energy-efficiency/transportation-alternative-fuels/resource-library/electric-vehicle-charging-infrastructure-for-canada/25756 [Accessed 19 September 2024].

[74] Government of Canada. *Zero Emission Vehicle Infrastructure Program* [online]. 2024. Available from https://natural-resources.canada.ca/energy-efficiency/transportation-alternative-fuels/zero-emission-vehicle-infrastructure-program/21876 [Accessed 19 September 2024].

[75] Government of Canada. *Electric Vehicle and Alternative Fuel Infrastructure Deployment Initiative* [online]. 2024. Available from https://natural-resources.canada.ca/energy-efficiency/transportation-alternative-fuels/infrastructure/electric-vehicle-alternative-fuels-infrastructure-deployment-initiative/18352 [Accessed 19 September 2024].

[76] Canadian Infrastructure Bank. *Charging and Hydrogen Refuelling Infrastructure Initiative* [online]. 2022. Available from https://cdn.cib-bic.ca/files/Investment/EN/CHRI-Initiative-Overview_2024-02-27-201535_vbws.pdf [Accessed 19 September 2024].

[77] The Canadian Transportation Alliance. *Canadian Regulatory Report* [online]. 2021. Available from https://www.transportationenergy.org/canadiantransportationalliance/wp-content/uploads/2021/11/CTA_Canadian-EV-Regulatory-Report_October-2021.pdf [Accessed 19 September 2024].

[78] Jarrat E. *Exclusive: NRCan holding early consultations on EV-ready recommendations* [online]. 2024. Available from https://electricautonomy.ca/news/2024-03-05/exclusive-nrcan-holding-early-consultations-on-ev-ready-recommendations-for-national-building-code/ [Accessed 19 September 2024].

[79] McEwen B. *'EV Ready' requirements for new buildings: A best practice guide for BC local governments* [online]. 2021. Available from https://www.bchydro.com/content/dam/BCHydro/customer-portal/documents/power-smart/business/programs/ev-ready-requirements-for-new-buildings.pdf [Accessed 19 September 2024].

[80] Gouvernement du Québec. *Québec's electric vehicle charging strategy* [online]. 2023. Available from https://www.quebec.ca/en/government/policies-orientations/quebec-electric-vehicle-charging-strategy [Accessed 19 September 2024].

[81] Jarratt E., and Yakub M. 2024 *EV charging network report: Canada's public charger installations up 33 per cent in 12 months* [online]. 2024. Available from https://electricautonomy.ca/news/2024-03-07/2024-canada-ev-charging-networks-report/ [Accessed 20 September 2024].

[82] Electric Autonomy Canada. *What's going on with EV charging in Canada? We've got it covered* [online]. 2024. Available from https://electricautonomy.

ca/charging/2024-07-30/whats-going-on-with-ev-charging-in-canada/ [Accessed 20 September 2024].

[83] Sokic N. *Monitoring, maintenance, mission criticality: How four Canadian EV charging networks work to provide reliable charging* [online]. 2024. Available from https://electricautonomy.ca/charging/2024-05-24/charging-network-reliability-canada-2/ [Accessed 20 September 2024].

[84] Agro C., McCann M., and McDonald J. *Same car, same charge, different prices? EV drivers face inconsistent, unreliable charging network* [online]. 2024. Available from https://www.cbc.ca/news/business/electric-car-charging-network-canada-marketplace-1.7094656 [Accessed 20 September 2024].

[85] Canada Energy Regulator. *Market Snapshot: Zero emission vehicles now account for over 10% of all new vehicles in Canada* [online]. 2024. Available from https://www.cer-rec.gc.ca/en/data-analysis/energy-markets/market-snapshots/2024/market-snapshot-zero-emission-vehicles-now-account-for-over-10-percent-of-all-new-vehicles-in-canada.html?=undefined&wbdisable=true [Accessed 20 September 2024].

[86] Jackson L. *Australia struggles to ditch SUV habit even as electric vehicle sales hit record* [online]. 2024. Available from https://www.reuters.com/business/autos-transportation/australia-struggles-ditch-suv-habit-even-electric-vehicle-sales-hit-record-2024-01-04/ [Accessed 21 September 2024].

[87] Zapmap. *EV market stats 2024* [online]. 2024. Available from https://www.zap-map.com/ev-stats/ev-market#:~:text=In%202023%2C%20the%20last%20full,%2C%20to%2016.8%25%20in%202024. [Accessed 21 September 2024].

[88] Cardinal E. *These are the most EV-friendly states in Australia* [online]. 2024. Available from https://www.drive.com.au/caradvice/how-many-ev-chargers-in-australia-2024/ [Accessed 21 September 2024].

[89] Australian Government. *National electric vehicle strategy* [online]. 2024. Available from https://www.dcceew.gov.au/sites/default/files/documents/national-electric-vehicle-strategy.pdf [Accessed 21 September 2024].

[90] Australian Government. *Driving the nation fund* [online]. 2024. Available from https://www.dcceew.gov.au/energy/transport/driving-the-nation-fund [Accessed 21 September 2024].

[91] New South Wales Government. *NSW electric vehicle strategy* [online]. 2024. Available from https://www.energy.nsw.gov.au/nsw-plans-and-progress/government-strategies-and-frameworks/electric-vehicle-strategy [Accessed 22 September 2024].

[92] New South Wales Government. *EV public charging master plan* [online]. 2024. Available from https://www.energy.nsw.gov.au/business-and-industry/programs-grants-and-schemes/electric-vehicles/EV-master-plan-map [Accessed 22 September 2024].

[93] Queensland Government. *Queensland's zero emission vehicle strategy* [online]. 2024. Available from https://www.qld.gov.au/transport/projects/electricvehicles/zero-emission-strategy#public-charging [Accessed 22 September 2024].

[94] Queensland Rural and Industry Development Authority. *Electric vehicle charging infrastructure scheme guidelines* [online]. 2022. Available from https://qrida.qld.gov.au/sites/default/files/2022-08/Guideline%20-%20Queensland%20Charging%20Infrastructure%20Scheme%20%28PDF%2C%20894KB%29.pdf [Accessed 22 September 2024].

[95] Watson Farley & Williams. *The future of EV charging infrastructure: Spotlight on Australia* [online]. 2024. Available from https://www.wfw.com/articles/the-future-of-ev-charging-infrastructure-spotlight-on-australia/ [Accessed 22 September 2024].

[96] Electric Vehicle Database. *EV market stats 2024* [online]. 2024. https://evdb.nz/ev-stats [Accessed 23 September 2024].

[97] New Zealand Government. *Charging our future: National electric vehicle charging strategy for Aotearoa New Zealand 2023-2035* [online]. 2023. Available from https://www.transport.govt.nz/assets/Uploads/EV-Charging-Strategy.pdf [Accessed 23 September 2024].

[98] New Zealand Government. *Charging our future: Government EV charging work programme (2023–2026)* [online]. 2023. Available from https://www.transport.govt.nz/assets/Uploads/EV-Charging-Work-Programme.pdf [Accessed 23 September 2024].

[99] NZ Transport Agency. *Enabling a nationwide network of public EV charging infrastructure* [online]. 2024. Available from https://nzta.govt.nz/planning-and-investment/planning/transport-planning/planning-for-electric-vehicles/national-guidance-for-public-electric-vehicle-charging-infrastructure/enabling-a-nationwide-network-of-public-charging-infrastructure/ [Accessed 23 September 2024].

[100] NZ Transport Agency. *EVRoam* [online]. 2024. Available from https://www.nzta.govt.nz/planning-and-investment/planning/transport-planning/planning-for-electric-vehicles/evroam [Accessed 24 September 2024].

[101] New Zealand Government. *Government to boost public EV charging network* [online]. 2024. Available from https://www.beehive.govt.nz/release/government-boost-public-ev-charging-network [Accessed 24 September 2024].

[102] Energy Efficiency & Conservation Authority. *Low emission transport fund – Public EV charging: Urban infill* [online]. 2024. Available from https://www.eeca.govt.nz/co-funding-and-support/products/public-ev-charging-urban-infill/ [Accessed 24 September 2024].

[103] NZ Transport Agency. *Guidance for EV charging infrastructure* [online]. 2025. Available from https://nzta.govt.nz/planning-and-investment/planning/transport-planning/planning-for-electric-vehicles/guidance-for-ev-charging-infrastructure/ [Accessed 24 September 2024].

[104] Wellington City Council. *Charged up capital: EV chargers* [online]. 2024. Available from https://wellington.govt.nz/your-council/projects/ev-chargers [Accessed 24 September 2024].

[105] Waikato Regional Council. *Guidelines for EV charging stations* [online]. 2023. Available from https://www.waikatoregion.govt.nz/assets/WRC/7388-EV-Charging-Guideline-WR.pdf [Accessed 24 September 2024].

[106] Energy Efficiency & Conservation Authority. *Public EV charger dashboard* [online]. 2024. Available from https://www.eeca.govt.nz/insights/data-tools/public-ev-charger-dashboard/ [Accessed 24 September 2024].

[107] Energy Efficiency & Conservation Authority. *Plugging into the future: How New Zealand is electrifying its roads: An overview of public EV charging infrastructure in New Zealand* [online]. 2024. Available from https://www.eeca.govt.nz/insights/eeca-insights/plugging-into-the-future-how-new-zealand-is-electrifying-its-roads/ [Accessed 24 September 2024].

Chapter 5

Demand and supply modelling including predicting utilisation

Andrew McGordon[1], *Stuart Bradley*[1] *and David Trousdale*[2]

This chapter outlines the methods for predicting public charge point energy supply, demand and utilisation. The chapter starts with a discussion on the provision of electrical power from the grid, historical changes in the structure of the grid and the complexities in terms of costs and carbon dioxide emissions as a result of these changes. The geographical spread of public electric vehicles (EV) charge points in the United Kingdom is highlighted with a focus on potential appropriate measures to define EV charge point density. The chapter notes the different charge point locations and use cases, with a brief comparison of charging power and costs. The chapter concludes by outlining the methods of modelling EV charge point utilisation, brought to life through two real-world case studies.

5.1 Electrical supply

The structure of the energy grid is in a state of flux. The previous structure of large suppliers (coal-fired power stations) and large users (industrial demands) that was prevalent particularly in industrial areas of the country has changed considerably over the last several decades. The nature of supply has changed such that although there are some significant power stations that remain in the present grid, there are now many smaller suppliers that are more geographically distributed [1].

Recent United Kingdom Government figures published in 2023 for the year 2022 show that the total installed capacity on the grid was 97.3 GW, as shown in Table 5.1, with interconnector capabilities shown in Table 5.2.

There is a relatively small number of large producers and only one of these is a recent large wind farm. Within the United Kingdom, there were only 18 plants with

[1]Warwick Manufacturing Group, University of Warwick, United Kingdom
[2]Principal Advisor - Climate Change and Sustainability, Newcastle City Council, United Kingdom

Table 5.1 UK grid installed capacity in MW by power producers in 2022

Generator type	Fuel	Installed capacity (MW)
Major power producer	Coal	3,341
Major power producer	Natural gas	30,684
Major power producer	Sour gas	1,380
Major power producer	Diesel/gas oil	932
Major power producer	Nuclear	6,848
Major power producer	Biomass	3,381
Major power producer	Waste	471
Major power producer	Hydro	1,466
Major power producer	Pumped storage	2,900
Major power producer	Onshore wind	11,965
Major power producer	Offshore wind	26,678
Major power producer	Solar	7,258
Major power producer	Total	97,303

[Source: Department for Business, Energy & Industrial Strategy (DESNZ). *Digest of United Kingdom Energy Statistics (DUKES): electricity Table 5.11* [online]. 2024. Available from https://assets.publishing.service.gov.uk/media/66a7daa4fc8e12ac3edb068e/DUKES_5.11.xlsx. This information has been licensed under the Open Government Licence v3.0. To view this licence, visit https://www.nationalarchives.gov.uk/doc/open-government-licence/ [Accessed 26 February 2025]]

Table 5.2 UK grid interconnector capacity in MW

Interconnector	Connecting country	Capacity (MW)	Year of first operation
Interconnecion France-Angleterre (IFA)	England–France	2,000	1986
North–South	Northern Ireland–Republic of Ireland	540	1996
Moyle	Scotland–Northern Ireland	500	2002
BritNed	England–Netherlands	1,000	2011
East–West	Wales–Republic of Ireland	500	2012
NemoLink	England–Belgium	1,000	2019
Interconnecion France-Angleterre 2 (IFA2)	England–France	1,000	2021
North Sea Link (NSL)	England–Norway	1,400	2021
ElecLink	England–France	1,000	2022
Viking	England–Denmark	1,400	2023

[Source: Department for Business, Energy & Industrial Strategy (DESNZ). *Digest of United Kingdom Energy Statistics (DUKES): electricity Table 5.11B* [online]. 2024. Available from https://assets.publishing.service.gov.uk/media/66a7daa4fc8e12ac3edb068e/DUKES_5.11.xlsx. This information has been licensed under the Open Government Licence v3.0. To view this licence, visit https://www.nationalarchives.gov.uk/doc/open-government-licence/ [Accessed 26 February 2025]. Note that notes indicating further detail in the table have been removed.]

an installed capacity of greater than 1 GW. This was broken down into 11 gas, 1 biomass, 5 nuclear and 1 wind.

In 2022, there were 1,313 recognised individual producers. The smallest producer included had a capacity of 50 kW; there were 60 producers between 50 kW and 1 MW; 699 between 1 and 10 MW, and 450 between 10 and 100 MW, with 83 between 100 MW and 1 GW. In addition to the growing numbers, these producers are becoming increasingly more dispersed geographically as shown in Figure 5.1. It is apparent that these producers are not necessarily located near urban areas where the demand is high, meaning there is a requirement for greater distribution and transmission infrastructure. Since the minimum producer recognised by the grid is 50 kW, a potentially large number of domestic producers are not accounted for. Many home solar photovoltaic (PV) installations will be of the order of less than 15kW with the primary aim to provide electricity only for the house. However, the presence of the Smart Export Guarantee [2] means that these producers can be an additional generation, or considered as a demand reduction.

The grid recognises different types of generation known as 'dispatchable' and 'non-dispatchable'. Dispatchable generation is generation that can be externally controlled, for example a gas power station or a pumped hydroelectric plant. These can react quickly to changes in demand when in operation, although start-up from 'cold' is much longer for gas plants. Non-dispatchable generation covers wind and solar, which, although it can be predicted to a large extent, cannot be fully controlled. Given the requirement to generate electricity producing minimum carbon dioxide, these intermittent generators are generally allowed to fluctuate to maximise their generation, and the dispatchable fossil fuel plants modulate their output accordingly. If agglomerated into a country-wide perspective, then average predictions for offshore and onshore wind combined are likely to be reasonably accurate in practice. However, variations in generation at a more local level can be significant. This causes challenges for balancing the grid at individual transformer levels.

Figure 5.2 shows how the contributions to the electrical energy generation mix have changed from 1950 to 2023. In 1950, coal provided more than 97% of the feedstock for electricity production. From the 1960s, nuclear provided a growing share of the generation capacity peaking at approximately 25% in the 1990s, which also coincided with the increasing use of gas for electricity production. In 2000, the generation mix was approximately 31% coal, 38% gas, 22% nuclear, with the remaining 10% coming from wind, oil, hydro and imports. By 2010, the generation mix had not changed much, but gas was now more dominant, providing 46% of electricity, with coal and nuclear reducing slightly to 28% and 16%, respectively. It is only in the last decade in the United Kingdom that the grid has seen a considerable reduction in carbon intensity, driven primarily by increased wind generation and an elimination of use of coal for electricity production (this occurred in 2024). In 2023, the mix was broken down into: gas 32%, wind 26%, nuclear 13%, other renewables 11%, imports 7% and solar 4%. As is clear from above, the modern grid is very different from

Figure 5.1 *Location of major power producers in the United Kingdom in mid-2024 [Source: Department for Energy Security and Net Zero. Digest of UK Energy Statistics (DUKES): electricity – DUKES chapter 5: statistics on electricity from generation through to sales [online]. 2024. Available from https://assets.publishing.service.gov.uk/media/66a7da1bce1fd0da7b592f0a/DUKES_2024_Chapter_5.pdf. This information has been licensed under the Open Government Licence v3.0. To view this licence, visit https://www.nationalarchives.gov.uk/doc/open-government-licence/ [Accessed 26 February 2025]].*

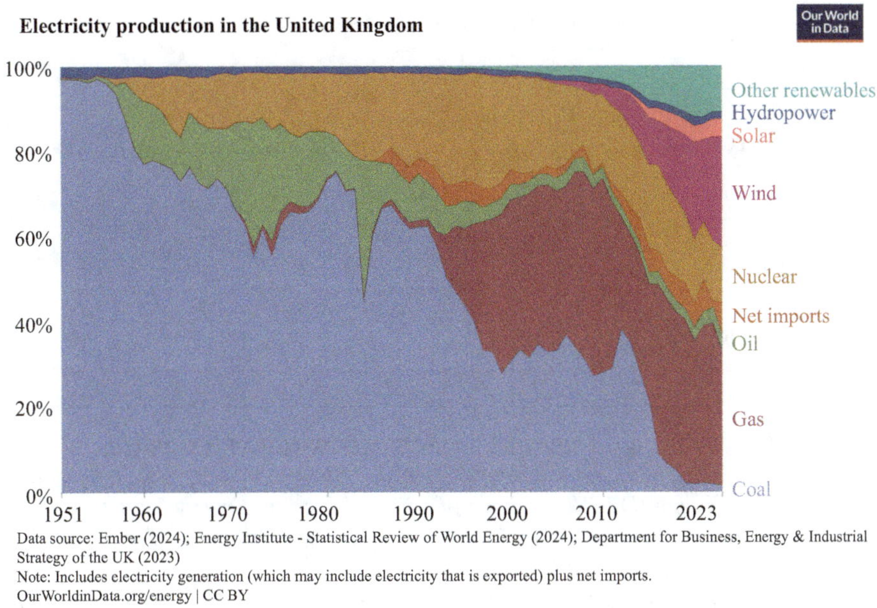

Figure 5.2 Electricity production in the United Kingdom by source, 1950–2023 [Source: Ember 2024, Energy Institute – Statistical Review of World Energy 2024 and Department for Business, Energy & Industrial Strategy of the UK (2023); Electricity production in the United Kingdom, Published online at OurWorldinData.org. Available from: https://ourworldindata.org/grapher/electricity-mix-uk?time=1951..latest [Online Resource], Licence: CC BY [Accessed 26 February 2025]].

the grid of only 25 years ago in terms of diversity of generation; geographically, temporally and spatially.

The change in electricity source is reflected in the carbon dioxide emissions intensity of the electricity generated, as shown in Figure 5.3. The reduction in emissions per unit of electricity produced when coupled with the increasing penetration of EVs is the driving force behind the decarbonisation of road transport. If we take a typical vehicle with an efficiency of 300 Wh/mi, then over a 100 mi journey, it will use 30 kWh. Ignoring charging efficiency, the carbon dioxide emissions from generating 30 kWh of electricity at 171 g/kWh in 2023 were 5.13 kg. Taking an average passenger car at 160 g/km of carbon dioxide emissions, this results in 25.7 T, a factor of five higher than the EV. However, using the 2013 figure of 450 g/kWh, the EV would 'emit' 13.5 T of carbon dioxide for the same journey. The same journey generates 2.5 times lower emissions in 2023 compared with 2013 solely due to grid decarbonisation.

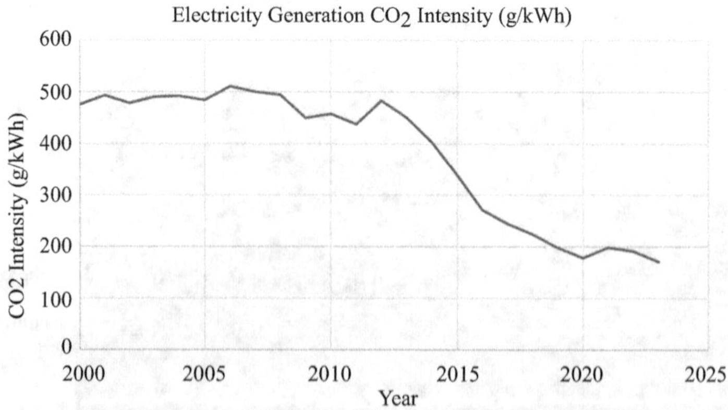

Figure 5.3 Carbon dioxide intensity of electricity generation in g/kWh, 2000–2023 [Source: Department for Business, Energy & Industrial Strategy (DESNZ). Digest of United Kingdom Energy Statistics (DUKES): electricity Table 5.14 [online]. 2024. Available from https://assets.publishing.service.gov.uk/media/66a7dacaa3c2a28abb50d91e/DUKES_5.14.xlsx, This information has been licensed under the Open Government Licence v3.0. To view this licence, visit https://www.nationalarchives.gov.uk/doc/open-government-licence/ [Accessed 26 February 2025]. Numbers generated from source by author].

5.2 Electrical consumption

In the United Kingdom, over the last 20 years, there has been a significant decline in electricity demand, reducing from a peak of 349 TWh in 2005 to 269 TWh in 2023, as shown in Figure 5.4. Within this decline, the consumption across the three sectors has decreased by approximately the same amount, at around 20%. Energy efficiency initiatives have reduced demand across the board as rising costs of electricity have become more important to consumers.

Additionally, the temporal nature of traditional domestic demands shows usage peaks in the morning and evenings, specifically for working days. There is also a noticeable reduction overnight, as shown in Figure 5.5. This represents a typical house in 2014. Home charging of EVs overnight (from 00:00 to 06:00) is considered a particular opportunity. However, it must be noted that an overnight home charger for an EV is likely to be rated at 7 kW. Figure 5.5 represents a demand ten times that of the current 18:00 peak. In the future when EV penetration is high and most houses feature overnight charging, the demand on the grid could become more levellised, or the peak on the grid could actually be overnight. In the United Kingdom, there are currently several domestic electricity tariffs on offer that incentivise use of electricity overnight. These are often marketed as 'EV tariffs' and are in addition to the more longstanding 'Economy 7' tariff. With these tariffs,

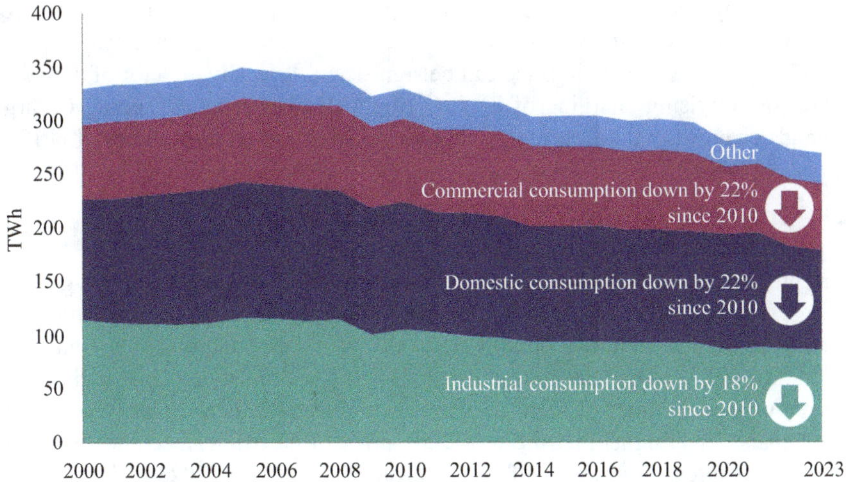

Figure 5.4 UK electricity consumption by major sector in TWh, 2000–2023 [Source: Department for Energy Security and Net Zero. Digest of UK Energy Statistics (DUKES): electricity – DUKES chapter 5: statistics on electricity from generation through to sales [online]. 2024. Available from https://assets.publishing.service.gov.uk/media/66a7da1bce1fd0da7b592f0a/DUKES_2024_Chapter_5.pdf. This information has been licensed under the Open Government Licence v3.0. To view this licence, visit https://www.nationalarchives.gov.uk/doc/open-government-licence/ [Accessed 26 February 2025]].

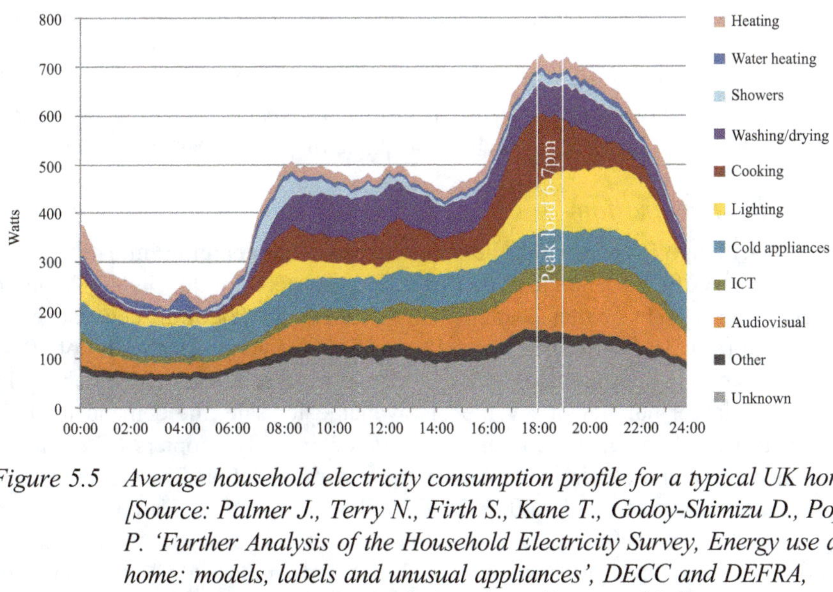

Figure 5.5 Average household electricity consumption profile for a typical UK home [Source: Palmer J., Terry N., Firth S., Kane T., Godoy-Shimizu D., Pope P. 'Further Analysis of the Household Electricity Survey, Energy use at home: models, labels and unusual appliances', DECC and DEFRA, London (2014), p2. Available from https://assets.publishing.service.gov.uk/media/5a7ee7ffed915d74e6227361/Electricity_Survey_2_-_Models__labels___unusuals_180214.pdf [Accessed 26 February 2025]].

customers pay two different rates for their electricity depending on time of day, for example Economy 7 offers a discounted rate of 7 hours overnight between 00:30 and 07:30. The difference in price can be significant, typically a factor of two, thus clearly incentivising charging of EVs overnight. However, as the timing of large demands changes in the future, this could have an effect on these types of tariffs.

5.3 Interplay of power demand and supply

Currently, the price of electricity generation across the United Kingdom's grid, on average, varies by time of day, linked to the dominant source. The general approach is to utilise the maximum amount of renewables (offshore and onshore wind, solar and hydro) followed by nuclear, biomass and finally combustion plants such as gas. It is not unusual at times of high renewable energy generation (primarily wind) and low demand (overnight) that the wholesale price of electricity becomes negative, with the historical low being −£77/MWh on 25 September 2023 at ~03:00 [3]. There can be considerable fluctuations in the price of electricity during the day which can be taken advantage of by customers who are able to use electricity at different times. EV owners are able to charge their vehicle when electricity is cheap, acting as both a sink and a source of power on the grid. More on this is noted in Chapter 11 where we discuss vehicle to grid.

5.4 EVs

In the context of this book, EVs are defined as battery EVs. There are a considerable number of plug-in hybrid vehicles, but these have an internal combustion engine as an alternative source of propulsion, and the battery packs are considerably smaller than those of pure EVs. Examples of currently available EV battery pack sizes range from 28 kWh (Dacia Spring) to 118 kWh (Mercedes EQS), with maximum charging rates varying from 22 kW (Smart EQ) to 350 kW (Lotus Eletre), with many 'mainstream' vehicles capable of charging at over 150 kW.

5.4.1 EV penetration

Another challenge for predicting the required future EV charging infrastructure is the penetration of EVs. The number of new EVs registered in the United Kingdom continues to grow. In quarter 2 of 2024, EVs formed 17.5% of all new car registrations, up from 16.5% in 2023 and 1.1% in 2019. However, by the end of 2024, EVs made up only approximately 4% of the total number of cars on British roads. Despite the larger number of EV models available for sale, the demand for EVs appears to be softening, due in part to saturation in early adopters (likely to be business users) and more limited uptake in the wider population. The Government's EV mandate stipulates that by 2025, 28% of all new cars sold must be EVs, rising to 80% by 2030 and 100% by 2035, with the 100% date 5 years later than originally proposed. Therefore, at the current time (2025), EVs remain a small minority of vehicles on the road, but due to the EV sales mandate, these figures will increase considerably over the next decade, needing to be matched by a similar growth in public charge points.

It is also predicted that the percentage of EV owners with domestic charge points will decline from the 85% in 2022, to around 65–70% in the future as more people without off-street parking adopt EVs [4].

5.5 Charge point infrastructure

What represents the 'best' measurement of charge point availability? From a user's perspective, their definition of charge point availability is that a charge point is always available when they want to charge. However, this can be considered an idealised case, similar to the case of driving on a motorway with no congestion. We know that it is impractical to install the same number of charge points as EVs on the road, so what is the 'correct' number of vehicles per charge point? Another consideration is how localised should this measure be calculated? From a user perspective, it is not particularly helpful for the country to have a low number of vehicles per charge point, if in their particular area the number of vehicles per charge point is very high and there are considerable challenges in finding available charge points.

The future required number of charge points for the United Kingdom varies; a medium estimate value is 500,000 total charge points by 2030 to service between 7 and 18 million EVs [5]. This equates to approximately 137 charge points per 100,000 of the population if assumed this is equally spread. In terms of EVs per charge point, this stands at approximately 17:1 compared with the estimated optimum of 10:1. It is important to consider figures in the United Kingdom from a national, regional and city perspective. For example, figures for the United Kingdom as a whole show 89 charge points per 100,000 of the population [6]. However, there are clear regional discrepancies across the United Kingdom, as shown in Figure 5.6, highlighting the lower density of charge points in the middle of the country, with London having a charge point density of approximately 2.5 times that of the next densest region (Scotland). The West Midlands region has 81 charge points per 100,000 of the population. If we consider the West Midlands itself, it is evident from Figure 5.7 that there is an increased density of charge points in Coventry where there are 500 per 100,000 of the population [7]. The reasons for this could be multiple and various, including incentives to install charge points, local priorities and density of EVs. To expand on the regionality perspective, Zapmap provided figures for mid-2024 per local authority. This breaks down Figure 5.6 to a more granular level as shown in Figure 5.8; however, the same scarcity of charge points is apparent in the middle of the country from approximately the Mersey to the Humber.

5.5.1 Charge point costs and emissions

It is not only the availability of public charge points that is important, but their cost. Current overnight domestic rates for homes offer costs as low as 8p/kWh compared

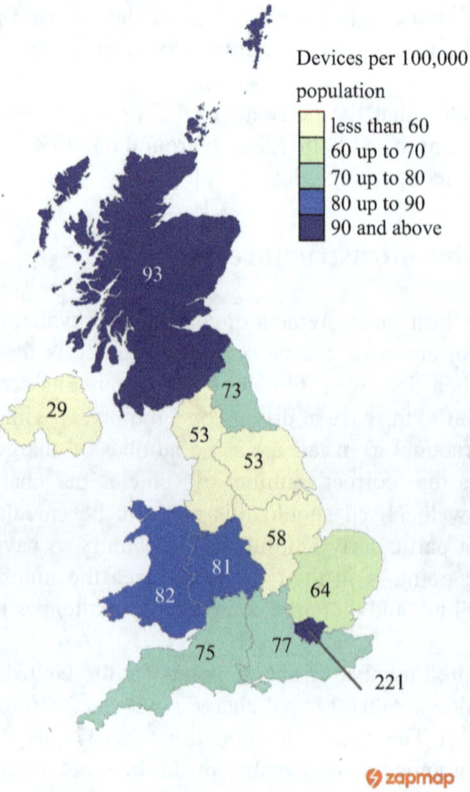

Figure 5.6 Charging devices for 100,000 population in the United Kingdom in 2024 by region [Source: Department for Transport. Electric vehicle public charging infrastructure statistics: April 2024 – Map 1 [online]. 2024. Available from https://www.gov.uk/government/statistics/electric-vehicle-public-charging-infrastructure-statistics-april-2024/electric-vehicle-public-charging-infrastructure-statistics-april-2024#maps1and2. This information has been licensed under the Open Government Licence v3.0. To view this licence, visit https://www.nationalarchives.gov.uk/doc/open-government-licence/ [Accessed 27 February 2025]].

with public rapid charging costs of 81 p/kWh on average [8]. A 50 kWh recharge of an EV battery could therefore cost £4.00 at home or £40.50 at an expensive public charge point. This difference reflects the additional cost of installing such infrastructure. It should be noted that a home charge point will typically be 7 kW, whereas a public charge point may typically be up to 150 kW. For overnight charging, an EV would receive an adequate charge from a 7 kW charge point – thus in residential areas to service EV charging overnight, 7 kW charge points may be preferred.

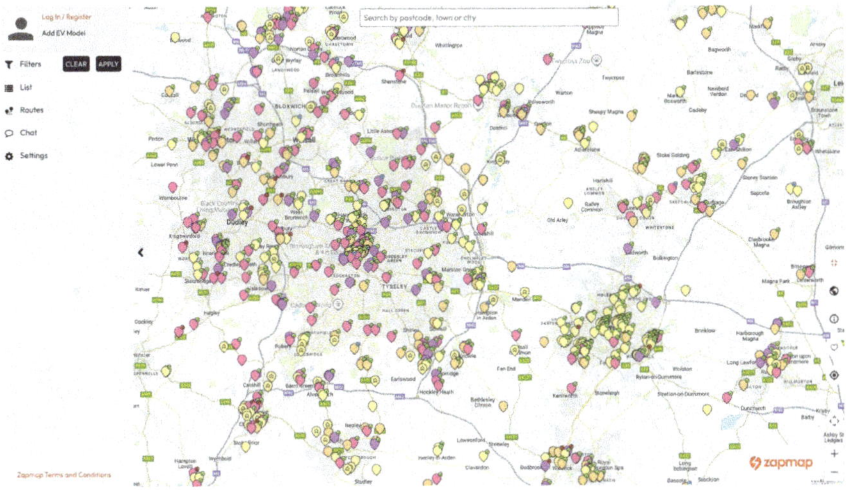

Figure 5.7 Locations of charge points in the West Midlands, centred on Coventry showing increased charger density [Source: Zapmap. Map [online]. 2025. Available from https://www.zap-map.com/live/ [Accessed 27 February 2025]].

Many drivers may not inherently want to rapid charge – they do it out of necessity and therefore it needs to be as painless as possible. Providing the vehicle has the necessary range they may want to charge when they reach the end of their journey, regardless of its length. Charging in this instance is known as destination charging, and is serviced by charge points suitable for this use case. Table 5.3 shows how charge point power and dwell time combinations may result in a relatively consistent energy delivery. This is a very generalised table, but reflects the fact that in the future, especially for those owners without home charging, there could be considerable choice over how and when to charge depending on the charging use case, and how much it will cost.

Table 5.3 demonstrates that it is important to separate charge point demand into several categories depending on the charging use case. The most common separation is between residential, destination and en route. Examples of these are: typical residential on-street parking areas where home charge points may be inconvenient; destination, for example a cinema, shopping centre, sports venue, and en route which covers both motorway/trunk road service stations, and traditional service stations where the opportunity exists for additional purchases, for example with an attached coffee shop. The interesting point to note is that the energy absorbed is comparable for all three of the cases covered in Table 5.3, but the costs are very different, even for public charge points.

Currently, there is little thought by charge point operators (CPOs) regarding the emissions of the electricity used to charge. For example, if there is a journey on the M6 from the east of Birmingham to the border of Scotland, according to carbon intensity calculations provided by the National Energy Systems Operator and focusing on 22 November 2024 as an example, there is a factor of 4 difference in

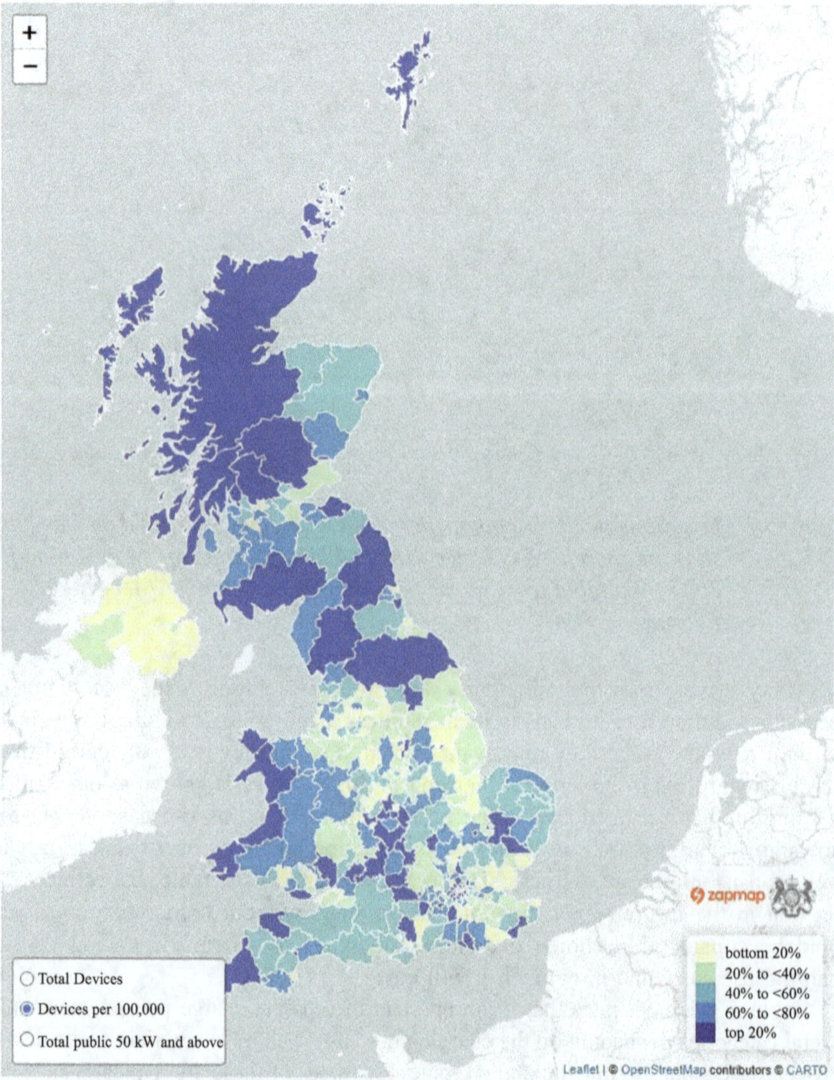

Figure 5.8 Charging devices per 100,000 population by local authority [Source: Department for Transport. Electric vehicle charging devices by local authority [online]. 2025. Available from https://maps.dft.gov.uk/ev-charging-map/index.html. This information has been licensed under the Open Government Licence v3.0. To view this licence, visit https://www.nationalarchives.gov.uk/doc/open-government-licence/ [Accessed 26 February 2025]].

Table 5.3 Typical dwell times and charger speeds

Charging type	Dwell/charging time	Charger power requirement (kW)	Typical energy absorbed (kWh)	Typical price (p/kWh)	Total cost (£)
Domestic (home)	7 hours–overnight	7	49	8	3.92
Residential (street)	7 hours–overnight	7	49	53	25.97
Destination	2.5 hours–daytime	20	50	53	26.50
En route	20 minutes–daytime	150	50	81	40.50

Table developed by Author based on source data
[Source of data: Zapmap. *Zapmap Price Index* [online]. 2025. Available from https://www.zap-map.com/ev-stats/charging-price-index [Accessed 27 February 2025]]

carbon intensity if a driver recharged in the West Midlands compared with the North West (West Midlands 226 g/kWh; North West England 57 g/kWh) [9]. This could only be a difference of tens of miles apart between charge points, but entirely different supply regions.

5.5.2 Grid electrical context

Bringing the strands of previous sections together, we now consider EV charge point demand predictions. If all the cars on British roads were electric, the additional power demand would amount to approximately 8.4 GW assuming an average annual distance of 7,500 mi driven and a typical EV energy efficiency of 300 Wh/mi [10]. 8.4 GW sounds a lot, but typical grid demands range from around 25 to 50 GW. In addition, electricity peak demand has gradually fallen from a peak of 62 GW in 2002 to approximately 52 GW in 2023 [11]. Table 5.1 in addition shows that the current installed capacity of the grid is approximately 97 GW; thus, the grid has the clear capability to meet these additional demands. Fortunately, the decarbonisation of the grid is also progressing at pace, thus allowing the potential for the majority of EV charging to be met by low-carbon sources.

5.6 Translating grid demand and supply into charge point demand and supply

Electricity demand and supply is an important factor when determining the best place to deploy charge points; however, these should be considered alongside other factors that also heavily influence effective deployment. This section outlines these factors and the methodology which should be followed to aid in the deployment process.

When deploying charge points, the overarching methodology remains the same regardless of the three en route, destination and residential use cases. The only aspects that change is the data that is used. The methodology is broken down into four parts. Demand analysis is the first step which is then used to create utilisation projections and financial models. This all then feeds into determining real-world locations for charge point deployments. The outputs of this analysis all feed in to the business case and planning of deployments at a local level step, further details of which can be found in Chapter 7.

5.6.1 Demand analysis

Demand analysis focuses on understanding how current and future demand for charge points in a specific area translates into potential locations for deployment. Put simply, demand is ultimately the number of EVs that may use a charge point now and in the future. Simplistic analysis at a localised level takes the number of current and projected EVs and divides by the number of current and projected charge points, giving an EV per charge point figure. Based on this simple analysis, high demand areas are where the EV per charge point figure is higher than in low demand areas where the EV per charge point figure is lower. The challenge, however, with this analysis is that it fails to take into consideration how different people charge based on the different charging use cases and the various other factors that influence demand based on these use cases. For example, an area with a high registered number of EVs and low number of charge points might indicate high demand specifically for residential charge points. However, if 98% of drivers in the area have access to their own charge point at home, the EV per charge point number plummets and may potentially result in little to no demand for public charge points. For each use case, the data required to conduct this analysis differs slightly, however, the approach stays the same and involves analysing the data, usually via a geospatial tool, and then identifying potential sites based on this data. At this stage, these sites are often streets or areas. The specific locations on these streets or in these areas are further refined as part of the deployment at a local level step, as noted in Chapter 7, and then finalised through detailed designs as part of the build stage, as noted in Chapter 8. We will now outline how the demand analysis process differs based on each use case.

5.6.1.1 En route demand analysis

As en route charge points are deployed on the strategic or major road network, including feeder and trunk roads, demand analysis focuses mainly on traffic flow along the road network and the proportion of which are EVs that require charging now and in the future. Regarding the former, the starting position is determining traffic flow for all vehicles along the route, usually based on traffic count data captured by highways authorities and projecting this into the future. Regarding the latter, this can be based on calculating the total proportion of EVs in an area or geography's roads and then applying this directly to traffic count data to give an indication of the potential number of EVs travelling on the road. Although this is not an exact science, it gives a potential indication. Following this, further analysis needs to be conducted as not every EV along the route will need to charge. To determine this, some assumptions need to be established. It is incredibly difficult to settle on exact numbers however a good rule of thumb is that anything between 5% and 15% of EVs travelling along the route may need to charge at some point. To refine this further, the number of current and future charge points along the route must be considered as those EVs needing to charge en route may charge at other locations. The analysis should also be cast further afield to consider charge points within a mile or so of key junctions as drivers may choose a small detour if the location is more attractive to charge. Taking all of these points into consideration will likely result in a reduction in the number of EVs which need to charge along a route. This final number forms the basis for en route charging demand in a specific location and should be used to right size the number of charge points required at a site, to satisfy immediate demand as well as future potential demand.

5.6.1.2 Destination demand analysis

Destination demand analysis differs to en route demand analysis because the potential users at a site might charge there for a variety of reasons, thus meaning that the data required to determine demand can be derived from a variety of sources. That being said, the first step is to determine the primary reason for demand which is derived from the type of site. For example, destination demand at a hotel will be determined by those staying overnight, whereas demand at a shopping centre, supermarket, leisure or cinema complex will primarily come from those visiting and spending a couple of hours charging while they do something else. Of course, destination sites may also serve a secondary reason for demand, especially in locations where multiple services coalesce, for example, at large out of town shopping complexes. In these locations, the primary reason might be a hotel, whereas the secondary reason might be a shopping centre. Furthermore, where these locations are close to residential areas, a tertiary reason for charging may exist, making it all the more difficult to determine the demand for charging. Any of these three reasons can combine at a destination site to push up demand. Reflecting on all of this, a structured process determining destination demand can be followed based on the primary, secondary, and in some cases, tertiary reasons for charging.

Regarding the primary reason for charging, the starting position is the potential number of users at a destination site. This can be based on the size of the car park and be perfected based on car parking usage data. If a destination services a local area, the current and projected future EV adoption rates in that area can be applied to the car park usage data to determine the number of vehicles using it which could potentially be EVs. If the destination services a wider area than just the immediate community, the EV adoption rates will need to reflect this. If the destination is a hotel, the number of rooms cross-referenced against the number of hotel car parks should be considered, with national current and future EV adoption rates applied to the figure to determine the potential number of EVs. Similar to en route charging, it cannot be assumed that everyone will charge, so analysis needs to settle on the percentage of vehicles which are likely to charge which is related back to the destination use case. For shopping centres, leisure centres and other facilities where users may spend a couple of hours, for a vast majority of drivers it is unlikely that they are reliant on charging there and are probably more sensitive to charge point pricing. This means that a lower percentage (possibly as low as, or lower than, 5%) of all EVs visiting the site will want or need to charge, impacting on the number of charge points needed now and in the future. For hotels, the percentage of vehicles which may need to charge could be higher than 50%, meaning a potentially larger number of charge points would be required.

A secondary reason for destination charging might not always apply but where it does, this will push up demand as combining the different destination use cases will increase the number of vehicles that may need to charge. Once this analysis is done, further analysis is required to adjust demand based on the charge points that currently exist and may exist in the future. If the sites are already well serviced, or will be in the future this may push demand for charge points down further. For hotel destination charging, en route charge points within the vicinity of the hotel should also be considered as this may reduce demand if users have stopped for a break, especially when travelling via the strategic/major road network. This may drive down demand. Taking all of these factors into consideration enables the estimation of the current and potential future charge points to meet current and future demand.

A tertiary reason (or secondary reason depending on the combination of services at the site) for destination charging may result from those currently living in (or planned to live in) residential areas close to the destination. Although this does not always apply, there are various shopping centres, leisure facilities and hotels that exist immediately next to residential areas. To understand the demand in this instance, the residential demand analysis process should be followed as outlined in Section 5.6.1.3 and in Section 7.4 in Chapter 7. Combining the demand across all destination use cases in an area enables the right sizing of the number of charge points.

5.6.1.3 Residential demand analysis

Residential demand analysis requires a different mix of data sources to determine charge point demand in a local area. The starting point is determining the granularity of the data at a local level. For example, is data available at street level or the next level up (in the United Kingdom, this would be an output area, then lower layer super output area and so on)? Once this is known, the key factors that determine demand for public charge points at an area level include current and projected EV adoption levels at a post code or house level and socio-demographic data including that derived from credit agencies. The latter focuses on population, number of households, income per household, propensity to own an EV and travel behaviours. These factors all heavily influence EV uptake. Another factor is access to an off-street driveway for private charging. Combining all of this data will give an indication of the areas where current and future EV charge point demand is high or low. This is often represented as heat maps which can then be used to target deployment. For example, areas with high current or future projected EV uptake but low off-street parking availability will present as high demand areas for charge points, whereas areas with high off-street parking will be low demand regardless of current and future projected EV uptake. Socio-demographic data is more nuanced and has to be handled with care as the assumption that more affluent areas will be higher demand fails to recognise that the emergence of lower cost EVs is starting to positively impact on adoption across all income levels. Nevertheless, certain subsets of this data (like propensity to want to own an EV) provide a vital snapshot which, when viewed alongside other data, helps in planning. Based on this analysis, demand for residential charge points is potentially high, with a useful rule of thumb being that 50% or more of residential EVs without access to an off-street charge point will want to charge locally. This is because the vehicle remains stationary in this location for the greatest length of time. Existing and future planned charge points should, of course, also be considered as these may currently be satisfying demand. Like en route and destination charging, assumptions need to be applied to adjust this 50% figure further. In most cases, it is assumed that those charging residentially do not have access to a charge point at work or elsewhere so the bulk of charging will take place at home. Cost to charge needs to also be considered and it is assumed that most charging is cheaper residentially so drivers will opt to do this over other use cases. There is only one use case that might influence demand for residential charging and that is destination charging, especially if this is clustered around a residential area or is provided in locations where local residents often visit. That being said, the cost to use these would need to be comparable to residential charging and that even if this is the case, it is unlikely that residential demand would solely be met by destination charge points. As a result, demand for residential charge points is based mainly on 50% or more of residential EVs needing to charge with little or no adjustment based on assumptions. This means that in areas with current or high EV uptake and low off-street car parking availability, a large number of charge points might be needed.

5.6.2 Utilisation analysis

Whereas demand is about the potential vehicles which might charge based on various factors and adjusted based on assumptions, utilisation analysis is about predicting how demand translates into actual and predicted use. Utilisation can be measured in many different ways and differs based on the three charging use cases. A paper focusing on demystifying utilisation was produced in 2023 by the Green Finance Institute to determine what is meant by utilisation and how it can be measured [12]. The paper recognises that there are two main ways in which utilisation can be measured. This can be time-based or energy-based. Regarding the former, this measures the amount of time vehicles are plugged in to a charge point and translates this into a percentage. For example a vehicle might be plugged in for 3 hours over a period of 24 hours and no other vehicle then plugs in. This would result in utilisation of 12.5%. Regarding the latter, this measures the energy supplied by a charge point in relation to the potential maximum energy that could be supplied over the same period. For example, if a 7 kW charge point delivers 14 kWh in total over a 24 hour period, this would result in a utilisation of 8%. When applying these in the real world, both have their value. Time-based utilisation gives an idea of when a charge point is used and cannot be used by others, regardless of the power delivered. This is an important measure as a charge point might be in use because the vehicle is plugged in but is not charging. The challenge with this measure, however, is that from a charge point energy throughput perspective, the measure cannot then be used to understand revenue from the charge point as that can only be calculated based on charge point energy usage. Energy-based utilisation is very useful from a financial calculation perspective as the cost per kWh can be multiplied by the total kWh drawn per day resulting in a daily income. This then feeds in to charge point income projections over a period of time. The challenge with this approach, however, is that not considering how long a vehicle is plugged in might mask periods when a vehicle is plugged in but charging at low speeds because it is near to reaching its maximum state of charge. Multiple incidences in a day where vehicles are charging much more slowly may be hidden by a higher average utilisation figure across a day. For a CPO, they want to see every charging session taking place at the maximum charge rate per charge point as this generates higher revenue. Based on the comparison of the approaches, the most accurate way to model utilisation is a combination of both.

Predicting utilisation is vital in understanding the viability of sites as predicted low time and energy-based utilisation would indicate a site which is not suitable for deployment, unless utilisation is predicted to grow from a low level over time. If that is the case, the utilisation ramp-up period would need to be considered as part of the site's viability. Following a structured methodology to predict utilisation is heavily reliant on various potential factors; however, these can be defined and detailed in such a way to develop realistic hypothetical predictions per each charge point. Although the details of each factor differ across the three use cases, the below factors themselves stay the same:

- time of day charging takes place,
- anticipated length of stay based on the use case,
- number of potential charging sessions per day,
- charge point speed, and
- anticipated vehicle state of charge on arrival.

Using these factors helps develop a maximum potential utilisation for each use case. This can then be amended based on localised features derived from the demand analysis to develop a realistic localised potential utilisation. That figure then helps in understanding the utilisation viability for each potential site, playing an important role in site selection.

5.6.2.1 En route charging

The vast majority of en route charging occurs during the time period 07:00 to 19:00 as this is when most journeys on key routes are completed. Of course, some charging will take place after 19:00 and before 07:00 but the number of these sessions will be much smaller. Within this period, there are peaks in weekdays during key commuting times of 07:00 to 09:00 and 17:00 to 19:00. Length of stay at en route charge points tends to be around 30 minutes to 1 hour.

The number of potential charging sessions can be as many as 2 per hour during the 07:00 to 19:00 period. The charging speed in these settings varies but is often 100 kW or greater and is provided by a level 3 DC charge point. Vehicles arriving at these locations will likely start with a low state of charge and will charge at higher speeds. However, in order to charge at the fastest speeds, batteries often need preconditioning to be able to accept charge at higher rates. Additionally, the rate of charging reduces as state of charge increases. Taking into consideration the above factors, the maximum potential time-based utilisation would be around 50% and the maximum potential energy-based utilisation would also be around 50%. When adjusting this based on real-world factors, the realistic utilisation for both would be somewhere around 30–45% due to time lost per charging sessions as a result of driving to the charge point, plugging in, unplugging and driving away, and energy efficiency losses due to charge points never delivering full charging rates for a range of reasons. If demand for the site is not consistent and traffic flow is low, the potential utilisation rate may fall further. Reflecting on all of these points, predicting en route charging utilisation needs to project trends forward based on future potential changes in factors like EV traffic flow and charge point competition. This could drive utilisation up or down and will have a major impact on the viability of the charge point in its local setting.

5.6.2.2 Destination charging

Destination charging occurs in two distinct charging time periods depending specifically on the use case. For a hotel most charging will take place between 17:00 and into the early hours of the following day, assuming those who are staying plug in when they arrive and leave the vehicle plugged in through the entire evening and into the following morning, only unplugging once they start their journey the next day. For a supermarket, shopping centre, leisure centre or similar location, charging will take place throughout the day, potentially starting from 08:00 and finishing at 20:00. For a cinema, charging is likely to take place from 11:00 onwards up until 23:00 potentially. As with the en route use case, some charging will take place outside these time periods, though this only makes up a small number of charging sessions. Length of stay in non-hotel destination use cases is usually around 2–4 hours with the number of sessions per day being 4–6. For hotel stays, this is often over 6–10 hours, with there being a maximum of two sessions. The most common charging speeds in these settings is between 7 and 11 kW provided by a level 2 charge point. For hotel charging, vehicles will likely arrive with a low state of charge because it is assumed they have travelled a longer distance to get there, whereas for non-hotel destination charging sessions, vehicles are not reliant on charging so will likely arrive with a medium to high state of charge. Taking into consideration the various factors, the maximum potential time-based utilisation for hotel destination use cases is 58% and 50% for non-hotel destination use cases. The maximum energy-based utilisation for hotel destination use cases is 58% and is based on an assumption of two sessions lasting 7 hours using a 7 kW charge point. For non-hotel destinations, this would be 50%, again based on a 7 kW charge point.

When adjusting this based on real-world factors, the realistic time and energy-based utilisation for a hotel would be about 29% assuming that a visitor would plug in their vehicle when they arrive and leave it plugged in all night and into the next day when they continue on with their journey. The maximum power delivered would be around 50 kWh in one single session, assuming the vehicle charged to full and did not arrive with an empty battery. For non-hotel destination use cases, the realistic time and energy-based utilisation is about 20–30%, assuming a 7 kW charge point. This also assumes that there are gaps between charging sessions throughout the day as a result of users not being reliant on charging in these locations, which pushes down demand. For lower demand sites, time- and energy-based utilisation could be anything around 15% or less. This is also the case because charging at these locations may result in vehicles reaching full charge and therefore not drawing any power whilst still plugged in. Projecting these calculations forward is more difficult in destination settings due to the range of different factors playing a different role in each use case; however, local and national EV uptake will either drive utilisation up or down as will the availability of competing charge points. Both impact on the viability of these sites. This is of greater importance than in en route settings because the realistic time and energy utilisation figures are starting from a lower potentially unviable position depending on the specifics of the site (see Section 7.3 in Chapter 7 regarding business case viability).

5.6.2.3 Residential charging

Residential utilisation is less complex than destination utilisation because the factors are much more predictable. The vast majority of residential charging starts in the early evening and runs overnight until the early morning. Length of stay can be as much as 14 hours, covering the period when a driver arrives home from work and then goes to work the next day. As a result of this prolonged stay, residential charge points tend to see 1 to 2 sessions per day, with the second session only taking place in the daytime as a result of those charging who work from home or who are not working. Residential charge point speeds can be anywhere between 3 and 7 kW. Drivers charging will almost always rely on residential charge points and may choose to either charge fully from a low state of charge or top up from a medium state of charge. Taking into consideration the above factors, the maximum potential time-based utilisation would be around 58% and the maximum potential energy-based utilisation would also be around 58%, assuming a level 2 7 kW charge point that is used to charge over two sessions delivering about 49 kWh per session. When adjusting this based on real-world factors, the realistic utilisation for both would be about 29% as in most cases only one charging session will take place overnight for about 7 hours to fully charge a vehicle and that vehicle will remain plugged in for several more hours after being fully charged, blocking the charge point for other users. Predicting utilisation in residential settings is reliant on local EV current and future uptake, availability of off-street charging and charge point competition. Projecting EV uptake forward and combining this with projected charge point competitor growth will give an indication as to whether utilisation will rise or fall (it is assumed off-street charging stays relatively constant). Like destination charge points, utilisation over 15% is classed as a good starting point. One caveat, however, is trailing cable charging using domestic power. If this is allowed in an area, this will greatly reduce demand for public residential charge points and may make them unviable altogether.

5.6.3 Financial modelling

The outputs from demand and utilisation modelling can then be used to financially model the potential revenue generated at sites which then can be compared with the capital, development and operational costs of deploying and running a charge point. Unlike demand and utilisation modelling, the financial modelling methodology is identical regardless of the use case. The starting position is to understand the potential gross revenue of a charge point. This is calculated by multiplying the annual predicted maximum potential energy-based utilisation when represented in kWh (and not as a percentage) by the cost charged to the customer per kWh. Once this is done, the costs to install a site must be considered so as to understand when a charge point will break even and then start making a profit. The capital costs to install a charge point relate to the charge point itself. Level 2 charge points are cheaper to buy than level 3 DC charge points, especially when compared with the higher powered 150 kW plus models. The development costs to install a charge point include the grid connection costs that usually relate to connecting the entire site but can be apportioned to each individual charge point, the civils costs in terms of labour and materials, any licensing costs for planning and traffic regulation orders and any overhead costs like project management. The operational costs relate to the energy import cost and the charge point back-office management costs including any payments made for processing charge point payments (usually as a percentage of each transaction). Other costs include on-going charge point maintenance costs and overhead costs including operational management of charge points and provision of a customer service (a helpline for an example). Subtraction of projected costs from the projected gross revenue will result in a net revenue which can be either positive or negative. Most CPOs expect a payback period for a charge point, though some may be happy for a charge point to be loss leading if it serves another purpose, for example, in securing access to a site which is expected in the future to become profitable. Over the payback period, the net revenue is usually a negative at the start because the CPO is paying back the initial capital and development outlay. The payback period differs based on the site and use case with most CPOs expecting a payback period of around 2–4 years. After this, the charge point becomes profitable. Most charge points have expected lifespans of 7–10 years provided they are properly maintained, though partway through this period the charge point may be upgraded or replaced. High utilisation sites may even see payback within the first few years, though this is not common. Predicting the financials of potential sites is critical to their viability as some sites may never become profitable either because of high development costs or low gross revenue due to poor utilisation. It is in the interest of the CPO to understand this as early as possible in the planning stage. They must feed this into the business case for charge point deployment (see Section 7.3 in Chapter 7) as the result of this analysis may mean that the business case does not stack up and that funding cannot be secured to deploy charge points. Other stakeholders, including local authorities and regional transport bodies, should also take financial modelling into consideration. This is because pushing the deployment of financially non-viable sites, which the CPO must operate as part of a larger portfolio of sites, may result in the CPO either refusing to bid for the contract to install them or, if successful, struggling financially throughout the duration of the concession. This may also result in the CPO having to push up prices across their charge point network, pushing utilisation down in profitable sites and denting the impact that charge points have in encouraging the adoption of EVs. Financial modelling is not only important for determining potential sites. It can also be used once a site is live to track financial predictions against actuals which then enables the CPO to take action if the site is not performing as expected. This could take the form of lowering the price per kWh to drive up utilisation or encouraging drivers to charge at their sites through other incentives (e.g., discount shopping vouchers).

5.7 Translating this into real-world locations

The outputs of demand, utilisation and financial modelling can be combined with power supply data in a localised setting to determine the most viable sites based on profitability and practicality. This is normally achieved by inputting the data for demand and power supply into geographic information system (GIS) software, then using this to plot potential practical sites as pins on the map with associated local data like the latitude, longitude and street which the charge point would sit on. Utilisation and financial modelling is normally completed for the potential sites in a spreadsheet which determines those which are potentially profitable. This data can then be fed into the GIS software to show where the sites are in relation to other sites. The outputs of this analysis feeds into two important steps in the planning stage, the development of the business case and the planning of deployment at a local level.

5.7.1 Real-world examples

We have highlighted two case studies that show how demand analysis, utilisation and financial modelling have been used in the real world to determine charge point sites. There are multiple use cases that exist across the industry; however, the examples demonstrate key aspects of the broader site selection methodologies. It should also be noted that these real-world examples are brought to life based on data available in the public domain.

Case Study 5.1: Residential charge point deployment: Oxfordshire County Council's use of Mind Foundry

Oxfordshire County Council (OCC) worked with Mind Foundry, a specialist in artificial intelligence (AI), to combine geospatial modelling with a number of differing data sources to model and predict EV charge point infrastructure requirements. Mind Foundry's platform uses machine learning and connects with both live and historic data sources to model different scenarios. The platform monitors the available electricity capacity which can then be used to inform where to deploy charge points. According to the Geospatial Commission which analysed the project, this has allowed OCC to visualise and understand immediate demand, local constraints and individual needs to make decisions about EV charge points. They are able to analyse potential sites and determine their suitability, working within the policies set out in their local EV infrastructure strategy, to quickly produce more accurate and targeted applications for charge point funding. According to Mind Foundry, the tool has also enabled OCC to 'increase profits from charging stations, better manage local and national electricity demand, ensure ethical consideration for local communities and optimise infrastructure budgets'. Although there are many tools that enable geospatial planning and modelling of charge points, Mind Foundry's use of AI is particularly interesting and will become an integral part of geospatial tools going forward.

[Source: Mind Foundry. *Case study – Optimising electric vehicle charging for Oxfordshire County Council* [online]. 2025. Available from https://www.mindfoundry.ai/resources/case-study/optimising-ev-charging [Accessed 27 February 2025]]

[Source: Geospatial Commission (Geospatial Commission became part of Government Digital Service in January 2025). *Getting to the point – Accelerating EV charge point rollout through geospatial data* [online]. 2022. Available from https://data.parliament.uk/DepositedPapers/Files/DEP2022-1031/Geospatial_Commission_Report.pdf [Accessed 27 February 2025]

Case Study 5.2: En route charge point deployment: Dublin

A recent paper published in 2023 for Sustainable Cities and Society explored the role of GIS in determining destination, en route and residential charge points in Dublin. Focusing specifically on en route charging the methodology used a free GIS tool called QGIS to integrate a number of data sources including existing en route charge point data, the major road network and existing car parks. The methodology also developed a process to select sites based on a range of criteria and an approach to prioritising charge point locations. The results of following the methodology were that for en route charging, 68 locations were identified and required by 2025, and 165 locations by 2030. These locations should be deployed along the motorway, primary, secondary and tertiary road network, situated 150 m from the road and 500 m away from existing charge points. The paper did not indicate the number of charge points required in en route locations. Nevertheless, the methodology is directly applicable to other geographies provided the data is available.

[Source: Charly A., Thomas N.J., Foley A., Caulfield B., "Identifying optimal locations for community electric vehicle charging", *Sustainable Cities and Society*, Volume 94, 2023, 104573, ISSN 2210-6707. Available from https://doi.org/10.1016/j.scs.2023.104573 [Accessed 27 February 2025]]

5.8 Conclusions

This chapter has covered the EV charge point supply and demand modelling approaches from several perspectives. At the time of writing, the EV car and van market, the EV charger provision industry and the electrical generation industry are in a state of flux, with considerable uncertainties going forward. This chapter has considered the changing face of the electricity generation industry with the diversification of supply into a greater number of more uncontrollable, smaller and more spatially distributed generators of greater feedstock types. In many cases, these generators are in locations where electrical demand is weak. The increasing diversification of the generation mix is leading to more variable carbon dioxide emissions, depending on the relative contributions of wind and solar. Thus, time of use for high demand users (EV charging in a domestic setting) is becoming increasingly important. Outside of household domestic charge points, the proposed location and required number of public charge points remain challenging to model accurately. Several methodologies have been outlined in this chapter based on demand, utilisation and financial modelling, which helps determine potential sites. This analysis is a pre-requisite for the specific EV charging technologies required at each location as described in Chapter 6.

References

[1] OpenInfraMap. *Power Plants* [online]. 2025. Available from https://openinframap.org/stats/area/United%20Kingdom/plants [Accessed 26 February 2025].

[2] The Office of Gas and Electricity Markets. *Smart Export Guarantee (SEG)* [online]. 2024. Available from https://www.ofgem.gov.uk/environmental-and-social-schemes/smart-export-guarantee-seg [Accessed 26 February 2025].

[3] Drax. *Electric Insights Quarterly – July to September 2024* [online]. 2024. Available from https://reports.electricinsights.co.uk/wp-content/uploads/2024/12/241202_Drax_24Q3_006.pdf [Accessed 26 February 2025].

[4] Longstaff T., and Healy A. *A model to forecast EV charging demand* [online]. 2024. Available from https://www.rolandberger.com/en/Insights/Publications/A-model-to-forecast-EV-charging-demand.html#:~:text=For%20example%2C%20EV%20penetration%20in,times%20between%202023%20and%202050 [Accessed 26 February 2025].

[5] Energy Saving Trust. *Forecasting public electric vehicle charging demand – a guide for local authorities in England prepared by the Local Government Support Programme* [online]. 2022. Available from https://energysavingtrust.org.uk/wp-content/uploads/2022/09/EST_LGSP-Guide_Forecasting-public-electric-vehicle-charging-demand.pdf [Accessed 26 February 2025].

[6] Department for Transport. *Electric vehicle public charging infrastructure statistics: April 2024* [online]. 2024. Available from https://www.gov.uk/government/statistics/electric-vehicle-public-charging-infrastructure-statistics-april-2024/electric-vehicle-public-charging-infrastructure-statistics-april-2024#annex-a-public-charging-device-by-100000-population-table [Accessed 26 February 2025].

[7] Mullen E. *Coventry declared UK's most electric car-friendly city* [online]. 2022. Available from https://www.coventrytelegraph.net/news/coventry-news/coventry-declared-uks-most-electric-23405240 [Accessed 26 February 2025].

[8] Zapmap. *Zapmap Price Index* [online]. 2025. Available from https://www.zap-map.com/ev-stats/charging-price-index [Accessed 26 February 2025].

[9] National Energy System Operator (NESO), Environmental Defense Fund Europe, University of Oxford Department for Computer Science, World Wide Fund. *Carbon Intensity API* [online]. 2025. Available from https://carbonintensity.org.uk/ [Accessed 22 November 2024].

[10] NKD Group. *How much additional power does the UK need to generate if every car in the UK was fully electric* [online]. 2024. Available from https://nkd-group.co.uk/2024/08/16/how-much-additional-power-does-the-uk-need-to-generate-if-every-car-in-the-uk-was-fully-electric/ [Accessed 27 February 2025].

[11] National Grid. *Can the UK grid cope with the extra demand from electric cars?* [online]. 2023. Available from https://www.nationalgrid.com/stories/journey-to-net-zero-stories/can-grid-cope-extra-demand-electric-cars [Accessed 27 February 2025].

[12] Green Finance Institute. *Demystifying utilisation* [online]. 2023. Available from https://www.greenfinanceinstitute.com/wp-content/uploads/2023/06/GFI-DEMYSTIFYING-UTILISATION.pdf [Accessed 27 February 2025].

Chapter 6
Charge point technologies

Stuart Bradley[1] and David Trousdale[2]

In this chapter, we focus on current charge point technologies in the United Kingdom and what contexts they are being deployed in. This includes the charge point itself, the customer front facing platforms and the back office systems required to manage the infrastructure. We start by exploring the key technological aspects of EVs and how they impact on charge point infrastructure.

6.1 EV architecture

EVs need to connect to an electricity source to add energy to the on-board battery energy store. This electrical energy can be from a fixed supply such as a grid connected charge point with a cable, or from a mobile source like an engine generator or a vehicle-mounted battery. The electricity transfer is controlled by a charging device and directed to the vehicle as dictated by the vehicle and grid conditions.

6.1.1 Vehicle battery types

EVs have a high-voltage battery, sometimes referred to as the main battery, operating at normally close to 400 or 800 V, and an auxiliary 12 V battery used for low-power circuits and to energise the main battery contactors that connect to the drivetrain. The 12 V battery is charged from a DC-to-DC converter or can be 'jump started' from an external power supply. Batteries are usually lithium ion with variations in chemistry such as lithium ferrophosphate and nickel metal hydride. They have a capacity ranging from 22 kWh for the earlier first generation vehicles like the Nissan Leaf, to over 100 kWh for luxury models such as those from Lucid, Tesla, Audi and Lotus. The trend is to have larger battery capacities than might be required to quell fears of needing to charge on a longer journey. As the public charge point network develops, the opportunity to down-size battery size, cost and weight will have an impact on vehicle adoption and their utility.

[1]Warwick Manufacturing Group, University of Warwick, United Kingdom
[2]Principal Advisor - Climate Change and Sustainability, Newcastle City Council, United Kingdom

6.1.2 Charging methods and on-board control

EVs frequently have two charging methods, both via a cable and connector mounted to the car bodywork, using an on-board charger (OBC) to convert the incoming AC power supply to the right parameters for the car battery. For example, the OBC will take 230 V 50 Hz electricity or other frequency and voltage combinations and convert to 400 V DC supplied to the battery management system (BMS) for the high-voltage traction battery. The BMS continually measures the main battery parameters, including the state of charge and the state of health, and determines the rate of charge acceptance. Battery acceptance rates will be lower if the state of charge is above 80%. This reduction is set to extend the battery service life.

Level 3 DC charge points supply directly to the battery via the BMS. This is commonly used for rapid and ultra-rapid charging. In this case, the electricity supplied to the car is between 400 and 1000 V DC, and is between 15 and 350 kW, with some future predictions to extend that upwards to 500 kW. A DC charge point is more expensive than an AC charge point because the conversion from AC to DC is done within the charger unit and not the car as with AC charging and the OBC.

An on-board DC-to-DC converter will convert the battery voltage to 12 V for the auxiliary battery, and to 5 V for ancillary devices such as low-power, low-voltage power supplies for instrumentation, cameras, control modules, sensors and low power lighting.

Analysis of the United Kingdom EV models shows that the overwhelming majority of cars are fitted with an OBC of 11 kW and lower, matching the power ratings that might be expected at domestic or workplace locations. The few exceptions are unusual models like the Tesla Model S Plaid – a very high-performance car, or the Smart ForTwo, which has AC charging capability only, and no DC.

6.2 EV makes and models – a UK context

Analysis of the United Kingdom EV fleet battery sizes shows that the majority have a lithium ion traction battery with a size from 40 to 80 kWh as is shown in Figure 6.1. If we split the fleet into car segments (A is mini, B is super mini, etc.), then we see a clear trend with smaller cars having smaller batteries. Nearly all cars have an OBC power rating of 11 kW and less to match domestic power supply ratings. This means that the time to charge is directly proportional to the battery size for the majority of cars when connected to a domestic charger).

However, the segment analysis in Figure 6.2 shows that a car's fast charging times are roughly consistent across the segments at circa 28 minutes when using their level 3 DC rapid charging capability, with smaller cars also more likely to have lower level 3 DC charging power ratings. There are several car models with improved charging technology, for example the Porsche and Audi GT architecture, or Hyundai group vehicles based on the E-GMP platform, both utilising an 800 V vehicle architecture. These architectures/platforms can charge at over 200 kW, meaning a 10–80% charge would take circa 15–18 minutes for this benchmark scenario in the most advantageous circumstances.

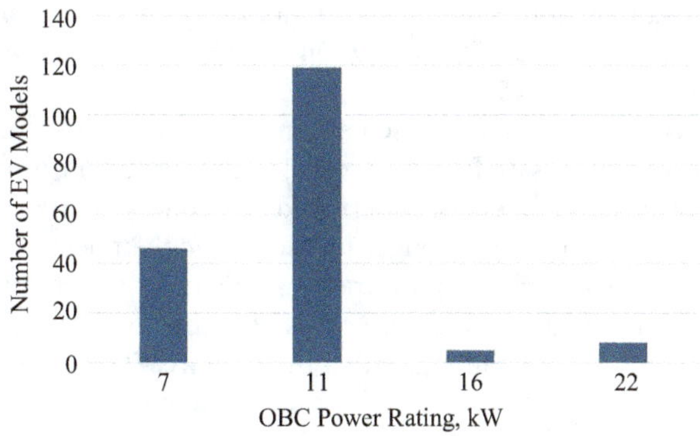

Figure 6.1 *Number of EV models with their OBC power rating. OBC power ratings are commonly 11 kW and lower [Source: Developed based on analysis conducted by the authors].*

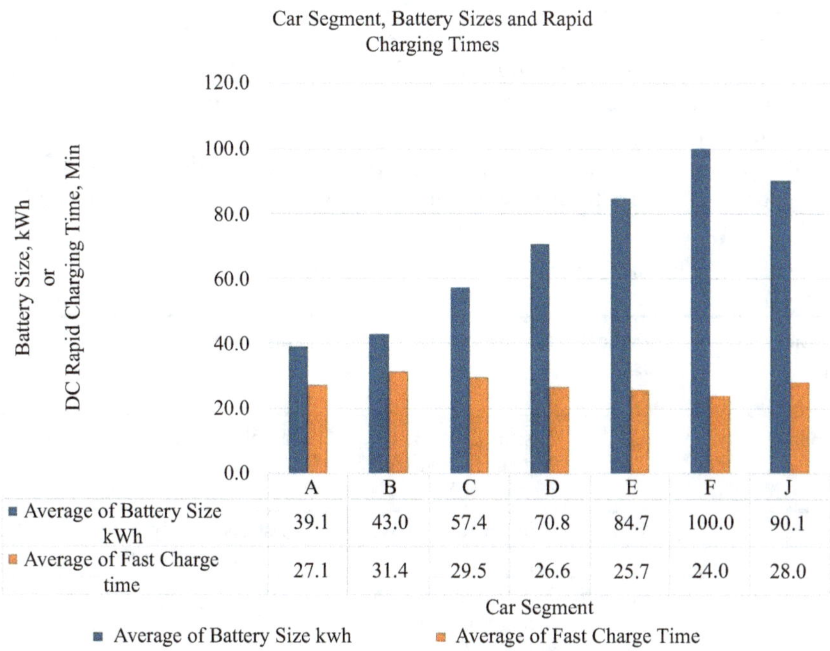

Figure 6.2 *The relationship between car segment, battery size and typical charge times when connected to a level 3 DC charger [Source: Developed based on analysis conducted by the authors]*

Figure 6.2 shows that the DC charging power across all car segments is proportional to battery size and that the allowable charge rate is a key limitation. It also shows that charge points could be sized to suit car segments and use cases, geographies and charging opportunity types. The typical charge point has a range of power ratings to suit the user's needs, and at motorway charge stations, we usually see level 3 DC charge points at rates between 20 and 350 kW, and AC charging at 11 kW and less for long-stays.

Typically, A segment cars are used around towns and cities, and their charging needs are normally satisfied by level 2 fast chargers in urban areas, and community areas such as cinemas, supermarkets, and city centre car parks. D, E and F segments are often used on longer journeys, and their needs are more likely to be met by level 3 DC rapid and ultra-rapid chargers at service stations and motorway junction stop areas. In all use cases, the time to recharge at level 3 DC rapid charging stops will be similar, with a time of circa 25 minutes from zero to fully charged. DC charge points take in high electrical power levels and distribute them to the connected vehicles.

6.3 Vehicle cables and connectors

The charge point connects to the vehicle by cable, with the male connection at the vehicle being one of the connector types listed below. The female connection joins to the charge point end for untethered connections. For tethered cables, the cable is permanently attached to the charge point. High-power level 3 DC cables are always tethered because of their high cost, weight and bulk. Table 6.1 shows the range of vehicle connector types.

A level 1 single-phase charge point is either one which uses a standard home socket or outlet to charge at very slow (1.4–3.2 kW) speeds or one built

Table 6.1 Describing the vehicle connector types

Connection type	No of pins	Typical use	Example cars	Observations
Type 1	5	AC	Nissan Leaf Mk1, Mitsubishi Outlander and i-MIEV	Largely obsolete
Type 2	7	AC	Mini to Rolls Royce	Most commonly used AC charging in Europe
CHAdeMO	10	AC and DC	Nissan Leaf Mk2, Mitsubishi Outlander and i-MIEV	
CCS	9	DC	Most modern EVs with DC rapid charging	2 large DC pins added to the Type 2 connector. Bulky and heavy
Tesla / NACS	5	AC and DC	Tesla, Ford and GM models in United States, Canada, etc.	A relatively small connector, widely used across the North American Supercharger network
CaoJi	7	AC and DC	HV connector for future, 1,500 V DC	

[Source: Developed based on authors analysis of freely available data]

Table 6.2 Setting out charge point descriptions

Charger	Frequency	Min power (kW)	Max power (kW)	Observations
Slow (Level 1)	AC	3	<6	Single-phase rated home charge point
Fast (Level 2)	AC	7	7.9	Single-phase rated home charge point
Fast (Level 2)	AC	8	<49	Three-phase AC charge point
Rapid (Level 3)	DC	50	<149	–
Ultra-rapid (Level 3)	DC	>150	–	–

[Source: Developed based on authors analysis of freely available data]

into a lamppost offering speeds of up to 5 kW. A level 2 single-phase home connection in Europe and the United Kingdom is limited to supply less than 8 kW to the car, usually 7.4 kW depending on the house's total power draw. Some domestic installations have three-phase supplies, in which case, higher charging rates are allowed. Many charge points have energy management systems checking the total home supply to monitor the total current draw and modulate the charge point power delivery. More sophisticated home energy systems can harvest, store and deliver energy from solar photovoltaic (PV) and battery systems.

Level 3 DC charge point systems provide power directly to the BMS, which controls the battery health, estimates state of charge and health, and other functions such as thermal management, switching and safety. In general, level 3 DC charge points allow higher charging power, from 15 to 350 kW, since the charging voltage is higher at 400 V, meaning that the cables do not handle high current. DC charging is also more efficient since the limited power OBC stage is omitted, along with associated losses.

Grid connections at home or work provide the opportunity for longer charging events, for example, overnight or during the working day. Quick, shorter charging sessions during long distance travel are usually from grid-connected level 3 DC charge points and are located at a wide variety of places such as motorway service stations, supermarkets or ferry terminals and airports. Table 6.2 shows the general characteristics of charge points.

6.4 AC charge points and charging

6.4.1 Power supply

For AC charge points, the incoming supply is from the grid voltage and frequency, so 230 V and 50 Hz for the United Kingdom and Europe, and 110 V and 60 Hz for North America for a single-phase supply. Three-phase supplies are needed for high-power charge points such as level 2 AC above 8 kW and high-power level 3 DC charge points, for instance, those found at motorway service stations. Most EVs have an OBC with a power rating of 11 kW. If charging via a level 2 AC charge point of a higher rating, for example, 20 and 43 kW, the vehicle will only accept 11 kW, which means that the battery charging time is the same as an 11 kW level 2 charge point.

The level 1 and 2 AC charge points consist of a control system, contactors and communication methods with the vehicle and potentially the home mains supply. Some charge points have unusual installations in lampposts (normally level 1) or underground in hidden chambers. In these cases, the public electrical supply is measured and converted according to the prevailing environmental conditions, for example, the lamppost might be severely limited in available power, whilst the underground installation might be thermally limited due to power supply losses being unable to be dispersed, or with a combination of incoming solar radiation and insulating ground conditions.

Government funding to support the installation of home-based AC charging in the 2010s produced many United Kingdom made charge point suppliers, with over 380,000 installations supported via the Office for Zero Emission Vehicles (OZEV) grant [1]. Typical makers included Ohme [2], Rolec [3], myenergi [4], Hypervolt [5], Pod Point [6] and Indra [7].

6.4.2 Public charging

Public level 2 AC charge points in destination settings can be those with standard power levels of up to 8 kW, and situated in places where the available charging time is more than 4 hours – for example at workplaces or community facilities. The charging facilities can be numerous and with mixed power levels to offer both standard and fast charging facilities to meet user needs. The EV parking spaces are separated from other vehicle spaces and the power supply to each space may be from a central power supply consisting of a transformer, control and monitoring system, distribution and switchboard. Power is sent to each charge point via a cable, with safety systems looking at current flow to each and centrally. Many level 2 AC public charge points are installed in car parks and community areas such as leisure centres, sports centres and supermarkets.

6.4.3 Home charge point costs

Level 2 AC home charging was supported by government grants, most of which are no longer available (in the United Kingdom at least). In the case of the OZEV grant, this was up to £300 for a first installation. Local authorities and other funding bodies may still support first installations by grant, depending on funding opportunities from national governments.

The cost of installing a home EV charge point depends on the charge point power rating, with most charge points in the United Kingdom rated at either 3.6 kW or up to 7.6 kW. Most are offered tethered or untethered, meaning with or without a cable permanently attached, and with a smartphone application to allow remote access and control.

Many specialist installation companies offer installation services along with supplying the charge point, and these services will include a survey in addition to the installation. The survey checks the domestic switchboard for capacity and safety, as well as potential cable installation routes. This initial survey can help the homeowner understand the implications of installing the EV charge point alongside alternative energy sources and storage systems such as solar PV panels and home battery energy storage systems. The EV charge point may include elements of a wider range of control and monitoring services to enable the efficient use of the home energy system, for example scheduling the charge events to coincide with low carbon intensity grid levels, or low cost.

The cost of a charge point installation ranges from circa £800 for a 3.6 kW charger and less than 10 m of cable, to £1,200 for a 7 kW charger and 15 m cable. This might rise by between £500 and £1,000 if significant improvements to the main domestic board are required. The charge point installation may need a local earth rod to provide a reliable return to earth in case of a fault and to prevent the car body becoming live under fault conditions. There are electrical installation guides and recommendations that describe the earth rod installation, including the IET 18th edition wiring regulations, forming part of the British Standard BS7671 series. These state that an earth rod shall be installed unless other methods are used such as continuous electrical system monitoring.

Few homes in the United Kingdom have 415 V three-phase power supply, the overwhelming majority being a single-phase supply at 230 V or so. To install a three-phase supply to a home is very costly, needing a dedicated cable from the local substation to the domestic board – which also needs to be significantly upgraded. As an example, the cost of upgrading the supply to a home, 200 m from the nearest substation, was quoted at £11,000 in 2023, including the necessary domestic board upgrades, installation and commissioning. This would allow much faster charging rates, although still limited by the vehicle OBC power rating of 11 kW, and would allow the EV charge point to be separated from the main domestic switchboard and main fuse. Most domestic use cases do not require charging speeds higher than 7 kW, and so can be adequately served by a single-phase connection and supply.

6.4.4 AC charge point system innovation

Most EVs have an AC OBC rating of 11 kW, with some unusual examples such as the Smart ForFour with a 22 kW OBC rating. With the vehicle charging system and battery being integrated within the domestic electrical system, there are opportunities to use the car battery as part of the wider domestic system. This integration means that the car battery can be used to store energy from solar PV and also return it to the house when needed. This return of electrical energy to the home is known as vehicle to load. If the domestic electrical system allows, the car battery might be able to return electricity to the grid, similar to the application of solar PV. This is known as vehicle to grid. Further details can be found in Chapter 11.

In both cases, the technology exists and has been successfully demonstrated by companies such as Indra Renewables. If vehicle to load can return 11 kW to the domestic system, that power rating is enough to run most heavy loads such as an electric shower (circa 9 kW for most shower systems), supplementing the grid draw. The times when this might be impactful to the domestic user is when the grid is expensive, for example during flexible pricing events, during a period of low renewables to the grid, or if local conditions limit the household draw. The impact of vehicle to load on the grid is to provide peak lopping services, thus supporting grid stability and local capacity by reducing the domestic load.

Vehicle to grid is a similar return of electrical power to the grid rather than the house system, and is limited by the power rating of the OBC in the same way. The reason to return energy to the grid is to generate an income for the homeowner. This is set by a solar PV export value, the Smart Export Guarantee. This rate varies according to the Smart Export Guarantee contract, the average is 5 to 10 p/kWh, with some being as high as 30 p/kWh for times such as high grid demands at peak hours (usually between 17:00 and 19:00). For distribution network operators, feeding power back to the grid can balance the load, as stated in Chapter 5.

6.4.5 AC versus DC charging

As we have seen above, level 1 and 2 charging normally takes place in residential or some destination settings, with power ratings matched to the needs of drivers who will leave their car for longer periods of time. The practical limitation to high-power delivery is often the car's OBC, which is almost always 11 kW. Level 3 DC charging is usually associated with higher power transfer, but is rarely seen as being appropriate for domestic car charging since the incoming AC has to be converted to DC, which entails higher capital costs for the rectifier assembly within the charge point. Level 3 DC charge points are therefore more likely to be used at higher power charging locations such as service stations on motorways and long-distance journeys, with AC being used at home and work. High-power charging may affect the battery life due to the resulting higher electrical and material stress within the battery cells and assembly, but with the adoption of more sophisticated and capable battery power management and thermal control, that risk is mitigated but not eliminated. A mix of AC and DC charging sources are most likely to be used in future, with a change to autonomous vehicles in the far future anticipated to demand more rapid charging as higher vehicle utilisation rates become more common.

6.4.6 Control and monitoring of charging operations for both AC and DC charging

Control of timing and operation is usually via a remote control system, using an internet connection and either a PC or Smartphone control software application. This handles safety and completion alerts too, allowing the user to decide if the car needs more charge or if the car can be moved.

The charge point might have the ability to reduce its power draw if it detects that the incoming supply is close to the rated power supply limit. This requires communication with a domestic smart meter, or a dedicated power management system using a current clamp, or other sensor to measure current. When DC charging, the level 3 DC charge point will constantly communicate with the vehicle BMS to control the power transfer, considering the vehicle battery conditions, the incoming power supply and the vehicle charge acceptance rate. For example, if a car is capable of accepting up to 230 kW of power and is connected to a 350 kW capable level 3 DC charge point, the maximum possible power transfer is 230 kW, but only if the car is ready and the battery conditions allow. The BMS may change the acceptance rate to a reduced level if the battery is too hot or cold, or if the battery state of charge is above 80%.

6.4.7 AC use cases

The extensive range of charge point technologies leads to a wide range of power ratings, public access models, billing methods and geographical availability. Home or work charge points, where the vehicle is immobile for hours, are able to take advantage of power ratings, from 3 kW (level 1) to 8 kW (level 2), and are able to charge most cars from state of charge levels of 20–80% over a period of a working day, or overnight hours. Other charging situations where fast rates are suitable include places of interest, overnight stops at hotels and long stay parking.

For long distance journeys when a quick charge time is usually preferred, this charge time acceptability ranges from 15 to 60 minutes. This might coincide with service station stops for rest and refreshment and will have minimal impact on the driver's journey experience and a positive impact on fatigue and general safety. These charging opportunities are best served by charge rates above 50 kW, and in some cases where large batteries such as in heavy good vehicles are being replenished, over 300 kW is required, allowing the driver to match their mandatory break periods with their vehicle's needs.

There are instances where choosing the wrong charge rate is costly or inconvenient, for example using a level 3 DC ultra-rapid charger at a cinema, where the car might only need a few minutes to finish charging before the film starts, or at a railway station car park which might charge a supplementary, punitive parking fee for the time that the charger and parking space is occupied, but not charging. These blocking situations can be remedied by using a more appropriate charging rate. The numerical distribution of chargers has changed significantly over time, with level 1 and 2 AC single and three-phase charge points being installed in public places, and representing over 70% of initial installations, with less than 0.5% for level 3 DC ultra-rapid charge points. In more recent years, as we can see in Figure 6.3, the main growth has been for level 2 AC fast and level 3 DC ultra-rapid charge points in public places. A key driver for the large number of level 2 AC charge points and their ongoing growth is the various government-funded schemes to support their rollout (see Chapter 4).

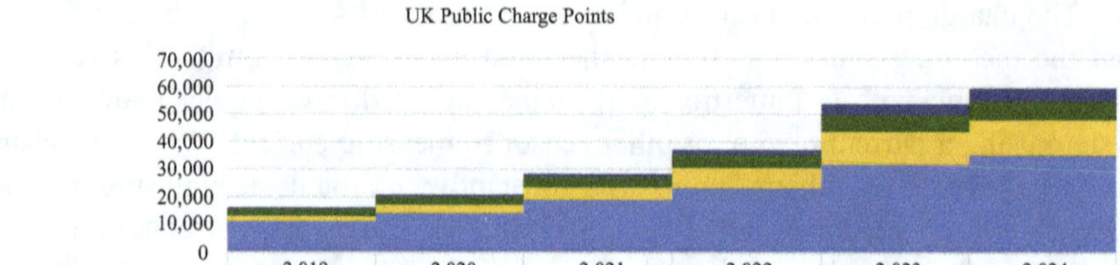

	2,019	2,020	2,021	2,022	2,023	2,024
Ultra Rapid (Level 3 DC)	476	562	998	1,731	3,824	4,667
Rapid (Level 3 DC)	2,899	3,309	4,089	4,976	6,294	6,942
Fast Chargers (Level 2)	1,658	2,867	4,293	7,092	11,710	12,715
Slow (Level 1 & 2)	11,250	14,226	19,080	23,464	32,037	35,266

Slow Fast Rapid Ultra Rapid

Figure 6.3 Number of charge points in the United Kingdom per year based on charge point types. Note that slow includes slightly faster charging speeds than defined in Table 6.2 [Source: Developed based on analysis conducted by the authors].

6.5 DC charging

The first EVs developed in the United Kingdom were all powered by lead acid batteries and charged from other cells, or from Edison DC distribution systems that were more common in the late 1880s (see Chapter 2). Today's EVs use level 3 DC charge points for high-power charging, but there is little fundamental difference between the reasons for using DC then and now. The adoption of level 3 DC charge points for high-power charging is to be able to electrically transfer energy from a fixed power system to a mobile energy storage system. The modern DC charging system uses power semiconductors within the charger assembly to convert the incoming AC power (50 or 60 Hz) to DC power which matches the battery needs.

6.5.1 DC charge points

Most DC charge points are rapid or ultra-rapid and have a power rating of more than 50 kW, and are situated in places with a suitable grid connection where vehicle users need high power transfer. The CCS connector and cable system allows a current up to 500 A (a 350 kW maximum power rating). Table 6.3 shows typical DC charge point manufacturers.

For level 3 DC public charge points, a range of payment and charging options are necessary, with many offering contactless payment using a debit or credit card, or using a smartphone app and QR code for charge point identification. The typical charging station arrangement is a series of parking bays arranged around a very high-power high-voltage grid connection and transformer with a local low-voltage distribution system to a series of DC charge points. A single charge point may serve one or two parking bays, with the cables being just long enough to reach a few metres from the charge point. The reason that the tethered charging cables are kept short compared with low-power AC charge points is because these cables are very expensive, are heavy and difficult to handle, and need to keep losses to a minimum

Table 6.3 Manufacturers of DC charge points

Maker	Model	Power rating (kW)	Maximum (kW)	Tethered connector
Tesla	Supercharger	120	250	NACS
ABB	Terra	50	180	CCS
Eaton	GreenMotion44/66	44	66	CCS
ProjectEV	EVD300D	50	300	CCS or CHAdeMO
Sevadis	Maxicharger	120	240	CCS
Tritium	PK350	N/A	350	CCS

[Source: Developed based on authors analysis of freely available data]

(and thus cable heating). Cables are heated by conduction losses. Heat generated is proportional to I^2R, where I is the current and R is the cable resistance, where the cable resistance is proportional to cross-sectional area and length, and conductor resistance per unit length. The CCS cable is rated at 500 A.

Load control and safety is managed by a local control system, optimising the power delivery to each charger according to need and charge condition, with information from the vehicle BMS being used to manage load across the charge points. Safety systems look at the electrical and mechanical conditions, assessing thermal conditions, insulation and damage on a continual basis. Depending on status, the safety system may suspend charging to one or more chargers, or to the whole local system. Remote monitoring and control is provided on DC charger systems, giving operators oversight and control of electrical delivery and billing.

The capital cost to build a level 3 DC charge point is proportional to the amount of civil engineering works including canopies, drainage, road surfaces, lighting and facilities buildings, the electrical infrastructure works, including the cost of electrical equipment and grid connection. The charge points themselves range from circa £15,000 to £80,000 per unit, depending on their power rating, number of cables and connectors, and their power electronics technology. Newer models with higher power ratings are more cost-effective per unit power than older models with a similar rating, with the cost reducing by 10% per doubling of the number of devices manufactured.

6.5.2 DC charging innovations

Since most domestic charge points use grid connections and are relatively low-power AC, it could be seen to be reasonable that a home DC network that incorporates solar PV, domestic storage batteries and vehicle batteries together with vehicle to home power systems would be a beneficial technology development. This type of home energy system would still have AC mains power to the home electrical system, but would have a parallel, connected DC power system that draws together power generation and storage. A bi-directional EV battery connection system would be at its heart since it would be the largest energy store on the DC distribution system. A single power electronics package that handles the conversion from grid voltage AC to DC and vice versa would be needed, but could be the same package that handles the EV charging duties.

DC charging relies on the conversion of high-power three-phase AC electricity to a suitable DC voltage for connection to the vehicle. There are a range of technology developments in this area in terms of power electronics, packaging and control. It is anticipated that future advancements will include an expansion of alternative applications of wide-band gap semiconductors such as gallium nitride and gallium oxide. This will allow smaller package sizes and higher efficiency, with the effect that DC charging system costs will drop. The higher temperature capability of these newer semiconductors will offer the opportunity for improved integration, and this may unlock the potential of having a single, small and efficient package within a domestic application.

DC charging may also become ubiquitous for low-power charging if the domestic DC system becomes popular since it will allow a reduction in the use of expensive materials such as copper and aluminium. As a result, AC charging could be replaced at home by DC charging.

6.5.3 DC charging use cases

As stated earlier in this chapter, DC charging tends to be used during longer journeys when a driving break is needed (either to charge or for other reasons) or when the driver has been called on an unexpected drive. In these cases, using level 3 DC charge points gives a quicker battery fill time than AC, and is more in line with normal break times when on longer journeys. For unexpected trips, the driver may be in a hurry, and might choose the highest power transfer rating to be able to get going as quickly as possible. Since DC charge points are more likely to be geographically located at en route locations, such as service stations, wayside restaurants and filling stations, the driver may be more willing to pay a higher price instead of waiting many hours at a lower power AC charge point. Table 6.4 shows

Table 6.4 Relationship between AC (light grey) and DC (dark grey) charging for three car types, Segments C (64 kWh), D (77 kWh) and J (85 kWh). Charging calculations are for 20 to 80%

Battery size, kWh	64	77	85
20%	12.8	15.4	17
80%	51.2	61.6	68
Energy to fill, 20%–80% kWh	38.4	46.2	51
Charge power, kW		Hours	
7.4	5.2	6.2	6.9
11	3.5	4.2	4.6
20	1.9	2.3	2.6
80	0.5	0.6	0.6
150	0.3	0.3	0.3
220	0.2	0.2	0.2

[Source: Developed based on authors analysis of freely available data]

the comparison between low power AC charging and DC charging for three different segment cars. The commonly found level 3 DC ultra-rapid power ratings that are often located at motorway services conveniently match with the typical en route travel stop time for food and services.

6.6 Platforms and back office systems

In Sections 6.1–6.5, we have focused primarily on the physical and electrical aspects of EVs and charge points, with a specific focus regarding the latter on level 2 AC and level 3 DC charge points installed in public charging settings. This section provides an overview of the platforms and back office systems required by a CPO or organisation installing charge points to effectively manage a charge point network. It should be noted that like the rest of this chapter, we are referring to platforms and back office systems in a United Kingdom context, though there is a major overlap with the rest of the Western world.

6.6.1 Platforms and back office systems today

The purpose of charge point platforms and back office systems is to manage a charge point network at scale over a broad geographical area. The platform is the gateway through which the driver interacts with the charge point and in a vast majority of cases is app and/or web based. The back office systems are those which sit as part of the platform and are used to manage processes and services behind the scenes. Key elements of a charge point driver facing platform include the ability to locate a charge point, charge point status, availability and price per kWh, the ability to start, monitor and end a charge and a summary of previous charging sessions. The platform often also includes a customer chat function that outlines current and previous conversation as well as the ability to raise and resolve issues, and a 'my account' area where the user can manage personal details as well as payment settings. Some front-facing platforms also have the ability to manage promotions as well as add details about a home charge point. Back office systems include a charge point management module showing live status and other details, a customer relationship management module to manage interactions with users and a payment module which processes driver payments, issues refunds and exports data to a CPO's accountancy software so as to track revenue. Both the platform and back office systems, providing they have been set up in this way, can exchange data with other platforms and systems (see Section 6.6.2). In terms of platforms operating today, each CPO has its own platform. In terms of back office systems, there are several popular solutions. Examples include ChargePoint [8], The Mobility House [9] and Virta [10]. In addition to back office system providers, there is a growing number of eRoaming providers who enable interoperability between charge points provided by different CPOs. Zapmap, although originally a charge point mapping tool, is increasingly becoming an eRoaming provider [11]. Other providers include Plugsurfing [12] and Hubject [13].

6.6.2 *Current technology standards*

There are various technology standards that underpin the operation of charge points. This chapter and Chapter 2 highlight the standards regarding the connector but this is only part of the picture. The other elements of standardisation include payments, plug and charge, roaming, charge point accessibility, open data protocols and cyber and physical security. This list, though not exhaustive, is the main standards which should be considered. This Section focuses on current standards in a United Kingdom focused context, whereas Chapter 11 highlights emerging standards.

Standardisation of payment, roaming and open data is determined by the Public Charge Point Regulations 2023 which states that new public charge points of 8 kW and above and existing charge points of 50 kW and above must offer contactless payment options to consumers. It also states that CPOs must enable consumers to pay through at least one roaming provider at their charge points and that data must be accurate and CPOs must use the Open Charge Point Interface (OCPI) to hold and share their data [14]. Though not specified in the regulations, the Open Charge Point Protocol (OCPP) is another standard which establishes a system of rules that allows EV charge points and CPOs to communicate with each other, sharing data. Plug and charge is where the driver connects to the charge point and automatically pays for the charging session and is enabled by ISO 15118, an international standard for charging EVs [15]. Charge point accessibility is specified by the British Standards Institute (BSI) Electric Vehicles Accessible Charging Specification PAS 1899:2022, further details of which can be found in Chapter 4 [16]. Cyber and physical security is covered by OCPP as all versions include security elements with later versions providing enhanced levels when compared with older versions.

6.6.3 *Platforms and back office systems – real-world use cases*

As stated in Section 6.6.1, there are various platform and back office providers, many of which provide information demonstrating how their systems work. As an example, Plugsurfing expanded its network of charge points on its platform when ChargePlace Scotland joined in 2024. In the United Kingdom, the platform enables drivers to find a charge point, start a charging session and pay for it in the app. Multiple CPOs in the United Kingdom have joined the platform, including Instavolt, Osprey charging, Ionity UK, Gridserve and MFG [17]. A further example is Virta who provide the Virta hub. This is a tool that enables charging station management through providing an overview of charge points, including remote monitoring, access management, and business insights. Pricing management is also an additional key functionality and enables price setting, cross-border VAT handling, regulatory compliance, loyalty pricing and alternate tariffs. A further key functionality of the hub is energy management. Through this the CPO is able to conduct dynamic load management, adaptive load management, congestion management and power capacity maximisation [10].

6.7 Other technologies deployed at charge points

Sections 6.3–6.6 primarily focus on the technologies related directly to the charge point; however, the deployment of a number of charge points in a single location may also be accompanied with other technologies that primarily relate to power. We have excluded battery swapping in this section as this is not a mass market technology in the United Kingdom at this moment in time, though it may have a future application (see Chapter 11). The most common technology accompanying a charging station is solar PV, often set up in a car port configuration where a PV canopy sits directly above a bank of charge points, feeding renewable energy to these when conditions are optimal. There are many examples of solar car ports in the United Kingdom and beyond, with local authorities like the City of York Council deploying these in en route and destination settings around the City's ring road [18]. It should be noted that even when solar PV is operating at peak energy generation, this is rarely enough to power all charge points, especially where they are level 3 DC. This means that solar car ports still need connections to the grid to effectively operate. In addition to solar car ports, battery energy storage systems are deployed. The reasons for this are varied and include storing solar energy during peak generation periods where not enough vehicles are using the charge points or as a way to lower the cost of energy usage by drawing energy when the grid is at its cheapest and then using this to feed charge points during peak pricing periods. The City of York Council has also opted to include battery energy storage systems at their en route and destination charging stations.

6.8 Conclusion

In this chapter, AC and DC charging systems have been described for EVs, including future potential innovations (e.g., local DC energy systems). Vehicle technology will change to use low-cost and highly efficient power electronics, but in all cases, the consumer's needs will continue to be met or even exceeded, especially in the case of ever faster ultra-rapid charging becoming increasingly common, making charging sessions shorter. Furthermore, the platforms and back office systems used by CPOs have been highlighted including the current technology standards, brought to life through two examples. The chapter ends by focusing on other technologies deployed in the vicinity of charge points, mainly solar PV and battery energy storage systems.

References

[1] Department for Transport. *Official Statistics – Electric vehicle charging device grant scheme statistics: April 2023* [online]. 2023. Available from https://www.gov.uk/government/statistics/electric-vehicle-charging-device-grant-scheme-statistics-april-2023/electric-vehicle-charging-device-grant-scheme-statistics-april-2023#:~:text=Headline%20figures,-Domestic%20Charging%20Devices&text=As%20of%201%20April%202023:,of%20339%2C775%20domestic%20charging%20devices [Accessed 11 March 2025].

[2] Ohme. *Charging the future* [online]. 2025. Available from https://ohme-ev.com/ [Accessed 13 March 2025].

[3] Rolec. *The UK's leading manufacturer of EV charging points* [online]. 2025. Available from https://www.rolecserv.com/ [Accessed 13 March 2025].

[4] Myenergi. *zappi – multiphase EV charger* [online]. 2025. Available from https://www.myenergi.com/product/zappi-ev-charger/ [Accessed 13 March 2025].

[5] Hypervolt. *Hypervolt Home 3 Pro* [online]. 2025. Available from https://www.hypervolt.co.uk/products/home-3-pro/overview [Accessed 13 March 2025].

[6] Pod Point. *EV home charging* [online]. 2025. Available from https://pod-point.com/solutions/driver/home-charging [Accessed 13 March 2025].

[7] Indra. *Join the smart charging revolution* [online]. 2025. Available from https://www.indra.co.uk/ [Accessed 13 March 2025].

[8] ChargePoint. *The leading platform for EV charging* [online]. 2025. Available from https://www.chargepoint.com/en-gb/businesses/software?srsltid=AfmBOooEepguAzfMsxQsidPPKQYxB2Ew6mJwvqgvEh0vLnRUDry2NlnT [Accessed 13 March 2025].

[9] The Mobility House. *Offer your customers the best smart charging solution* [online]. 2025. Available from https://www.mobilityhouse.com/int_en/charging-solutions-for-operators [Accessed 13 March 2025].

[10] Virta. *Virta Hub* [online]. 2025. Available from https://www.virta.global/charging-solution/virta-hub-cpms [Accessed 13 March 2025].

[11] Zapmap. *EV charging made simple* [online]. 2025. Available from https://www.zap-map.com/ [Accessed 13 March 2025].

[12] Plugsurfing. *Open your network to carmakers, fleets, and over 2 million EV drivers across Europe* [online]. 2025. Available from https://plugsurfing.com/charge-point-operators-solutions/ [Accessed 13 March 2025].

[13] Hubject. *Join the largest eRoaming network for borderless electric mobility* [online]. 2025. Available from https://www.hubject.com/ [Accessed 13 March 2025].

[14] Department for Transport, Office for Zero Emission Vehicles. Public Charge Point Regulations 2023 guidance [online]. 2023. Available from https://www.gov.uk/government/publications/the-public-charge-point-regulations-2023-guidance/public-charge-point-regulations-2023-guidance [Accessed 13 March 2025].

[15] Poultney L. *The EV charging nightmare could soon be over – universal Plug-and-Charge protocol due for US rollout in 2025* [online]. 2024. Available from https://www.techradar.com/vehicle-tech/hybrid-electric-vehicles/the-ev-charging-nightmare-could-soon-be-over-universal-plug-and-charge-protocol-due-for-us-rollout-in-2025 [Accessed 13 March 2025].

[16] British Standards Institute. *PAS 1899:2022 – Electric vehicles – Accessible charging – Specification* [online]. 2022. Available from https://www.bsigroup.com/siteassets/pdf/en/insights-and-media/insights/brochures/pas_1889_final.pdf [Accessed 13 March 2025].

[17] Plugsurfing. *Travel with your EV on the biggest charging network in the UK* [online]. 2024. Available from https://plugsurfing.com/blog/electric-vehicle-charging-network-news-uk/ [Accessed 13 March 2025].

[18] City of York Council. *Electric Vehicle HyperHubs* [online]. 2025. Available from https://www.york.gov.uk/hyperhubs [Accessed 13 March 2025].

Chapter 7
Planning for deployment

David Trousdale[1]

In Chapter 3, we outlined the high level end-to-end process relating to the deployment of charge points. This chapter, and subsequent chapters, goes into further detail regarding the process, with this specifically focusing on planning for deployment. This comprises five key steps focused on the 'why', 'where', 'what' and 'how' behind the deployment. Beginning with the initial decision to deploy charging infrastructure (the 'why'), this is then followed by determining the geographic area to deploy (the 'where'). After this comes the business case including seeking funding (the 'what') and then planning for deployment at a local level and the governance and approvals step (the 'how'). Completion of all of these steps then enables the deployment to move on to the build phase.

Throughout this chapter, each of the steps is brought to life through ten case studies including further context regarding the different stakeholders involved in the process at different steps and their perspectives regarding deployment planning. Furthermore, the chapter is supplemented by my own experiences having planned deployments as part of multiple projects. It is written primarily from the perspectives of a charge point operator (CPO) and local authority.

7.1 The initial decision to deploy the infrastructure

The initial decision to deploy charge point infrastructure is the first step in understanding the 'why' behind charge point deployment. Case Study 7.1 demonstrates that there are three key stakeholder groups who play a role in determining the 'why' which directly relates to the outcome that charge point deployment seeks to achieve. Understanding the outcomes at the start is critical to ensuring a successful deployment. In the Case Study example, London demonstrates the importance of government-driven deployments, though in a vast majority of deployments each stakeholder group has some role in shaping the outcomes. Where deployment is government-driven the overarching outcome tends to be that deployment of charge points will improve air quality through supporting the transition to electric vehicles (EVs) (see Case Study 7.2). This is often linked to a broader transport strategy that

[1]Principal Advisor - Climate Change and Sustainability, Newcastle City Council, United Kingdom

Case Study 7.1: The 'why' behind deployment – settling out the business model for London

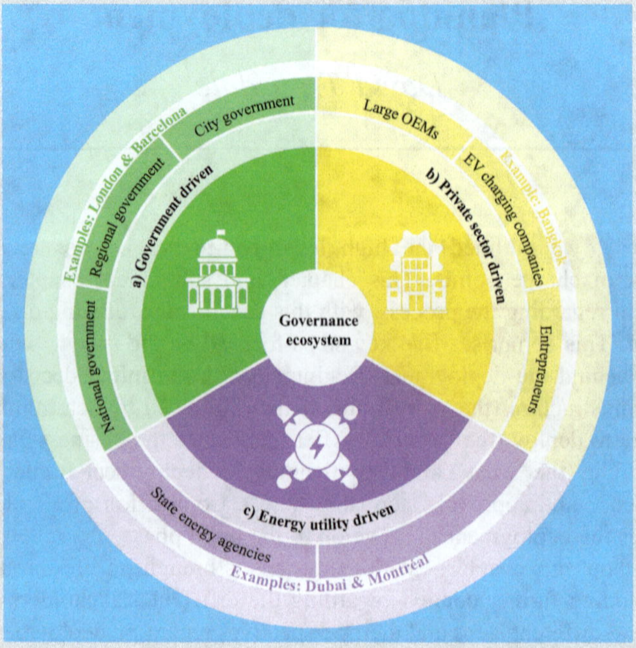

London is made up of 32 boroughs. The overarching strategic direction is set by the Mayor of London and London Assembly, with each borough adopting a charge point approach specific to their location. The driving force, however, behind the different approaches is broadly the same across Greater London. There are different business model drivers for charge point deployment though deployment across London is largely government driven with a heavy focus on satisfying the residential charge point use case with a smaller focus on en route and fleet charging use cases. The boroughs have taken advantage of the government-funded On-Street Residential Charge Point Scheme (ORCS) to deploy 4,300 level 2 residential-focused charge points. Transport for London (TfL) has also developed the level 3 DC Rapid Charge Point Framework to satisfy the en route and fleet use cases with a specific focus on light goods vans and taxis. TfL estimates an investment between £15 and £48 million is required to deliver between 150 and 570 additional level 3 DC charge points by the end of 2025. By the end of 2020, London had installed 300 charge points.

Source: C40 Cities Climate Leadership Group and E-Mobilita. *EV charging infrastructure: Business models and city case studies* [online]. 2023. Available from https://www.c40knowledgehub.org/s/article/EV-charging-infrastructure-Business-models-and-city-case-studies?language=en_US [Accessed 03 October 2024]

Case Study 7.2: The 'why' behind deployment – North Tyneside's EV Strategy

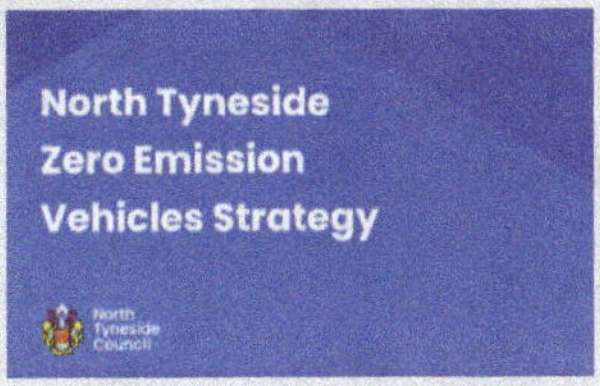

North Tyneside Council, like many other local authorities in the United Kingdom, developed an EV strategy to outline the reasons why the Council is promoting uptake and the actions that must be taken to drive adoption. The Strategy aligns with the Council's vision of 'moving to a green, healthy, dynamic and thriving North Tyneside' and notes that the Council has a leadership role in promoting and enabling EV adoption. Core to the Strategy is the fact that EVs reduce carbon emissions and promote clean air. The Council recognises as part of the Strategy that there are a number of challenges preventing adoption. This includes the lack of publicly available charge points across North Tyneside and that there is demand for on-street charging for people who do not have a driveway or private off-street parking, for example, at terraced properties. The Strategy sets out eight actions which the Council will take to deliver its aspirations. The key actions include:

- A phased replacement programme for North Tyneside's fleet to introduce EVs and supporting infrastructure.
- Upgrading and expanding the existing network of EV charge points in the Authority's car parks and premises.
- In areas of terraced streets where a substantial number of houses do not have private off-street parking, North Tyneside will review opportunities to seek external funding and work with commercial operators on potential charge point provision.
- Work with the NECA and Nexus (the transit operator) to expand their network of EV charge points at the Council's public transport interchanges.

The Council is making good progress in delivery, having recently secured through NECA, part of £15.8 million of LEVI funding to roll out charge points.

Source: North Tyneside Council. *North Tyneside Zero Emission Vehicles Strategy* [online]. 2024. Available from https://northtyneside.gov.uk/sites/default/files/web-page-related-files/1b%20Appendix%201-North%20Tyneside%20Zero%20Emission%20Vehicles%20Strategy.pdf [Accessed 25 April 2025]

includes a hierarchy of transport modes where public transit and active travel sit higher than journeys by private vehicle, with EVs being the preferred vehicle type if travellers must use their car over other modes. Furthermore, the need to improve air quality is closely linked to a local authority's climate change or net zero strategy. Further outcomes related to government-driven deployment include ensuring a just transition to EVs is achieved through making sure deployments are equitably spread across areas to give residents fair access to charge points and also that deployments generate an income that feeds into council budgets to support delivery of other council services. In addition, local authorities might seek to deploy charge points to deliver reduced motoring and fleet costs, satisfy resident and business demand and/or achieve noise reduction.

The outcomes that private sector stakeholders (primarily CPOs but also landowners) want to see in any deployment is that there is demand for charge points and that this translates into profitability. Analysis completed by the McKinsey Center for Future Mobility demonstrates that there are two interlinked factors that relate to profitability; utilisation which is the number of active charge point sessions over any period of time, and price charged to the end-customer. Assuming the CPO has built its cost base per kWh correctly the latter should be higher than the cost to buy energy, supply it to the user and pay for any other supplementary systems (card processing fees, back office systems) and tax (value added tax, sales tax). Subtracting the cost per kWh from that charged to the user generates a per kWh net revenue. How both interact is important as a site might have lower utilisation but a higher per kWh net revenue which may be sufficient to breakeven. Conversely, a higher utilisation but lower per kWh net revenue may also result in the same outcome. Looking at a typical fast public charging station in California, the CPO would break even if utilisation increased from 15% to 20%, or if the price for charging customers increased from \$0.45 to \$0.53 per kWh [1]. As a general rule of thumb, utilisation anywhere above 15–20% is an indicator of a good site. The criticality of utilisation and pricing in profitability is the reason why CPOs conduct in-depth modelling of the demand, utilisation and pricing of sites (see Section 7.2) before progressing to deployment.

The outcomes that energy operators, suppliers and governing agencies want to see are that deployment of a charge point network does not put strain on the power network and that charge point and EV innovations can provide grid stability by feeding power back into the grid and through smart charging systems which only draw power from the grid when there is capacity during off-peak periods or when renewable energy is generating too much power. In the United Kingdom, distribution network operators (DNOs) manage capacity through granting connection approvals for each site based on the maximum power demand required and costing these connections based on upgrade costs.

What is challenging in any area or region is trying to triangulate these different outcomes to ensure that all stakeholders support deployment, as in many cases the outcomes might not be compatible. For example, a local authority might insist that a deployment ensures a just transition. This implies deployment in areas that currently have poor demand and utilisation as a result of less affluent communities (usually high indices of multiple deprivation) not having transitioned to EVs due to high vehicle or charging costs. This is not compatible with the CPO outcome that charge point deployments should be profitable. Failing to address this at the start of the deployment programme might result in a poor response to any concession procurement exercise run by a local authority. Overcoming this challenge would likely involve ensuring that when proposing sites, a portfolio of high utilisation and low utilisation sites are packaged together so that the overall portfolio is profitable even if certain sites are not. The grid stability outcome required by the DNO is largely addressed in planning the deployment at a local level step but there might be indicators at the very start of the planning step that connection requests for potential sites might be turned down based on local knowledge regarding planned developments, capacity constraints and existing or planned charge point networks impacting on capacity. Ensuring this outcome is still achieved would involve considering known power constraints during the analysis of the geographic area to deploy step (see Section 7.2). This data can be analysed to either identify gaps in capacity where charge points can be deployed or ensure that planned deployments are only partly reliant on the grid through solar and battery storage. As long as the organisation that is planning to deploy charge points can clearly articulate the outcome it is trying to achieve and demonstrate how this aligns with and interacts with the outcomes of key stakeholders, the deployment programme should start off on a good footing. A further core part of determining the outcomes that are to be achieved is understanding in detail the use case that a charge point deployment is seeking to satisfy.

There are three use cases for public charging which are the focus of this chapter:

- **En route charging**: this is where a driver is travelling en route to a location and wants to stop for a short period of time (anywhere between 15 and 45 minutes) to charge their vehicle so as to continue their onward journey. This use case lends itself to level 3 DC rapid (50–100 kW) or ultra-rapid (100 kW plus) charge points which tend to be more expensive to install due to faster charging equipment, larger grid upgrades and surrounding charge point features like canopies and other support services (e.g., toilets). Due to the higher cost to install, price per kWh is higher than in the other two use cases. En route charge points tend to be deployed on or near to strategic road networks and their connecting roads. The en route use case tends to result in charging hubs of six or more charge points. If deployed in the right place, demand and utilisation are high.
- **Destination charging**: this is where a driver is travelling to a location to do something else. This might, for example, be a shopping centre, gym, cinema or hotel. The driver will be there for anywhere between 1 hour and 12+ hours and will likely not be reliant on the charge to continue their onward journey, though if they have travelled to a hotel, they might need the charge depending on the distance travelled. Due to the length of stay, this use case lends itself to level 2 charge points which can be anywhere between 7 and 22 kW though some destination locations may opt, depending on the average length of stay, for level 3 DC rapid charge points. Because charge points are charging at lower speeds, this means the charge point itself and the cost to install it, as a result of requiring a smaller grid upgrade, are much lower than en route charging. This

means a lower per kWh cost for the driver. Charge point numbers installed vary based on the location. If deployed in the right place, demand and utilisation are high, though these charge points are much more susceptible to competition especially if drivers do not need to charge.

- **Residential charging**: this is where a driver charges on-street at home because they do not have access to a private driveway with their own charger or cannot charge at work. Their vehicle will be stationary for several hours meaning that a slower charging speed can meet their needs. In most instances, level 2 charge points with speeds anywhere between 3 and 11 kW are installed using a myriad of technologies including raised bollards, lamppost charge points and, in some cases, charge points which raise from the ground. The slower charging speeds tend to mean less expensive charge points and grid upgrades. As a result, cost per kWh charged to the driver tends to be less than en route and destination charge points. This is also the case because residential drivers are acutely aware of those who charge privately on their own drives and can take advantage of cheaper charging tariffs offering lower per kWh rates during off-peak periods. Charging costs for off-street private drivers tend to be less because the tax is lower on domestic energy use. This means that most residential deployments try to ensure fair and affordable access to charge points. Residential charge points tend to be deployed in their hundreds over larger areas to ensure residents can access them near to their home. Additional complexities relating to deployment include approvals and regulations governing who can park in residential charging bays and for how long.

It is critical that whoever is installing and operating the charge points understands the use case/s they are seeking to satisfy from the very start of the deployment project as this determines a myriad of factors. This includes the type and number of charge points to be installed, the types of locations to be installed in, the likely demand and utilisation, the business model, the commercial model and the funding mechanisms available. It also includes the governance process and planning rules, the size of grid upgrade required, the capital, development and operational expenditure (OpEx) and the profitability of the sites based on the utilisation versus price per kWh charged to the user. The process by which these details are understood is covered in Sections 7.2–7.5. However, at least developing a high-level understanding of these during the initial decision to deploy infrastructure step means that the 'why' behind deployment can be clearly articulated. The project is also less likely to fail at subsequent steps in the planning, build and operation stages.

Looking in detail at Case Studies 7.2 and 7.3, both North Tyneside and Newcastle City Council set out early in the deployment process to understand the use case the charge point deployments would satisfy, with both targeting residential charging. Both focused on on-street residential charging and sought to secure funding to enable rollout. North Tyneside recognised that its car parks could also play a role in satisfying a residential use case and sought to deploy charge points there. Both North Tyneside and Newcastle secured Local Electric Vehicle Infrastructure (LEVI) funding through the North East Combined Authority (NECA).

Case Study 7.3: The 'why' behind deployment – Electric Vehicle Charging Infrastructure Strategy for Newcastle

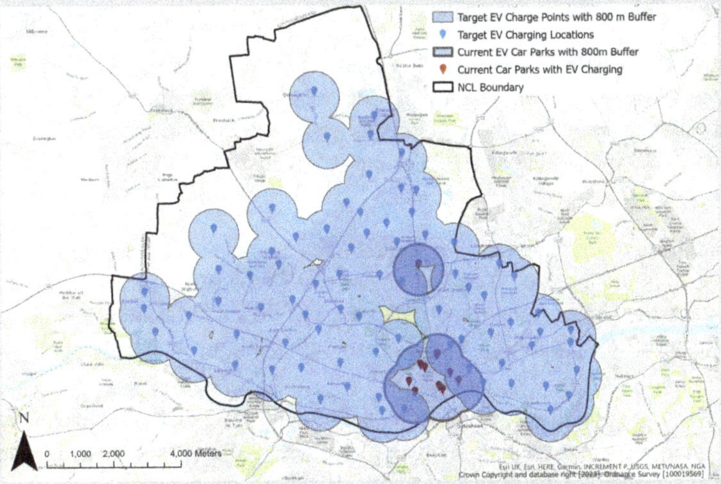

Newcastle City Council, in their deployment of charge points across the city, demonstrated the importance of determining the key drivers behind deployment at the start of the process. Having identified five priority actions, the Council highlighted the different use cases relevant to the city and how these should be satisfied through charge point deployment. Two of the key actions included delivering an inclusive neighbourhood charge point rollout and establishing a network of publicly available destination charge points. To determine potential target locations, the Strategy analysed several factors including:

- Proximity of public car parks or suitable Council land holdings.
- Availability of off-street parking in residential properties.
- Presence of lampposts at the front of pavement (subject to technical and cost viability assessment).
- Car ownership levels – this was used to determine whether pairing EV charging infrastructure with an EV car club may be more appropriate for an area.
- Width of road and footway and ability to integrate EV charging bays without unacceptably restricting footway access.
- Presence of disabled parking spaces.
- Density of Council fleet operations.

This analysis resulted in the identification of 77 target locations for residential charge points (see picture above). Furthermore, analysis of a range of factors including existing Council car parks and sites with high demand helped determine the need for 53 destination charge point locations across the city. Steady progress is being made in deployment. Recent LEVI funding of £15 million, secured by NECA in February 2025, is supporting the rollout of charge points across Newcastle and its six surrounding local authorities.

Source: Newcastle City Council. *Electric Vehicle Charging Infrastructure Strategy for Newcastle* [online]. 2023. Available from https://greenernewcastle.newcastle.gov.uk/assets/assets/EV%20Charging%20Strategy%20REV%20Final%20Oct%202023.2.pdf [Accessed 17 April 2025]

In addition to understanding the outcome/s and use case/s driving charge point deployment, it is important to understand as part of the 'why' which business model will be sought as this can set in motion a range of different tasks further along in the planning, build and operation stages. Broadly speaking, there are three different business models:

- **Owner operator**: this is where an operator (usually a CPO) owns and operates the charge point, and also owns the land where the charge point is placed.
- **Owner operator leaser/renter**: this is where an operator (usually a CPO) owns and operates the charge point but rents or leases the land off a landowner.
- **Concession**: this is where a contract is granted by one body (usually a local authority or a landowner) to a CPO to usually (though not always) install, operate, maintain and often upgrade the charge point infrastructure. Operating risk is shared between the concession management organisation and concessionaire.

Each model tends to be funded in a different way with the first two tending to be private sector funded, whereas a concession is often established by a local authority so might draw on grant funding. Each approach has a different timeline and requires different routes to secure funding. Determining the model at the start helps to set expectations in the organisation planning deployment and for key stakeholders who, for example, may have differing levels of interest in the process depending on the model. A CPO will be very interested in a concession model and would want to influence the contract early in the process. It also helps to set out a rough timeline for delivery and the amount of funding required. Failure to understand the business model in detail at the start of the process might result in a change in approach later down the line, resulting in lost time and, potentially, the need to start again. Focusing on Case Study 7.4 as an example of determining the business model at the start of the process, NECA adopted a concession model. Through the second round of the British Government's Levelling Up Fund to deliver a transport decarbonisation package of investment, NECA appointed Zest as a CPO to provide, install, operate and maintain charge points for four North East local authorities. The same approach is anticipated to be taken to award £15.8 million of LEVI funding which NECA was successful in securing in February 2025.

Case Study 7.4: The 'why' behind deployment – NECA's appointment of Zest via the NEPO framework to deliver charge points for four local authorities

In 2023 NECA secured £19.5 million from the second round of the British Government's Levelling Up Fund to deliver a transport decarbonisation package of investment. Within this package, 92 EV charge points were planned for installation. Deployment was delivered via a joint contract for four Tyne and Wear Authorities (Gateshead, Newcastle City Council, North Tyneside Council and Sunderland City Council) alongside other contracts within Durham, Northumberland, Nexus and South Tyneside. This contract was procured via the NEPO procurement framework which was open to a number of charge point suppliers including Amp EV, Apcoa, Believ, Connected Kerb, Equans, Swarco UK, Ubitricity and Zest. As part of the supplier selection, it was agreed that a negotiated direct award presented the most viable option for the four local authorities. As CPOs on the framework were ranked when it was established, officers tendered to the first ranked supplier. Where no response was received, a further tender opportunity was issued to the second ranked supplier, Zest. An Invitation to Tender exercise was completed and the returns assessed by a panel of evaluators from the participant local authorities. This ensured a quality submission and value for money. Zest was appointed as the successful supplier as part of this process. The value of the contract was £1.2 million, inclusive of Levelling Up Funding and funding through the On-Street Residential Chargepoint Scheme (ORCS) for the delivery of residential-based charge points in North Tyneside.

NECA successfully negotiated match funding of 10% from Zest, resulting in savings valued at approximately £124,000. The contract was developed utilising the NEPO framework terms and conditions. Roles and responsibilities were established around the design and development of the infrastructure including ensuring compliance with Health and Safety and Construction (Design and Management) Regulations 2015. The procurement approach delivered value for money, enabled economies of scale and resulted in a more attractive private sector investment opportunity. NECA is looking at all options to deploy charge points across the region using £15.8 million of LEVI funding secured in early 2025 and City Region Sustainable Transport Funding announced in the same year.

Source: North East Combined Authority. *Zero Emission Vehicle Strategy* [online]. 2025. Available from https://www.northeast-ca.gov.uk/how-we-work/transport/zero-emission-vehicle-strategy [Accessed 26 April 2025]

Source: North East Combined Authority. *Delegated Decision Report* – 8 November 2024 [online]. Available from https://www.northeast-ca.gov.uk/downloads/3242/241108-dd-report-contract-for-electric-vehicle-charging-units.pdf [Accessed 26 April 2025]

Image provided with permission from Zest.

At the end of this step, the organisation deploying charge points should have established an overview of why charge points are being installed through outlining the outcome/s to be achieved, the use case/s targeted and the business model/s sought. Following this, they can then determine the geographic area to deploy.

7.2 Determining the geographic area to deploy

Determining the 'where' in relation to charge point deployment follows on from the 'why'. At this step in the process, a local authority or CPO does not necessarily need to know the exact location to deploy charge points. What they must do, however, is adopt an analytical approach that aligns with the target use case as this will determine what information is needed to be analysed to identify deployment locations. This is usually achieved through a geographic information system that gathers information from a number of sources and can then display this through different perspectives depending on target use cases. Transport for the North, for example, analyses the various data streams via their EV charge point infrastructure visualiser which is available to all local authorities and key stakeholders in the North of England [2]. The tool, as shown in Case Study 7.5, analyses data at a macro level to understand demand for charge points and gaps for residential and en route charge point deployments across large areas. The tool can also be used for destination charging, though it is not necessarily designed for this. The tool and others similar to it enable analysis through three steps; ingestion of data, visualisation and decision-making.

Case Study 7.5: The 'where' behind deployment – mapping the various factors which influence the deployment of a charge points at a macro geographic level across the North of England

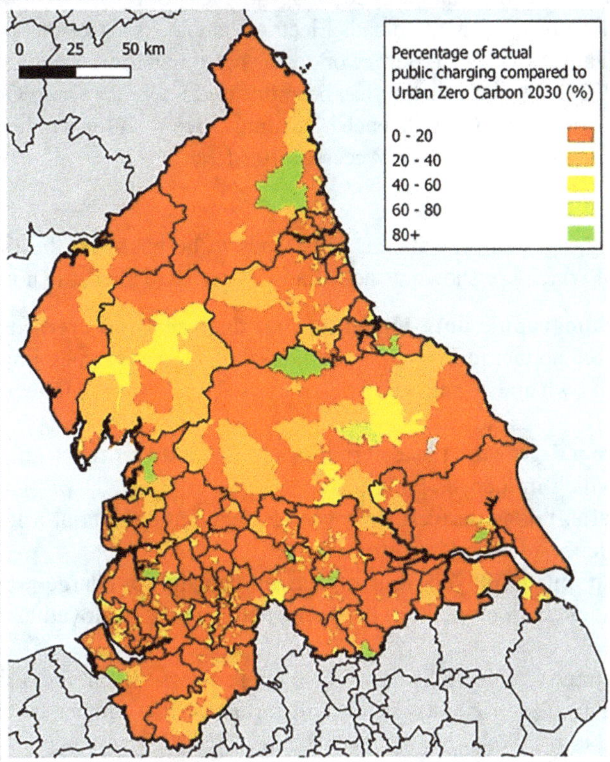

As part of Transport for the North's strategic transport role spanning from the edge of the Midlands all the way up to border with Scotland, the organisation developed an interactive EV charge point infrastructure visualiser tool which enables local authorities, CPOs and other key stakeholders to understand the demand for level 2 residential and level 3 DC rapid charge points. The tool can be viewed regionally; by combined authority; by local authority; or by MSOA. Demand can also be viewed along the strategic and major roads networks. This includes major roads operated by National Highways and local highway authorities. The tool draws on a range of data sources including demographics, housing, employment and jobs data, existing and projected EV adoption rates and existing and projected EV infrastructure (see example above showing percentage of actual public charging against the projected 2030 'Urban Zero Carbon' scenario). Other data sources include available grid capacity, demand for charge points and traffic flow data across the strategic and major road networks, and local roads. The tool is flexible enough to ingest other data sources and has been rolled out across England via the other sub-national transport bodies like the West Yorkshire Combined Authority.

Source: Transport for the North. *Transport for the North Electric Vehicle (EV) State of Play Report* [online]. 2024. Available from https://www.transportforthenorth.com/wp-content/uploads/Electric-Vehicle-EV-State-of-Play-Report-4.pdf. Contains STB EVCI Framework data shared under the Open Government Licence, 2024. This information has been licensed under the Open Government Licence v3.0. To view this licence, visit https://www.nationalarchives.gov.uk/doc/open-government-licence/ [Accessed 10 March 2025]

Regarding ingestion of data, analysis tends to be based on the following data sources, all of which are shown as actual and projected. This list is not exhaustive:

- **Socio-demographic data including that derived from credit agencies**: this focuses on population, number of households, income per household, propensity to own an EV and travel behaviours. These factors all heavily influence EV uptake.
- **Access to a driveway**: this determines whether a driver relies on public charge points, which influences EV uptake.
- **EV uptake**: this is derived from vehicle registration data at a localised level and influences demand on charge points.
- **Charge points deployed**: this impacts on demand for charge points and indicates the level of competition between charge points deployed currently and in the future.
- **Grid capacity**: this indicates where there are gaps in the grid and what available headroom exists which will influence the number of charge points that can be installed.
- **Traffic flow**: this will give an indication of potential demand on certain routes.
- **Travel demand**: this indicates the modal split in an area and can help identify opportunities to transition to EV where cars occupy a high modal share and other modes are not available.
- **Charge point demand**: this is based on the various data sources listed above and is usually calculated as the number of potential vehicles who might charge at a charge point. CPO centric models will tend to show this data.
- **Charge point utilisation**: this is based on the various data sources above and is usually calculated as a percentage of time across an entire day where a charge point will be in use charging. This is then translated into a total kWh drawn per charge point per day. CPO centric models tend to show this data.

When the data is then visualised, this is where the different approaches between residential and en route analysis start to appear. For residential charge points, data is visualised at middle-layer super output area (MSOA) or lower-layer super output area (LSOA) level, usually covering areas of either 400 to 1,200 households (LSOA) or

2,000–6,000 households (MSOA). This is because charge point demand is displayed for whole suburbs or conurbations where residents live and also because multiple streets must be considered to ensure those which are most viable are recommended for further analysis. Visualisations are usually heat maps of an area where various factors are combined to identify locations with high, medium and low demand, or where power capacity is good or poor. Other data can then be layered on top of these maps to identify the best streets where charge points could be deployed. For en route charge points, these are normally visualised using strategic and major road networks as the base to which all other data is viewed against. This starts with static data like service stations, existing charge points, junctions and available land if known. To this is then added traffic flow data (including numbers of EVs if known) and power capacity. The result is a visualisation showing narrow corridors with slightly broader catchment areas at junctions where drivers can leave the road network in search of a charge point. The target areas might be narrowed further if a service station or fuel station operator has tasked a CPO to determine the best locations based on their existing network. However, even if this is the case, the operator might be looking for new locations not currently run by them. Using the available data, proposed sites can then be pinpointed on the network to identify the best possible locations for deployment.

Once both residential or en route data is visualised, a further step is development of different scenarios based on different parameters which then can be displayed to help decision-makers understand different potential futures. This is key to decision-making. For residential charge points, scenarios might include a fall in income in an area combined with a fall in EV adoption, resulting in fewer suitable areas for deployment. The reverse (higher income, higher adoption) might also be visualised, identifying more areas. Other scenarios considered might be deploying based on a just transition where there is equitable access to charge points regardless of demand or adoption, and one where deployment is based purely on demand alone, resulting in areas that are likely to be the most commercially viable. For en route charge points, different scenarios primarily relate to higher or lower adoption rates, higher or lower charge point competition and slightly more complex scenarios based on average range of EVs being higher or lower, impacting on the need to charge en route. Once these are visualised, a most likely scenario, which may draw on other scenarios, is developed and will usually form the basis for decision-making in terms of the long list and then feed through to planning deployment at a local level in Section 7.3. What should be apparent from this process is that there is no right answer to determining the geographic area to deploy. The various parameters will influence potential locations and are ultimately determined by key stakeholders like local authorities, for example, who require a certain outcome to be achieved as stated in Section 7.1.

Determining the geographic area to deploy for destination charge points is slightly different to en route or residential charge points because the CPO will have entered into a contractual agreement with either a landlord or a company with an existing estate to deploy on. Case Study 7.6, where Mer has entered into an agreement with Blue Diamond, a garden and living centre specialist, to deploy charge points at garden centres across the United Kingdom, is a good example. In this instance, Mer will have worked with Blue Diamond to analyse destinations with the

Case Study 7.6: Mer's deployment of destination and en route charge points in the United Kingdom

Mer are a European CPO who provide destination and en route charge points. In this case study, Mer entered into an agreement with Blue Diamond, a garden and living centre specialist, to install over 300 charge points at garden centres across the United Kingdom. As part of the rollout, Mer analysed the demand for charge points at different garden centres, taking into consideration various factors including the charge point use case to be targeted, size of garden centre and the position of the garden centre in relation to the strategic road network. Mer also looked at other factors like local grid capacity and the position of the garden centre in relation to other points of interest like being in close vicinity to a national park. This analysis enabled Mer to determine the right infrastructure to meet the needs of users at each site. For example, due to its location near to Junction 22 off the M5, Mer chose to install 1 level 2 22 kW and 4 level 3 DC 150 kW chargers so as to satisfy both en route and destination charging use cases. In a different location 16 mi from Peak District National Park and next to the A38 Mer chose to deploy 1 level 2 7 kW charge point and 4 level 3 DC 150 kW charge points again to satisfy destination and en route charging use cases.

Source: Mer. *Mer Helps Blue Diamond with Destination EV Charging* [online]. 2024. Available from https://uk.mer.eco/news/case-study-blue-diamond-destination-ev-charging/ [Accessed 06 October 2024]

highest possible footfall and therefore demand for charge points. The visualisation they will likely have developed would have started by plotting garden centre sites across a broad geography or country. On top of this would have then been added localised data including car park size and usage, visitor numbers and existing charge points and utilisation if known. To this they then will likely add EV adoption rates within a catchment area and also the location of the site in relation to the strategic road network or other points of interest. Amalgamating this data should then indicate sites that show the most promise for charge point deployment and also the types of charge point. Mer's analysis of Blue Diamond's estate demonstrates the importance of blurring the boundaries between use cases as certain garden centres, due to their location near the strategic road network, were suitable for a mix of level 2 destination and level 3 DC en route charge points. Like en route and residential deployments, different destination scenarios are considered. These tend to focus on, for example, high versus low utilisation derived from footfall, high versus low EV adoption rates in an area, and high versus low charge point competition, all impacting on demand. The most likely scenario is then chosen to base further analysis on, supporting the decision to finalise a list of sites to develop further.

The output of determining the geographic area to deploy is a long list of potential sites which feed into the process to develop the business case and planning for deployment at a local level.

7.3 The business case including seeking funding

Completing the business case is primarily about clearly setting out the justification for deployment, drawing on information already gathered as part of Sections and 7.2, and supplementing with further analysis to build the case. As stated in Chapter 3, there are several different approaches to writing business cases; however, a commonly used structure, in the United Kingdom at least, is the five case model. The five cases are the strategic case that sets out the case for change and demonstrates how a deployment provides strategic fit, the economic case which sets out how the deployment will deliver value for money versus other proposals and the value to society, and the commercial case which establishes how the deployment is commercially deliverable. The financial case demonstrates the affordability of a deployment and the management case sets out that robust arrangements are in place for the delivery, monitoring and evaluation of deployment, including feedback into the organisation's strategic planning cycle. This section brings to life, through case studies, how a business case would be structured in each specific deployment context.

Beginning with the strategic case, Case Study 7.7 demonstrates how the West Midlands Combined Authority set out the strategic fit and subsequently the requirement to deploy 10,000 charge points across the West Midlands. The rationale was that without a structured approach to deployment, guiding public and private investment in charge points across the region, there was the risk that only profitable sites focused on level 3 DC charging would be developed, leaving behind

Case Study 7.7: The strategic case for deployment – the West Midlands Combined Authority

In 2020, the West Midlands Combined Authority (WMCA) began the process of establishing a new approach to the delivery of charge points and enabling infrastructure across the West Midlands. In a paper issued in 2020 to the WMCA Board, the Authority set out the strategic case for a regional strategy for EV infrastructure deployment and the requirement for further development of the economic, commercial, financial and management case underpinning the overall business case. The strategic case involved setting out the rationale, drivers and objectives of the deployment. In the paper, the Authority set out the requirements for charge point infrastructure across the region based on projected adoption rates. This stated that by 2040, 10,000 level 2 7 kW chargers would be required, mainly in long stay car parks targeting residential, workplace and destination (hotel) use cases. In addition to this, a number of level 2 22 kW and level 3 DC 50–350 kW charge points were required at short stay car parks, local hubs and transit station. The board paper noted that if the private sector was solely responsible for deployment, the various future gaps in provision would emerge. This anticipated reduced coverage in less affluent areas and in areas of electrical grid constraints, less on-street charging in preference for high power destination charging, delay in rollout due to needing to ensure profitability and a potential lack of interoperability. To address this, the paper stated that WMCA should invest in assets across the aforementioned use cases and that there would be benefits in doing this including:

- Mitigating the risk of land banking, where the charge points may not be available for years until the private sector feels that the market is mature enough to be profitable, losing the benefit of stimulating demand.
- Allowing a more holistic view of the economic benefit of investment, including the provision of a large, de-risked, electrical network upgrade to catalyse renewable energy, storage and other commercial developments.

The report recommended that the strategic case for deployment should be based on development of a regional strategy. The overarching benefit of this approach was that this would ensure efficient deployment of infrastructure, provide investment confidence for the private sector and central government and enable coordination via a forum and steering group. It would also result in coordinated engagement with DNOs and, ultimately, coordinated delivery of resources to deploy charge points. The strategic case also highlighted the need to develop the investment case for charge points, facilitate and coordinate delivery and secure funding to assist the investment case for local delivery. To drive forward delivery, the report highlighted three next steps which were the development of a more granular evidence base, agreement to develop a regional delivery plan and secure investment for specific charge point deployment opportunities.

Source: West Midlands Combined Authority. *Electric Vehicle Charging and Enabling Energy Infrastructure: A West Midlands Approach* [online]. 2020. Available from https://governance.wmca.org.uk/documents/s3953/Report.pdf [Accessed 08 October 2024]

communities in urban areas that required on-street charging. This would be further exacerbated by the fact that across the region, EV growth is expected, meaning the demand for charge points would increase. Therefore, the strategic case for deployment across the West Midlands was that this would ensure charge points were deployed ahead of, and to encourage, demand and that these would be deployed in a range of locations to ensure an equitable distribution. Furthermore, the Combined Authority noted the importance of deploying infrastructure to encourage adoption of EVs so as to enable air quality improvements across the region. A CPO, in structuring their strategic case, would most likely adopt a case for change based on the identification of a gap in the market where charge points can be deployed profitably or where the market is congested. Deployment would be based on offering a unique selling point versus competitors. Regarding identifying a gap, this might be locations for level 3 DC charging on a strategic road network or in terms of destination charging through an agreement with a retailer or hotel that does not already have charge points in place. Regarding a unique selling point, this might be the ability to offer smart charging, lowering the cost to the customer or a unique technology enabling the ability to use their domestic home tariff to charge out and about. Regardless of whether the deployment is publicly or privately driven, the strategic case establishes the bedrock on which the rest of the business case is built.

Focusing on the economic case, Case Study 7.1 highlights how the deployment of charge points alongside a fleet transition would deliver value for money for North Tyneside Council. The economic case was based on the cost of installing multiple charge points at depots versus the operational savings which would be achieved through cheaper per mile costs based on electricity versus diesel or petrol. North Tyneside, in their investment, will no doubt have considered the costs of ‘business as usual’ (based on ICE vehicles) and compared this with a ‘do minimum’ of a partial transition to EV and an ‘all-in’ approach where the entire fleet was transitioned. The ‘business as usual’ would deliver no operational savings. The ‘do minimum’ would deliver some savings and start the Council on the transition. The ‘all-in’ option, though delivering the greatest operational savings, would also require the greatest investment and might also not be practical at the time the Council embarked on its journey due to the availability of EVs on the market.

Based on this analysis, it is clear why the economic case for 'do minimum' formed the basis for the way forward.

For a CPO, the economic case would primarily focus on the value for money of one deployment approach over another. For example, if deploying on-street chargers, this might focus on the cost of deploying in areas with high anticipated demand but higher grid upgrade costs to enable the charge points versus an area with lower demand but lower upgrade costs. The level of detail required in the economic case is therefore greater due to needing to consider various different deployment options.

Looking at the commercial case, Case Study 7.4 brings to life how this part of the five case approach sets out the deliverability of a charge point deployment in terms of the procurement and supply chain. In spending the Levelling Up Fund, NECA identified that a joint procurement spanning across four of the seven local authorities (Gateshead, Newcastle City Council, North Tyneside Council and Sunderland City Council) would deliver value for money, enable economies of scale and result in a more attractive private sector investment opportunity. The North East Procurement Organisation (NEPO) framework was established with input from transport officers from NECA and across the region in anticipation of the EV investment. Through this approach, Zest were appointed to deploy 92 charge points to the value of £1.2 million. The procurement secured savings as a result of successful negotiations with Zest to match fund approximately £124,000 (10%) of the contract value. Within the mechanisms of the contract, the constituent contracting authorities are entitled to a revenue share to cover programme management and delivery support. This was initially 2% of all revenue generated with the rate rising should usage exceed expectation. The procurement via the NEPO framework highlights the importance of ensuring that there are suitable providers who can bid for the contract and that a well-prepared client should result in a contract that meets its needs while delivering value for money. The commercial case for a CPO would be entirely different and would focus on two key elements. The first is whether, if bidding for public sector contracts, they are on the correct procurement framework to bid for an opportunity to deploy infrastructure. The second is relevant regardless of this and relates to whether the CPO has the supply chain to deploy the number and type of charge points in the timescales required to meet the expectations of the client. If the answer to either of these is 'no' the commercial case collapses; however, in the real world, the commercial case is not quite as binary and will outline how the CPO will use a range of providers in its supply chain to deliver a deployment. Regarding a framework the CPO is not on but is keen to access, the commercial case might outline how a partnership would be sought with a provider on the framework to open up the opportunity.

The financial case and its importance in a real-world deployment is demonstrated in Case Study 7.8. In this example, analysis conducted by the International Zero Emissions Vehicle Alliance, Cenex and ICF demonstrates that, across Canada, China, India and the Netherlands, affordability and profitability of charge points is central to the financial case for CPOs. Profitability is primarily

Case Study 7.8: Understanding the business case for EV charge point infrastructure – a modelling approach

The crux of the business case for a CPO is the financial case, focusing specifically on the affordability and funding for a preferred deployment. Core to this is ensuring there is a detailed understanding of the capital, revenue and whole life costs of the proposal and its impact on the balance sheet, income and expenditure of the organisation. The analysis conducted by the International Zero Emissions Vehicle Alliance, Cenex and ICF highlighted that across Canada, China, India and the Netherlands, there is a consistency in terms of the key factors that determine charge point profitability. Modelling of utilisation, energy costs, charge point installation costs (CapEx, OpEx), ancillary income (e.g., retail and advertising), incentives and subsidies determines projected profitability and is critical to securing investment and delivering the deployment. To enable this, the Zero Emissions Vehicle Alliance, Cenex and ICF developed a Charging Infrastructure Business Case Assessment Tool (CIBCAT). CIBCAT is based on a discounted cash flow model, with the key result being expressed as a project Net Present Value (NPV). The user inputs to the tool are used to both calculate project initial set up costs (CapEx) and forecast annual operational cost. The model is based on several assumptions:

- The technology lifespan is 10 years, with NPV calculated over the same duration.
- A single utilisation profile is used and applies to all charge points.
- Utilisation events within utilisation profiles are from a uniform distribution.
- Charge point OpEx applies to all charge points equally.
- Grid upgrade costs are applied on a per kW basis.
- Electricity prices are constant throughout each day and year. This was chosen partly to keep inputs simpler for the user, but also to exclude energy price volatility risk from the model. Such risk can be managed by a fixed price energy contract.
- Charge point utilisation has no seasonal or within week variation.
- Projected solar yield and battery storage power can be included if relevant.

Utilisation is key to the business case. The modelling across the geographies found that the payback period, or the total time needed for the project to generate sufficient cash inflows to recover its initial costs, can be shortened by half when the average charging station utilisation increases from 10% to 20%.

Source: International Zero Emissions Vehicles Alliance, Cenex and ICF. *Understanding the Business Case for Electric Vehicle Charging Infrastructure* [online]. 2024. Available from https://zevalliance.org/wp-content/uploads/2024/01/Understanding-the-Business-Case-for-Electric-Vehicle-Charging-Infrastructure_jan24.pdf [Accessed 08 October 2024]

related to two factors: net revenue per kWh and utilisation. Regarding the former, net revenue is the income per kWh to the CPO after energy and other costs have been factored in. Higher net revenue per kWh will reduce the payback time for capital expenditure (CapEx). Regarding the latter, higher utilisation reduces the payback period, with this being shortened by half when utilisation increases from 10% to 20%. The ideal situation for profitability is a high net revenue and high (15% plus) utilisation, though this is difficult to achieve as the former may impact on the latter because a high net revenue implies a high cost charged per kWh which may result in less competitive charge points and lower utilisation as a result. Affordability primarily relates to the CapEx of the charge points and associated infrastructure costs to install them. As a general rule, the greater the charge point speed, the more expensive the charge point unit and the higher the grid upgrade cost. This means that installing hundreds of level 2 charge points over a wide area as part of a residential charge point deployment will likely cost much less than installing two level 3 charge point hubs, each comprising six charge points offering speeds of 350 kW. Comparing the cost to install with the profitability of charge points will make or break the financial case based on the payback period required by the investor. Some may be happy with payback periods in excess of 7 plus years where a long term site lease of 25 years has been agreed, whereas other will want to see payback and profitability as quickly as possible. For a local authority building a business case, although their primary focus is not on profitability, it is important to consider when going to market for a CPO that the sites proposed at least reflect a degree of profitability based on cost to install versus utilisation and anticipated net revenue per kWh. Regarding the latter point, the local authority, if setting the price, should propose realistic options for the successful CPO; otherwise, the procurement exercise will not result in competition because CPOs will be put off. The worst-case scenario is a failed exercise because no CPO bids due to anticipated poor profitability and affordability. That being said, a local authority should consider the financial case alongside the strategic case as the aspiration to ensure a just transition (including poor demand sites) will need to be balanced with the need for CPOs to make a profit so as to ensure a successful deployment.

The management case is demonstrated through Case Study 7.6 as this part of the process focuses on whether robust arrangements are in place for the delivery, monitoring and evaluation of a charge point deployment. In Case Study 7.6,

Mer's rollout of destination charge points across Blue Diamond's garden centre network would have required Mer to have the capacity and capability to plan for and manage deployment as well as monitoring the success of each site once in place. This is highlighted in the way in which the deployment programme gradually expanded to include additional sites. From experience, capacity and capability to deploy charge points should not be underestimated. Staff must be employed and trained to carry out various activities, including site selection, site plan production, DNO connection application completion and monitoring, sourcing of charge point equipment, engaging with contractors to install charge points, managing the project financials, seeking electrical and safety approvals for each deployment and then managing the charge points once in place. Mer, like many CPOs, have the necessary teams to carry out these various steps, as evidenced by the fact that they continue to organically expand their network. The management case is ultimately about ensuring these various elements are in place (or there is a plan to put them in place) so that charge point deployment occurs as smoothly as possible.

Once all five cases are outlined and documented in the business case, the final parts of the process are gaining approval for the business case and then seeking funding to deploy the charge points. If public sector funded, what will likely follow approval is a procurement exercise to select a CPO or CPOs to deliver the deployment. Payment by a local authority is normally based on delivery against a pre-determined timeline. If private sector funded, the CPO will likely need to provide the business case to its investors to raise the necessary capital to fund the deployment. As is clear in the financial case, they will be very interested in both the profitability and return on investment.

7.4 Planning deployment at a local level

Planning deployment at a local level takes the long list of sites developed in Section 7.2 and focuses on determining specific locations at a granular street level so as to finalise the final deployment locations. It should be noted that during the build stage, detailed construction phase plans are completed including detailed designs. This stage does not cover this level of detail but does gather a range of data sources to ensure analysis considers real-world opportunities and challenges impacting on the viability of potential locations. Case Study 7.9 demonstrates this through Birmingham's EV Charging Strategy Local Charge Point Deployment. There are various tools available to local authorities to enable them to carry out deployment planning at a local level. This can be achieved through entering information into a spreadsheet and ranking locations based on various factors though, for a number of local authorities and CPOs, the use of a geospatial mapping tools enables various data sources to be analysed and layered to identify the best and most viable locations. The most common data sources used for residential on-street charging analysis include:

Case Study 7.9: Birmingham City Council's EV charging strategy local charge point deployment

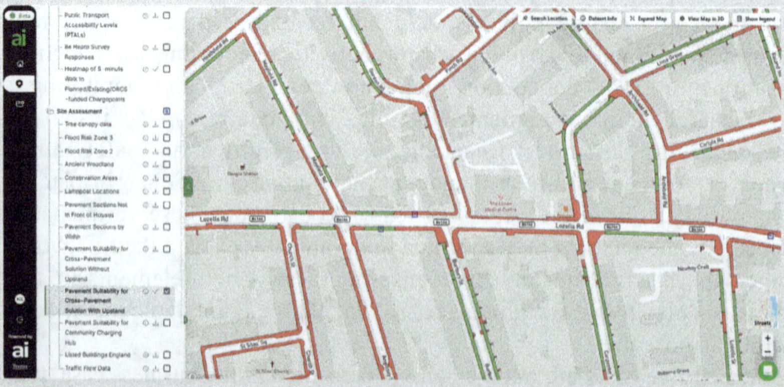

Birmingham City Council, in their deployment of charge points across the city, demonstrated the importance of planning for deployment at a local level. Through the EV Strategy and subsequent deployment plan, the city documented the various locations where charge points should be deployed. In the Strategy, they highlighted that en route and destination charge points are key to Birmingham's deployment and that specific park and ride sites for site users, taxi ranks and supermarkets/amenity sites were suitable for destination charge points. For en route charging, key routes into and out of the city (M6, A45, A456, A4540) were suitable for level 3 DC charging hubs. For on-street deployments, Birmingham recognised that a more granular localised approach was required. Working with Advanced Infrastructure limited, a spatial analytics and software company, the Council created an interactive EV charge point mapping tool and bespoke street-level geospatial datasets for all 69 wards of Birmingham. Project partner Element Energy was also engaged to develop the Residential On-Street EV Charging Implementation Plan using the outputs from the mapping tool and data. The tool incorporated various data sources key to planning for localised deployment, including pavement suitability for on-street charge points, heat maps of 5 minute walking distance to an existing charge point, existing charge points, public transport accessibility levels, potential for off-street parking, car ownership statistics and transformer demand and generation headroom. Combining these various data sources enabled the city to identify the most viable sites and build its deployment plan around these.

Source: Birmingham City Council. *City-Wide Electric Vehicle Charging Strategy* [online]. 2021. Available from https://www.birmingham.gov.uk/download/downloads/id/20988/city-wide_electric_vehicle_charging_strategy.pdf [Accessed 13 October 2024]

Source: Advanced Infrastructure Technology. *Empowering Birmingham City Council's EV Charging Strategy with LAEP+ and Granular Data* [online]. 2024. Available from https://www.advanced-infrastructure.co.uk/case-study/empowering-birmingham-city-councils-ev-charging-strategy-with-laep-and-granular-data [Accessed 13 October 2024]

- **Proximity to nearby substation, including total capacity and available headroom**: this determines whether there is any available power capacity locally to accommodate additional charge points and, depending on the speed, the number that can be installed in any location. A deployment might choose to connect to a different substation with greater headroom to be able to deploy.
- **Existing on-street parking including what traffic regulation orders or rules exist:** this enables planning of the number of bays which could be used for EV charging and, depending on the type of bay, whether traffic regulation orders (or their equivalents outside the United Kingdom) are needed to govern parking usage. This might limit the ability to deploy large numbers of charge points.
- **Existing off-street parking at street level (current and planned)**: understanding this helps determine demand in an area, as a high percentage of off-street parking households will mean less demand as drivers will charge cheaply at home.
- **Street level EV uptake (both current and projected)**: this helps determine current demand for charging (factoring off-street parking) and future demand.
- **Projected utilisation**: this is developed based on the various factors influencing demand, with high utilisation (anything over 15%) being preferable to CPOs as these locations are more profitable than lower utilisation sites.
- **Nearby car parks**: in locations where there is a lack of on-street parking these might offer alternative locations to deploy charge points. This data will usually be accompanied with walking distance data helping to indicate how far a driver might walk to charge their car. In Case Study 7.9, this is no more than 5 minutes.
- **Existing residential and destination charge points**: these are mapped to understand the level of competition for charging in an area. A dense cluster of charge points will likely indicate high competition and a saturated market.

- **Pavement width and street furniture/lamppost location**: this relates to the practical aspects of deployment as narrow pavements or lampposts distant from the edge of the kerb are unsuitable for charge point deployment, meaning locations with these features would be discounted, or only developed alongside redesign of the street layout.
- **Nearby points of interest**: these might influence demand for charge points, especially where these drive up demand for on-street parking in an area.

For en route charge points, the data is different and focuses on:

- **Traffic flow data including proportion of EVs (current and projected)**: this is an indicator of likely demand at a site, with high traffic flow locations suggesting high demand and are therefore a preference.
- **Projected utilisation**: based on traffic flow, higher projected utilisation sites which are more profitable are favoured over lower utilisation sites.
- **Existing level 3 DC charge points (current and planned)**: this indicates likely competition which will impact on utilisation.
- **Available land and land ownership**: this is important, especially when a CPO needs to understand who should be approached to lease or rent land from.
- **Proximity to nearby substation, including total capacity and available headroom**: as already stated, this impacts the number of charge points that can be deployed.

For destination charge points, the analysis is primarily driven by the customer seeking to deploy charge points and includes:

- **The customer's current estate including type of location**: this determines what sites are available for deployment.
- **Car park size and layout**: this is a useful indicator of the number of charge point bays that can be currently installed and the potential for the site to expand in the future.
- **Traffic flow data around potential deployment locations including proportion of EVs (current and projected)**: this indicates the demand for charge points.
- **Existing level 2 and level 3 DC charge points (current and projected)**: this gives an indication of the competition surrounding the potential charge points.
- **Nearby points of interest**: this helps understand whether there will be additional demand coming from outside the site.
- **Projected utilisation**: this is based on the above factors and indicates the potential profitability of proposed charge points.
- **Proximity to nearby substation, including total capacity and available headroom**: as already stated.

Once this data is layered and visualised, this enables the identification of certain locations suitable for deployment. This includes a prioritisation of different factors based on the practicalities of deployment, potential cost and alignment with the initial decision to deploy. This includes a focus on sites that align closely with the specific outcome being sought as part of the deployment. For example, if a local authority is seeking to ensure an equitable distribution, it may choose high-demand sites close to substations with headroom and mix those with sites in areas with little current or projected demand combined with low incomes. A CPO might include low-demand sites for en route charge points because they anticipate future growth and want to secure a site as a 'land-grab'. Completing this prioritisation will result in a short list of sites.

The short list is further refined by analysis of the aspects which will impact on deployment including virtual and physical site visits. This helps understand whether a site is suitable for deployment based on features at the site. For example, vandalism picked up on a visit might make the site less attractive. Based on these site visits, concept site designs can be drawn up demonstrating the number and types of charge point to be deployed, power diagrams showing connection points, location of the nearby feeder pillar containing metering, cabling layout and other notable features like street furniture. Engagement with key stakeholders then takes place to understand the practical and planning aspects of deployment at a site-by-site level and ultimately determines the viability of sites. This may result in further refinement of the short list. A CPO will want to engage with the landowner to understand whether proposed sites are available and whether the land can be purchased, leased or rented. Initial conversations with the relevant planning department will take place based on the short list to understand whether planning is required. This also includes engagement with the local highway authority regarding parking regulations to understand the requirements. The refined short list is then shared with the power provider/DNO and contractor/installer. The former will determine the cost to install at each site based on the number of charge points, proximity to the nearest power connection point and substation capacity. If the costs are too high, the CPO/installer might engage with an independent DNO. Regarding the latter, the contractor/installer will give an indication of cost based on the concept designs.

The final part of the process is the finalised short list collating the relevant data for each location. A detailed site design is developed at this stage. Alongside this, the development of a financial model and project plan is completed covering each site, as well as combining all sites into one overarching model and plan. These documents, alongside the short list and detailed site designs for each location, are critical to the governance and approvals process in Section 7.5.

7.5 Governance and approvals

The final step of the planning stage is securing approvals and determining the governance arrangements for the next stages. Where a CPO is driving deployment, this involves consolidating the outputs from Sections 7.1 to 7.4 and finalising a business case with an associated business plan to secure investment. Once this is done, the CPO can then set in motion the various approvals needed in place for the build stage including finalised quotes for power connections and upgrades if required and completed civils plans and associated quotes to carry out the work at each site. Planning applications may also be required for certain deployments that are submitted to the local authority. In the United Kingdom, for example, planning permission is required where a charge point is installed on-street, in a conservation area or on/next to a listed building, where permitted development rights have been removed or if more than one charge point is provided for a parking bay. Planning is also required if the charge point unit is taller than 1.6 m. Traffic regulation approval, if applicable, is required in on-street charge point deployments where bays need to be designated specifically for EV charging and is sought from the local transport authority. Other planning tasks completed in this final stage might include ensuring internet connectivity is in place at sites and whether a connectivity upgrade is required (e.g., boosting mobile signal). Planning to purchase, lease or rent land, including associated costs, should start at this point, including the drafting up of agreements and settling on a cost so that these can be finalised. If any of these changes result in significant changes to projected costs, the business case may need to be re-visited and adjusted to ensure the deployment still makes financial sense. Once these final tasks have been completed and funding secured for deployment, the CPO can move into the build stage. The CPO should then put in place the management structure required for the build stage including project managers and support teams. Further detail is outlined in Chapter 8.

In most deployments, local authorities are key to the governance and approvals process, as demonstrated in Case Study 7.10. In this example, Durham County Council's legal, procurement, planning, assets, highways and health and safety teams played a key role in the planning of deployment and delivery, especially where a CPO is deploying a network on the local authority's behalf. Analysing this in detail, the procurement team established the tender opportunity and ran the exercise to select the successful CPOs. The Council's legal team were key to drafting and signing contracts between CPOs and the Council. The planning team determined whether planning was required for deployments and if so, the approvals required. The assets team granted approval for the types of charge points that were being deployed on the Council's behalf. The highways team were involved in planning deployments on roads, approving requests for road closures or temporary traffic measures and ensuring the CPO returned the road back to its original appearance where deployments took place on the highway. The health and safety team were responsible for approving the final designs for deployment and approving the certification provided by the CPO once the charge points were installed and the site was ready to go live. Durham County Council in Case Study 7.10 demonstrated how having all of these decision-makers onboard at the start of the project ensured smooth delivery.

Case Study 7.10: Durham County Council's delivery and governance process underpinning charge point deployment

As part of Durham County Council's charge point deployment, ensuring strong governance was in place was key to delivery. Across County Durham, 43% of residents live in rural areas and 40% of houses are terraced with no off-street parking. The Council set out to deploy accessible charge points across the County to ensure residents have good access. Durham County Council's aim was that every Durham resident lives within a 5 minute walk of a charge point. Key to rollout was a partnership arrangement between the Council and parish and town councils, NECA, the local authority 7 group of local authorities in North East England, Durham University, Northern Powergrid and Charge My Street (a community benefit society that installs charge points). Deployment had support from all the key internal decision-makers within the Council to ensure the approvals process for deployment would be as smooth and efficient as possible. Legal, procurement, planning, assets, highways and health and safety teams were engaged and involved in preparation of key deliverables underpinning deployment. Furthermore, the Council set up an internal EV working group that met every two to three weeks to maintain pace of delivery. Senior buy-in also ensured that approvals were efficient when required. The Head of Environment and Corporate Director of Neighbourhoods and Climate Change were fully engaged throughout the process, unlocking internal funding if needed to ensure delivery progressed. A further key part of the governance process was ensuring that residents were engaged throughout deployment. This involved active engagement including partly basing deployments on requests from residents. For example, one resident asked for a charge point at the Tanfield Business Centre, where they spend most of their day. The Council installed one.

Source: Wignall F. *Durham County Council is bringing vehicle charging to every community* [online]. 2022. Available from https://ashden.org/news/durham-county-council-is-bringing-vehicle-charging-to-every-community/ [Accessed 17 October 2024]

To ensure all governance steps are taken into consideration, this involves mapping out the decision-making points, as demonstrated in Case Study 7.10. Ensuring the key stakeholders who are involved in approvals are aware of the deployments and understand what will be installed helps ensure that timelines are adhered to. The most common approvals required as part of the project plan, which were highlighted in Chapter 3 but have been re-listed here, include:

- **Approval of final site detailed designs (internal to the CPO)**: this either comes at the end of the planning stage or at the start of the build stage.
- **Approval to start the build (internal and external to the CPO)**: once gained the CPO can start the install.
- **Approval of build alignment with designs (internal and external to the CPO)**: the CPO will sign off that they are happy the design aligns with the built layout.
- **Safety approvals (external to the CPO)**: the CPO will seek approvals for electrical safety of the charge point site.
- **Charge point system and commissioning approvals (internal and external to the CPO)**: the CPO will seek approvals that the charge point operates as expected and has been properly commissioned to go live (e.g., not being left in standby mode). These approvals will also include go-live of the charge point.
- **Release of payment approval if applicable (external to the CPO)**: if the CPO is delivering for a customer (like a local authority), they may have to complete all of the outlined approval steps to receive funds.
- **Approval to change pricing if applicable (external to the CPO)**: a CPO, if delivering and operating charge points on behalf of a customer, will likely need approval if changes are made to agreed pricing.
- **Approval to upgrade charge points (internal and external to the CPO)**: a CPO will need approval if charge points become obsolete and require physical or software upgrades. If this is the case, the CPO may need to gain approvals again as upgrades progress.
- **Approval to decommission a charge point (internal and external to the CPO)**: the CPO require this where charge points are being taken out of action.

Planning for these ensures there are no surprises in the build and operation stages.

References

[1] Fröde P., Lee M. and Sahdev S. *Can public EV fast-charging stations be profitable in the United States?* [online]. 2023. Available from https://www.mckinsey.com/features/mckinsey-center-for-future-mobility/our-insights/can-public-ev-fast-charging-stations-be-profitable-in-the-united-states [Accessed 04 October 2024].

[2] Transport for the North. *STB EV Charging Infrastructure Framework* [online]. 2024. Available from https://evci-visualiser.transportforthenorth.com/transport-for-the-north [Accessed 06 October 2024].

Chapter 8

Deploying charging infrastructure

Myles Roberts[1] and David Trousdale[2]

The focus of this chapter concerns the deployment of charge points and includes analysis of challenges that may arise and other relevant factors. The experiences outlined and case studies shown are United Kingdom focused; however, we have attempted to illustrate the technological and economic challenges and how they apply to other Western geographies. By the time a project reaches the deployment stage (usually the build stage in the construction industry), the project manager should have clear guidelines, with the 'planning for deployment' stage having established that the project fits within the organisation's EV charging strategy, is commercially feasible and is deliverable within the proposed budget and timeline. During the delivery of this stage, the project manager will address already identified risks and issues and may identify new benefits (opportunities) that could be achieved, should certain aspects of the project be altered. It should also be noted that although this chapter is written from char.gy's perspective, their approach is fairly standard in the United Kingdom at least, and is similar to that adopted by other charge point operators (CPOs).

8.1 Project requirements established prior to deployment

When launching a new charge point build project, the project manager must ensure that they have as much detail to hand in relation to the proposed installation, with sufficient information about the location of the deployment and power supply to progress through the build process. To ensure that any unidentified risks or unanswered questions are addressed as soon as possible, the project manager must make sure that:

- The nominated site, with its given location, size and transport accessibility, aligns with the CPO's strategy.
- The proposed project is understood in detail, including the number of planned charge points and their power ratings, and any proposed additional on-site facilities.

[1]Off-Street EV Charger Project Manager, char.gy, United Kingdom
[2]Principal Advisor - Climate Change and Sustainability, Newcastle City Council, United Kingdom

- The proposed project design is suitable for the proposed site. This includes that the new required electrical demand (including the additional load after the installation of the charge points) has been considered and that the civils work (including digging for the installation of the charge point and/or ducting for the electrical/data service) has been planned. Civils work also includes the planning of required traffic management, marking of necessary parking bays and signage.

Identifying these elements prior to the build should ensure that the project is achievable within the allocated budget and expected timeline.

8.2 Stakeholder management planning prior to deployment

Crucial to project delivery is identifying all stakeholders involved including those who are directly engaged in the build (e.g., suppliers) and those with a more distant role but one that is nevertheless important in deployment (e.g., local authorities). The most common stakeholders are:

- **Charge point suppliers/other suppliers**: these provide the charge point/s and the associated equipment.
- **The local or state authority or landowner**: the CPO obtains the relevant documentation and permission/s from these organisations to carry out the works at sites/locations.
- **The electrical contractors**: they have the capabilities to install, electrically test and certify the charge points to correct standards.
- **Distribution Network Operators (DNOs)**: the project manager needs to engage with the DNO to determine any works which may be required for the installation of charge points, including the timeframe of when works are to be completed. Agreement needs to be reached with the DNO to supply the CPO with a Meter Point Administration Number (MPAN) so that a metred connection can be used to pay for the energy dispensed by the charge point when in public use.
- **Civils contractors**: these need to be engaged to plan ground works. Key elements in planning include seeking permissions to carry out the work (in the United Kingdom, this is a Section 50) and avoiding cable/services strikes while digging.

The CPO will also need to manage internal stakeholders when preparing for the deployment. The focus of the activity includes planning for required testing, inspection and correct certification for all parts of the electrical installation and commissioning the charge points on the CPO's platform. Regarding the latter this relates to the CPO's charge point management system (CMS) that monitors charge point operations in real time, checking their status, authorising individual charging sessions requested by drivers and measuring the resulting amount of energy delivered and finally billing the driver for the total cost of the session.

8.3 The role of project management prior to deployment

The project manager has a critical role to play in deploying charge points. Deploying public charge points requires planning to ensure the required permissions are in place at the start of the project, the required materials have been delivered to site and the necessary skills and capabilities are available at the right time to drive forward delivery. Checking all of these elements helps ensure that all is accomplished within the planned budget and timelines. The overall functions are expanded on in the following sections according to the standard roles set out in a charge point deployment project as outlined in Section 8.2.

8.3.1 Project planning

A project scope lists a project's high-level objectives, timelines and any limits to what is to be achieved by a project. It will outline what work is to be included and the objectives addressed. It also includes ensuring that all stakeholders have a clear understanding of the project's objectives so as to manage their expectations and prevent scope creep. A project design will be drawn up, specifying each project deliverable, how each is to be achieved and the quality standards to be met (including definition of complete). Related to the design is a plan detailing the tasks and timings thereof to meet the project design and budget, as well as any interdependencies between the tasks. Typically, certain deliverables need to be completed before others. For example, power must be secured at an EV charging site before charge points can be tested and commissioned. Resources (people and equipment with specific skills and capabilities) are required to fulfil each task in the plan, with a schedule planning when each resource is required to perform various tasks so as to meet relevant objectives at specific points during the project. To ensure all the resources are able to achieve their objectives in time, a budget is required to cover resources and materials costs. The cost of the land/premises, where required, is often not included in a project's deployment costs. This is because this will have been considered at the project feasibility stage and is beyond the project manager's direct control.

In order to protect the reputational risk of the organisation they are working for, the project manager should ensure that the designs that are being implemented are approved by competent people and that these comply with relevant national regulatory requirements, industry-specific and general/publicly accepted standards relevant to the technology being installed. Examples include local electrical grid performance, local traffic impact and noise or environmental consequences of a new EV charge point being installed. Typically, the project manager is responsible for submitting technical designs and associated studies or reports that are required by relevant authorities. These are reviewed, considered and, assuming the authorities are happy, approved. The project manager will use their industry experience to establish which stakeholders are relevant to the particular location and type of build, what artefacts are required by each stakeholder, how long each one takes to provide their views on the build and issue permits where required. Permitting costs are included in the project budget.

8.3.2 Contract management

Having established a plan with associated costs, all required materials and resources need to be secured and guaranteed to be available at the required time and in the necessary quantities. The guarantee comes in the form of a contract. This can be a purchase order to pay for a material or service at a quoted price or a contract with a specialist supplier to supply a unique service at a specific time in the project. Most contracts will establish specific standards and the consequences of not meeting these (including the impact on the timeline). It may also cover the consequences of non-delivery for the rest of the project. These consequences will have been quantified and penalties included in the contract with the supplier to cover the cost of adverse impacts on the project, should the supplier not be able to meet its commitments. In large and complicated projects, the project manager should have the support of a legal team (or at least a contract specialist) to help them with the legal intricacies related to contract management.

8.3.3 Change management

As the project plan evolves and matures, any changes to the project scope, design, plan and especially their impact on project budgets and timelines must be quantified and approved. Each individual project change should be recorded as a change request. These trigger a round of project re-planning, focused on the impact of, and possible mitigation of, consequences relating to the planned change. As these consequences may be material and unavoidable, individual change requests must be approved at the relevant level of management seniority, akin to the project approval originally. The details and impacts of the change on the project plan must be communicated in a timely manner to relevant suppliers and other stakeholders to minimise risks and allow as much time as possible to mitigate any adverse consequences.

8.3.4 Risk management

Throughout the above planning cycle and until project completion, risk management is a key function to help ensure that the project keeps to the original plan and budget. In this process, hazards are identified which could influence delivery of the project. The impact of the hazard occurring is multiplied by the likelihood of the hazard occurring (typically on a scale of 1–4 or 1–5), with the resulting answer quantifying the risk associated with a hazard. This is called a risk assessment. Should the risk be high enough, mitigating steps should be taken to respond to the risk, with a subsequent risk assessment after mitigation conducted to determine the residual risk. Typically, 'red' risks must be mitigated to 'amber' or 'green', and 'amber' risks to 'green'. If the residual risk is not appreciably less than the original risk, then alternate mitigating steps must be taken. This process must be started and repeated throughout the planning cycle, as planning will involve some element of trade-off between time, resources and budget, with planning choices made having pros and cons. Any risk that is deemed to be too high will require mitigating. This will likely require a change to the original plan. Once mobilisation and building starts, project progress and budget are continually monitored and risks evaluated on a routine basis. Original risks may be closed as their related tasks are completed. New hazards will likely be identified, assessed, mitigated and subsequently monitored until they too are either closed, or project completion is achieved. It is possible that certain risks may persist beyond the initial build of a charge point. An example is the risk that objections from local residents increase, possibly curtailing operating hours in the future once the site is open to the public.

8.3.5 Resource management

During the mobilisation and build steps, the ongoing tracking of project progress and any deviation from it will result in specific tasks being rearranged, requiring associated resources to be rearranged in tandem. Typically, once a plan is set in place and suppliers have committed to it, there could be cost/budget implications to reschedule work and resources. A materials supplier may raise prices to deliver a required material sooner than originally planned; similarly, they may offer cheaper prices to delay delivery. Or the opposite may occur. This will depend on the specifics of the situation at the time. Should no practicable option be available to reschedule a particular task, then that issue becomes a constraint around which other planning must revolve as the risk of that task being missed will have more serious implications for project completion. Effective resource planning involves continual scenario evaluation, assessing combinations of different timings of tasks and their associated requirement of materials and resources, and choosing the most appropriate option.

8.3.6 Budgeting and cost control

A CPO will have a finance system and accounting controls in place. These should be used to reduce the use of spreadsheets to track budgets and the amount of estimating of spend. There will likely be certain financial reports that cannot track to the level or category of spend required for specific resources as the project budget may not be set up in an appropriate way, or the finance department does not have the staff to provide reporting as frequently as desired. In this instance, a spreadsheet might be the only suitable tool.

8.3.7 Quality management

Each project deliverable must have a qualitative standard defining what must be achieved to mark the deliverable as complete. These project standards may refer to national regulatory requirements, industry- or client-specific standards. Contracts associated with each deliverable must have terms enshrined that ensure that not meeting specified standards will require rework at the supplier's expense, along with compensation for any consequential adverse costs or other impacts.

8.3.8 Progress tracking and reporting including scheduling and time management

The project manager must track task completion on a regular basis (between daily and ideally weekly, depending on the project size, complexity and particular project stage). This can be done by meeting with suppliers remotely or in person on- or off-site. Suppliers must provide evidence of completion to the required standard. Depending on the reliability of a supplier and any previous experience related to the current project, alternate means of verification may be sought, especially if the deliverable is technically complicated. This additional independent verification will likely cost, and if it is being introduced as a result of current poor performance, the cost thereof should be borne by the supplier. This will likely be contained as a term in the supplier's contract. Payment for completed deliverables must be tied to having achieved the prescribed quality as stated in Section 8.3.7. Reporting of project progress is typically done in a hierarchical manner, with the project manager reporting weekly to a programme manager on project progress and budget expenditure against the current approved plan, as well as any change in current or new risks or issues. Additional reporting is done on an exception basis as and when is required. Similarly, the programme manager will report to a programme board, with senior executive management reviewing progress, issues and other exceptions at a strategic level.

8.3.9 Communication and stakeholder management

It is good project management practice to establish a communication plan at the start of a project, with stakeholders identified and categorised by means of a RACI matrix as outlined below:

- **R – Responsible**: who is responsible for the delivery of the project and is typically the task owner.
- **A – Accountable**: who ensures that a task is complete and to standard.
- **C – Consulted**: who should be consulted before a decision or action is taken.
- **I – Informed**: who should be informed after a decision or action is taken.

All stakeholders must be assessed. This includes relevant internal staff, material and resource suppliers, energy suppliers, electricity network operators, relevant national, regional and local government bodies, civic bodies, and in the author's experience, local climate action groups and EV and motoring clubs. Such a matrix avoids ambiguity regarding each stakeholder's role and responsibility in the project. Additionally, relevant information can be planned and communicated at the appropriate time.

8.4 Construction (Design and Management) (CDM) regulations relating to EV charge point installation

Construction (Design and Management) Regulations 2015 (CDM) is the United Kingdom's primary legislation for managing health and safety in the construction industry [1]. It outlines the roles, responsibilities and duties of everyone involved in the construction industry (including charge point deployment) to ensure a safe working environment. The word 'industry' is used advisedly, as any building or construction work is included, from a small home renovation project performed by a single tradesman to the HS2 high-speed rail project (Britain's largest infrastructure project ever) with a workforce of 20,000 people [2]. The Health and Safety Executive (HSE) is the regulatory body for health and safety in Great Britain. It provides guidance and support to each of the roles named below. This section outlines the basics behind CDM which must be followed when deploying charge points.

8.4.1 The client

The client is defined in the CDM regulations as any person for whom a project is completed for. This could be a single person (e.g., a homeowner), a group of people, or a business. The client is usually the person instructing contractors and designers, and deciding what work will be done. Importantly, this is usually the person paying for the work to be done. At the start of a project, the client must provide relevant information available about the site including surveys or the results of other investigations, and the following types of project-specific information (pre-construction information) in a clear, concise and convenient form. This should contain information about:

- The project objective(s), including what success looks like and key dates.
- The project planning and management structure, including resources and time allocated at each stage.
- Project cooperation and coordination arrangements for all various project teams, including who will communicate with whom, by which medium and about which topics, ensuring each stakeholder is communicated with as appropriate.
- Health and safety hazards relevant to the site and planned project activities, including how design and construction hazards will be addressed.
- Any relevant information from an existing health and safety file.

The principal designer should help the client in collating this information and provide them with a 'pre-construction information pack'. This information will help the principal contractor plan the work and is the basis for developing the construction phase plan.

8.4.2 The principal designer

A principal designer should be appointed on all construction projects involving more than one contractor (including subcontractors). This requirement applies to any type or size of construction project. Key factors relevant to the role include that:

- There can only be one principal designer appointed at any time.
- The principal designer is required on any project with more than one contractor.
- The principal designer is responsible for health and safety during the pre-construction stage.
- The principal designer must have the right skills and experience to be able to perform the role. They should have:

- Technical knowledge of the construction industry, relevant to the project.
- An understanding of how health and safety is managed through the design process.
- The skills to be able to oversee health and safety during the pre-construction stage of the project and the ongoing design.

8.4.3 The principal contractor

The principal contractor has the overall control of health and safety during the construction step of the project. The client must appoint a principal contractor in writing before construction work starts on their project if more than one contractor is involved. The principal contractor must have the skills and experience to:

- Plan and manage the work so it can be done safely.
- Plan and manage cooperation between contractors.
- Produce the construction phase plan.
- Ensure contractors comply with legal requirements.
- Consider health and safety when scheduling work.
- Take into account the principles of prevention.
- Ensure other contractors apply the principles of prevention.
- Ensure other contractors follow the construction phase plan.
- Ensure a site induction is provided.
- Ensure welfare facilities are provided.
- Prevent unauthorised access to the site/s.
- Monitor the work.
- Consult and engage with workers.
- Liaise with the principal designer.
- Provide information for the health and safety file.

8.4.4 Construction phase plan

The construction phase plan is required to ensure CDM compliance and must be developed by the principal contractor. This describes how health and safety will be managed during the construction stage, with pre-construction information received and any client requirements provided being used to draft the construction phase plan. The construction phase plan should be proportionate, workable and realistic to the size and nature of the work and the risks involved. It must be sufficiently developed to allow work to start on-site and regularly reviewed and added to as new contractors start on the project. The following topics should be considered in the construction phase plan:

- Project information including the project description and key dates.
- Key members of the project team and contact information.
- Health and safety objectives and expectations.
- Site induction arrangements.
- Site rules.
- Arrangements for the cooperation and coordination of work.

- Provision of welfare facilities.
- Fire and emergency procedures.
- Control of site health and safety risks.

The principal contractor is responsible for preparing the construction phase plan before setting up the construction site and starting work. This plan should also be shared with subcontractors. Regular reviews and briefings should be undertaken. The principal contractor is also responsible for the following artefacts which should be covered in detail, including:

- **A method statement and risk assessment**: a risk assessment identifies and quantifies all the hazards associated with performing a task, with the associated method statement describing the process, steps and resources required to carry out the task safely and without risk to health.
- **Site works – public safety**: public interaction and safety must be considered at every stage. Methods include appropriate barriers and signage, avoiding trailing cables or trip hazards, keeping equipment secure when not in use, and minimising exposure to live work.
- **Accident reporting**: a near miss in safety terms refers to an event that could have resulted in injury, illness, or damage, but did not, due to chance or intervention. These are considered 'close calls' and are important to track and investigate as they can reveal underlying safety issues. Principal contractors and contractors should measure time on-site, accidents and near misses and specifically any RIDDOR (Reporting of Injuries, Diseases and Dangerous Occurrences Regulations) events [3]. The method and frequency of this reporting should be weekly or monthly and is often shared with the client.

8.4.5 Designer responsibilities

Designers are any individuals or organisations that provide or alter designs for a project. They might be architects, engineers, interior designers or surveyors. Their responsibilities include:

- **Ensuring the client knows their responsibilities**: a designer must not begin work unless they are satisfied the client is aware of the duties they have assigned to the designer, including appointing relevant roles and ensuring construction phase plans and health and safety files are prepared.
- **Considering pre-construction information**: when preparing or modifying designs, designers must take pre-construction information into account to eliminate foreseeable risks, which may impact design decisions. If it is not possible to eliminate such risks, the designer must take steps to reduce or control the risks, inform the principal designer and ensure the information is included in the health and safety file.
- **Provide information about the design, construction and maintenance of the charge point**: designers must take all reasonable steps to provide sufficient detail alongside the design to assist the client, other designers and contractors in complying with their duties as outlined in the CDM regulations.

8.4.6 Contractor responsibilities

Multiple contractors often work simultaneously on a project and must fulfil specific duties under the CDM regulations. Like designers, contractors must not commence work on a project before they are satisfied that the client knows the duties required under the regulations. Contractors must:

- **Manage their work**: contractors must plan, manage and monitor the construction work they carry out to minimise risks to health and safety. Contractors must also provide appropriate supervision, instructions and information to each worker under their control.
- **Comply with direction and the construction phase plan**: where there is more than one contractor assigned to a project, each contractor must comply with direction given by the principal contractor and follow the parts of the construction phase plan relevant to their work.
- **Prevent unauthorised access**: the principal contractor must ensure necessary steps are taken to prevent unauthorised access to the construction site. It is the contractor's responsibility not to begin work before this duty is fulfilled.

8.5 Translating the theory into practice: charge point power considerations

As part of any charge point deployment, it is important to understand the electrical power considerations of each charging power level (e.g., level 1, 2 or 3), as this is one of the main factors when designing a charging hub or a network of individual charge points. This is a key part of mobilisation, construction and commissioning. It should also be noted that this section is only concerned with publicly accessible charge points. All charge point installations must be registered with the relevant DNO before installation, in order that they can assess the impact on the local power network before approval is given.

8.5.1 Level 1 AC on-street residential charge points

In most urban settings, lampposts are individually powered by a 25 A rated DNO supply. These are used to obtain power along the public footpath by parked cars. A number of CPOs in the United Kingdom are able to use lampposts to power charge points, these being single-phase. Charge points are typically mounted on lampposts that are located at the front of the footpath, adjacent to the kerb. Should the lamppost be located at the back of the footpath, a charge point in the form of a free-standing bollard can be installed adjacent to the kerb and used in the same way as the front of footpath lamppost charge point. Dynamic load management is not normally a concern in this scenario. This is because the lamppost DNO head is rated at 25 A and fitted with a 25 A fuse. The charge point is set to operate at 22 A (5 kW) leaving 3 A of headroom to power the usually low power LED lantern. If the lantern is not a LED lamp, then the charge point must be further derated to ensure sufficient power for the lantern without blowing the DNO fuse or overheating the DNO head.

8.5.2 Level 2 AC on-street residential and destination charge points

Level 2 charge points (8–22 kW) typically require a three-phase supply. Cost-effective deployment is restricted to types of locations which are more likely to have a three-phase electricity supply. The related question of adequate spare capacity and dynamic load management to power the charge points is the same as per single-phase supply. In locations without three-phase supply, the grid needs to be upgraded to support this, which will add to DNO costs.

8.5.3 Level 3 DC en route charge points

Level 3 DC charging, with multiple units deployed at EV charging hubs, will have the same power requirements as that of a light industrial plant. The smallest sites in this category offer 2 50 kW DC charge points, installed in the late 2010s and early 2020s at British motorways services. Modern equivalent CPO offerings generally provide a minimum of two 120–150 kW rapid charge points at small en route hubs. Larger hubs offer more charge points in this rapid category, and busier routes will increase the power rating offered to the ultra-rapid category (greater than 150 kW, as much as 350 kW), ensuring a faster turnaround of EVs, provided the EVs can charge at this ultra-rapid speed. Presently in Britain, super hubs are opening with 40 or more ultra-rapid charge points installed, with rest and food facilities on-site as well as other amenities (see Case Study 8.5). The power capacity requested from a DNO will be very expensive (hundreds of thousands of pounds) and include significant grid reinforcement. The CPO will want to ensure that the capacity is fully exploited, optimising fewer ultra-rapid charge points or more rapid charge points. The more powerful the charge point, the more expensive it is. For example, an Autel 240 kW unit is listed at £51,149 (£213 per kW) [4], whilst a 120 kW unit is listed at £33,495 (£279 per kW) [5]. Interestingly, the faster the charge point, the cheaper the per kW of rated capacity. However, if a connected EV cannot benefit from this higher power, the additional capital expenditure for the faster charge point will have been wasted, not to mention the portion of the DNO supply capacity that has been allocated to enable these more powerful charge points.

8.6 Translating the theory into practice: the key steps when deploying charge points

Whereas Sections 8.1–8.5 set out the project management methodologies, regulations and power considerations that must be understood, this section analyses the specifics related to a charge point deployment. A charge point deployment begins with the appointment by the CPO of a project manager who will manage project delivery (as per Section 8.1). It should be noted that prior to deployment beginning, shortlisted sites should have been identified and the business case underpinning the feasibility of deployment should have been completed. Although Chapter 7 mentioned the key aspects of a business case, it is useful here to recap these as all aspects are revisited during deployment. An important factor underpinning this is cost.

From a project manager's perspective, the electricity grid connection, the charge point type and number directly affect the costs related to electrics. The more electrical power required per charge point, the more expensive the charge point dispenser is, and additionally the more expensive it will be to reinforce the electrical grid supplying the charge point with sufficient power to meet the rated output. Level 2 AC charge points are rated at a lower power rating, are cheaper to purchase and operate than higher-powered level 3 DC charge points, and require less (and cheaper) electricity grid reinforcement per charge point. As a result, they make a compelling option for residential and destination use cases. Conversely, level 3 DC charge points are more powerful and an order of magnitude more expensive to purchase and operate. When combined with the much higher grid reinforcement costs, this means they are best suited for en route use cases where utilisation is likely to be higher. There is a balance to be struck between the power rating (and increasing proportional cost) of the planned charge points and the speed at which EVs can recharge. CPOs earn a return on their investment as a margin on each kWh of energy provided. Investing more to install higher-powered charge points than the EVs are able to use in the specific location and use case will result in a sub-optimal investment of CPO capital. Aside from electrical costs, site geology and earthworks required to enable the designs to be built are relevant. Public access roads may need to be built or modified alongside proposed on-site facilities. Again, the CPO should have a good sense of what the costs will be based on previous experience. All of these factors should be detailed in the business case and reviewed during mobilisation and deployment.

Having been appointed to deliver the project, the project manager will become familiar with it, including reviewing key documents such as the business case. If, in their role, they start noticing issues that may conflict or challenge basic assumptions underpinning the deployment, these should be analysed in further detail. As a result, the project design may evolve, or in the worst case, the deployment may be cancelled. The business case should give the project manager enough information to draft a project initiation document (PID), detailing all the high level details the project manager needs to consider when working out the project plan. The project manager will review this PID with the project stakeholders to ensure alignment with the overall business objectives. This process may be cyclical, with the project manager raising questions that the business has not thought about in enough detail or any unintended consequences that the project manager sees. The project manager will draw on all their experience to ensure that any issues identified are addressed before planning is finalised and contracts committed to.

8.6.1 Mobilisation, construction and commissioning

Once the project's objectives are clearly understood by the project manager, they can then put in place the necessary project plans and infrastructure to deliver the project.

8.6.1.1 Planning, permissions and permits

Various permits are required from different stakeholders to enable the project to proceed. The most common required are those from political and government agencies (e.g., local government and highways agencies like National Highways in the United Kingdom).

DNO new connection cost and timelines

DNOs must be informed before any charge point is installed (regardless of the use case). The business case will have determined the number and size of charge points to be installed on-site. This requirement will determine the size of the new connection required to the electricity grid. The larger the size, the longer it takes for both permissions to be obtained and eventual works to take place, and the larger the corresponding connection costs. The DNO needs to assess capacity on the local network to confirm that the charge point, once installed, can be powered. When applying to the DNO to install charge points, the DNO completes a supply capacity assessment to assess whether the grid has the capacity to supply the additional energy requested at the location. The more charge points requested to connect, and the higher the rated power, the more power the DNO will need to supply. This may well require the DNO to strengthen the electricity grid, and result in the customer contributing to grid reinforcement costs. Since April 2023, DNOs in the United Kingdom use a pool of funds to cover these costs up to a relative local standard. However, costs in excess of this will need to be paid by the CPO, which can rise to hundreds of thousands of pounds for larger charging hubs.

Given that EVs arrive and depart at charge points at varying times throughout the day (and have varying states of charge on arrival), it is very unlikely that all charge points will be supplying their maximum rated power all the time. The CPO will likely want to install more total charging capacity than the electricity supply it requests from the DNO (the more electricity supply capacity requested, the higher the grid reinforcement costs are likely to be). The ability to share a lower electricity supply capacity with all installed charge points at a site is the function of dynamic load management (see Chapter 11 for more detail), with a central system monitoring all the charge points and modulating each charge point's charging in real time. This ensures that all the electrical power requested does not exceed the installed supply capacity and trip the local breaker. Given this issue, British electrical wiring standards [6] and DNOs themselves require that dynamic load management is used when charge points are installed.

Factoring in all of the above, the DNO will develop a connection cost and provide this to the CPO. This is made of two parts:

- **Statutory costs** relate to network planning, any network reinforcement of the existing grid to supply the required capacity at that point and a proportion of the network asset cost that the connection will use. Since April 2023, in the United Kingdom, the customer pays for grid reinforcement above £1,720/kVA, with the DNO paying for costs below this threshold from the Distribution Use of System Fund (contributed to by all electricity customers).
- **Non-statutory costs** are proportional to the construction work where the charge points are to be installed. The length of trenching between where the

cable joint with the electricity grid is made and the type of surface to be reinstated all depend on the specifics of the location in question. A roadway crossing with tarmac is twice as expensive as a tarred footpath. The cheapest is a soft roadway verge or undeveloped ground. Traffic management on the public highway varies from hundreds of pounds for pedestrian management with a footpath closure, to thousands of pounds for vehicular traffic management, and similar for bus rerouting and temporary bus stops.

Independent connection providers can be contracted to complete the cable jointing and laying. They, or an earthworks contractor, complete the digging and surface reinstatement. In some cases, an Independent DNO is appointed by the contractor, instead of the DNO, to install and operate a local distribution network. The substation they install and manage becomes the electrical interface with the regional DNO. For brevity, they are not referred to explicitly in this chapter.

Data requirements

Public charge points are more often than not remotely monitored and controlled via the CPO's CMS. Typically, the platform communicates over the GSM mobile telephone network with a server or controller at the charge point site, and each charge point is connected via a LAN to this controller. In the author's experience, this controller provides the dynamic load management capability, as well as being the communications conduit to the platform to monitor and control individual charge point statuses and each charging session. It should also be noted that the LAN is used to enable payments if the charge point has a physical terminal. The provision of LAN cabling at each site must be considered as part of the groundworks, with data cables installed in separate ducts to power cables so as not to interfere with data flow.

Land ownership

Prior to deployment, the CPO will have confirmed the ownership or use of the land where the charge points are to be deployed, with land rights lasting at least as long as the planned project lifespan.

Wayleaves for power and access

Wayleaves are access rights over adjoining land between the source of a service and the land where the service is to be used and grants access rights for the purpose of installation, operation, and maintenance of the service. Usually the landowner from whom the wayleave permission is obtained is remunerated on a periodic basis for this access by the service provider (e.g., the DNO, water or telecoms utility company).

Local government permission

Local government permission will depend on the site location. If a site is located in an urban environment, the local town or city council will have jurisdiction over issuing permits for that location. A rural location would normally fall to a county council to seek permission.

National government agency permissions (e.g., National Highways)

Should the site be located along major national road arteries, then the national body that regulates the national road network would need to be involved in the approvals process. An example of this would be National Highways which regulates and operates the motorways in England and Wales. Where this is the case, they check for any conflicts with other national agendas before giving consent to proceed.

Stakeholder consultation
Aside from governmental and technical authorities as discussed in this section, other relevant stakeholders include local residents who may be inconvenienced by the installation of the charge points. Their concerns must be addressed to mitigate ongoing opposition to the project once construction has completed. Similarly, stakeholders such as local 'action climate groups' (who are concerned with the local environment and generally promoters of low-carbon schemes) may likely have valuable local insights to help shape the design of the solution being installed.

8.6.1.2 Planning completion

Once the project manager has obtained all the necessary requirements outlined above, they then seek approval from their own company's management to commit to the project by establishing contracts with suppliers to provide the required services. Supplier timelines (and any permits not yet awarded) will typically define the project's realistic construction plan and resulting overall timelines (the baseline). From a CDM perspective, the pre-construction information will be completed by the principal designer.

8.6.2 Construction step

The principal contractor takes over delivery at this point. They refine the construction phase plan as defined by information given over by the principal designer following the mobilisation step. The principal contractor will henceforth manage the build step ensuring that materials and skills are delivered to site as and when required to enable the construction plan to proceed as scheduled. Depending on the size and complexity of the project, the construction step may take weeks to years, but from experience, the construction step is almost always shorter than the mobilisation and feasibility steps that come before. Construction will involve the groundworks, cabling and installation of the associated feeder pillar where charge point bollards are installed. The power distribution board and site load management and monitoring and control system, as well as power from the DNO connection point to the power distribution board, will also have been installed as part of the construction step. Foundations will be put in place for each charge point unit in preparation for installation.

8.6.2.1 Charge point installation

In practice, most site groundworks and construction will be nearing completion as charge point installation begins. The process of installing and testing is relatively quick, taking a number of days versus the weeks or months for the rest of the construction step. Planning for this to take place at the end of the construction step mitigates the risk of the relatively fragile and expensive units being damaged by other construction vehicles moving around on-site. Charge point installation consists of two parts. The first is the physical lifting of the charge point off the delivery truck to place it on foundations where it is secured in place. The second is the electrical works connecting power and data cables from the on-site power distribution board and site load management, monitoring and control system to the unit. The control system remotely communicates over the internet with the CPO's management platform.

8.6.2.2 Charge point testing, including site testing

With installation completed and electrical power provided to the charge points and on-site control systems, the network is then powered up in sequence and in accordance with BS7671, the United Kingdom's 'Requirements for Electrical Installations' standard [6]. Everything from the site lampposts to the high-power network supplying the charge points directly will be tested according to this standard. It is a regulatory requirement and the resulting Electrical Installation Certificate (EIC) must be added to the CDM's health and safety file. In reality, each system will be tested independently and a related EIC generated per system. In this way, if part of an already tested system is changed, then only that system needs to be re-tested and a new EIC generated, and not one for the entire site.

8.6.2.3 Charge point commissioning

Once all the electrical equipment on-site is powered and certified as safe to use, commissioning involves testing all functionality. This is completed by following the manufacturer's instructions. To mimic an EV connecting to the charge point, an 'EV tester' is connected to simulate an EV being plugged in. The charge point functionality and related site load management, monitoring and control systems are taken through all modes of operation to ensure they operate as expected. Related systems such as load management and on-terminal payment must also be tested. Testing documentation and results must be filed for project quality control purposes and as evidence that the installation step has been successfully completed.

8.6.3 Opening to the public

At the end of the construction step, the site is ready to be made available for the public to use and operation of the site is handed over from the project deployment team to the CPO's network management team, who will operate, maintain and periodically test all equipment throughout the operational life of the site. All surplus construction material, safety barriers and site hoarding will be removed, the site cleaned and user signage and branding material erected. One of the last steps in making the site ready for the public is to link up the site's details to the CPO's website and charge point finder tools as well as publishing links to third-party websites and apps such as Zapmap to enable EV drivers to locate, establish availability of and use the charge points.

8.7 Translating the theory into practice: real-world projects and case studies where charge points have been deployed

There are many types of projects that involve the installation of charge points. In this section, we look at the most common ones across residential on-street, destination and en route charge points. These are brought to life by case studies developed based on char.gy's experience. char.gy is an established CPO with a track record of deploying charge points in the United Kingdom. It should be noted that all case studies followed the processes outlined in Section 8.1–8.4.

8.7.1 Residential charging

On-street charge points in the United Kingdom tend to be installed in two ways. This is either through integrating into a lamppost or via a dedicated charge point bollard. There are other more niche deployments where the charge point is hidden underground until needed or where a lance is used to 'plug' in the charging cable into a recessed point; however, the deployment of these is not the focus of this section. Starting first with lamppost charging, the charge point is mounted on existing street lighting furniture, which has its own DNO supply, which significantly lowers the cost of providing power to a charge point in a public space. With DNO heads in a lamppost rated at 25 A, fitting a 25 A fuse provides 22 A to a charge point, enabling it to operate at 5 kW. Permission from the lamppost owner and operator, the local authority itself and the DNO are required before such an installation can take place. Where lampposts are installed at the front of the footpath, a charge point can be attached directly to the post by drilling holes into it. Where the lamppost is located at the back of the footpath, a power supply is taken from the lamppost to a bollard charge point located at the front of the footpath, adjacent to kerb in line with where cars park. Civils work is required to run the power from the lamppost to the base of the bollard charge point, and to complete the installation of the bollard charge point socket. Case Study 8.1 outlines how lamppost charge points were installed for the London Borough of Southwark by char.gy.

Case Study 8.1: On-street residential charging: the London Borough of Southwark

Between 2018 and 2023, char.gy supplied and installed almost 1,600 lamppost charge points for the London Borough of Southwark. The first contract in 2018 was for 50 charge points for Southwark to trial the then new technology of lamppost charging. After a year of evaluation, Southwark selected char.gy as their supplier. As a result, char.gy supplied, installed, operated, maintained and relocated a number of charge points as the urban road layout changed. Since the first 50 charge points, char.gy supplied charge points in tranches, each tranche being a mini-project in itself. A typical on-street residential charging deployment covers:

- **Site selection**: this considers any residential requests, local authority prioritised sites or other relevant local authority policies and strategies. char.gy uses a geospatial tool to predict EV uptake (and hence charge point usage) over various time horizons, at the street level. It uses existing vehicle registrations, commercially available socio-demographic information, land use data, local journey data and electricity network maps to forecast EV use over time. Once street-level locations have been identified, installers survey identified lampposts for suitability, with front-of-path lampposts selected to mount on post charge points, whilst back-of-path lampposts power a bollard charge point installed at the front of the footpath nearest to the road.
- **Site approval**: once viable lampposts have been identified for the project area, char.gy presents a short list of viable lampposts to the local authority for review and approval. Some authorities elect to conduct a Section 17 public consultation (designation and regulation of special roads which applies to a charge point deployment) to determine any resident opposition before approving sites.
- **Mobilisation**: Section 50 permits are completed for relevant groundworks. This permit enables CPOs to install, maintain or remove a charge point on public highways. Elements of the permit include installation of an earth rod for TT earthing, base installation and power ducting for bollards.
- **Groundworks**: once Section 50 permits have been issued, groundworks take place at the relevant locations.
- **Bay dedication**: some local authorities choose to conduct a Section 17 public consultation, as mentioned, prior to location approval to manage residents' objections ahead of installation, and then dedicate EV charging bays as EV only to help ensure availability. Southwark, however, decided not to dedicate EV charging bays alongside charge points as a way of mitigating resident objections. Southwark instead opted to install more charge points to ensure EV drivers are able to find an available charge point to use.
- **Installation, testing and commissioning**: it takes up to 90 minutes to install, test and commission a lamppost charge point.

Once commissioned, char.gy managed the charge points on Southwark's behalf. Over time, the urban streetscape may change, resulting in the need for some charge points to be relocated each year. Each relocation is treated in the same way as a new deployment and follows the process outlined above.

Source: Based on the author's experience while working for char.gy

A new lamppost might be required if there is concern about the existing post's age or suitability to mount a charge point on it. In such cases, a new and better-suited lamppost can be installed in the ideal location for on-post charge points, at the kerb side, with new DNO heads. Case Study 8.2 demonstrates how a new approach to charge point deployment has been adopted by the London Borough of Barnet.

Case Study 8.2: On-street residential charging: the London Borough of Barnet

In January 2025, the London Borough of Barnet awarded char.gy the contract to operate 1,000 charge points for 15 years across the Borough. What is different about this project is that Barnet will be installing new double door lampposts at the required locations for the project. This ensures that every lamppost is perfectly suited for the project and no on-street surveying is required, significantly reducing the amount of time needed to carry out this task during the mobilisation step. Additionally, earth rods will be installed at the same time as the lampposts removing the need for additional Section 50 permits. Key elements of deployment include:

- **Site selection**: this is driven by a mixture of Borough and char.gy chosen locations with the latter using its EV charging geospatial modelling tool to identify promising locations for future deployment. The Borough is undertaking Section 17 public consultations as part of the site approval process, reducing the risk of residents' objections after installation.
- **Installation**: the new lampposts will have new DNO heads and new DNO connections and are suited for char.gy's new 'on-door' charge point design where it attaches to the lamppost access door. This enables the lamppost switchgear to remain in the lower access chamber while the charge point switch gear is installed in the upper access chamber. The installation process will be optimised as a batch of lamppost access doors will be drilled and charge points attached off-site in a workshop, with the installers arriving on-site with all physical assembly completed, and only the electrical works being done locally. As a result, on-site activity is much reduced.

Source: Based on the author's experience while working for char.gy

Multiple local authorities prefer to install charge point bollards, a number of which are now 22 kW, requiring the provision of new three-phase DNO connections, with associated feeder pillars on the footpath. Installing pairs or more of charge points from each feeder pillar helps to reduce the overall installation cost per charge point. Case Study 8.3 brings to life how this approach was adopted by Coventry City Council.

Case Study 8.3: On-street residential charging: Coventry City Council

Between 2019 and 2020, char.gy supplied 400 charge points to Coventry City Council. With most being lamppost powered 5 kW charge points, approximately 50 were specified to be 22 kW charge points (in pairs), requiring a three-phase power supply per pair. With the City Council having identified the locations, char.gy worked with the Council, the DNO (then Western Power Distribution, now National Grid Electricity Distribution) and the contractor (Balfour Beatty) to design and install the required infrastructure per each location. The DNO was concerned about this on-street EV charging technology as it was new at that point and it was not understood what the potential impact on the electricity grid might be across the City. The DNO conducted an evaluation and network modelling exercise to understand the impact in more detail which then enabled the installation to take place. Following each site's approval, the MPAN for that site was provided.

The contractor and its independent connection provider completed the cable jointing and groundworks for the power cables and installed the feeder pillars and bollard bases. char.gy then arranged for its energy supplier's meter administrator to install an electricity meter in each feeder pillar. After this, char.gy installed the necessary switchgear, connected and tested all the cabling, and once safe and certified, commissioned each charge point on char.gy's CMS. The DNO was advised when each charge point was commissioned.

Source: Based on the author's experience while working for char.gy

8.7.2 Destination charging

The challenge when deploying destination charge points is providing power from the supply to the destination location where the charge points are installed, failing which a new DNO connection is required. Where the latter is the case this increases the cost of installation but it also provides an increased supply which enables multiple charge points to be installed. Case Study 8.4 outlines the process of installing destination charge points at a hotel in North Yorkshire.

Case Study 8.4: Destination charging: Ravenhall Hotel

char.gy was contracted to supply three charge points to Ravenhall Hotel located in Ravenscar, North Yorkshire, to enable the hotel to offer its residents and visitors with destination EV charging. Key elements included:

- **Design and planning**: this involved a site visit to start the project network design process and the completion of a load study to ensure that there was sufficient hotel electricity supply available to power the charge points. With sufficient power having been shown to be available, the project was deemed technically feasible. A high-level project cost model was developed and the business case put to char.gy's board for preliminary approval. With this having been obtained, char.gy approached the DNO (Northern Powergrid) for permission to install charge points at the hotel. char.gy then finalised the precise charge point locations and groundworks costs and related electrical network design to enable a high confidence budget to be calculated. Design and planning took several months to complete. The next steps involved:

- **Mobilisation**: with the project and its associated business case approved, char.gy mobilised. This involved planning the electrical works and groundworks. Relevant suppliers and the hotel owner were engaged.
- **Groundworks and electrical works**: these were each performed over two consecutive days, one week apart to avoid any overlapping work. During the electrical works step, the installers connected, tested and certified all the cabling and switch gear from the charge point network point of supply to each charge point.
- **Commissioning**: the charge points were commissioned on char.gy's CMS. The DNO was advised that the charge points had been commissioned and were now in operation.

Source: Based on the author's experience while working for char.gy

8.7.3 En route charging

En route charging sites invariably offer multiple level 3 ultra-rapid charge points to enable multiple drivers to charge at the fastest speeds possible, at the same time. A number of these charge points need a large and expensive grid connection, with the CPO optimising the number of charge points and speeds offered to match the size of grid connection required. Dynamic load management enables a number of charge points to operate at full power at the same time. As EVs connect the charge points start to de-rate to share the available power. Case Study 8.5 demonstrates how CPOs are taking steps to mitigate the need for an expensive grid connection with on-site battery storage and solar PV farms. En route charging sites offer modern rest and food facilities that can be used during the relatively short charging sessions, allowing drivers the opportunity to make the most of their stop.

Case Study 8.5: Public en route charging: Winchester's 'Super Hub'

InstaVolt, one of the United Kingdom's en route charging CPOs, opened in March 2025 the United Kingdom's largest en route 'Super Hub' on a 4 acre site, with 44 160 kW charge points and numerous novel features that indicate what en route charging could look like as the sector matures. The site is readily accessible off the main arterial routes to the coast at Winchester (M3 and A34) and offers on-site dedicated food and rest facilities. Another notable feature for drivers is the inclusion of 4 long bays for vans, and 4 'pull-through' bays where an EV towing a trailer or caravan will not have to unhitch or manoeuvre with the trailer behind to get into position to connect and charge. To enable a smaller grid connection than otherwise would be required for 44 ultra-rapid chargers, a 4 MWh battery storage system was installed on-site, complemented by a 500 kW solar PV array. It is clear from the design of the site that deployment was well structured and factored in a number of complex elements, including a large grid connection and a battery energy storage system. It is also clear that the deployment factored in the local demands of the site so as to meet the needs of its customers.

Source: Instavolt. *Now Open UK's largest ultra-rapid only EV charging superhub off the A34 in Winchester* [online] 2025. Available from https://instavolt.co.uk/instavolt-ev-charging-station-winchester-super-hub/ [Accessed 30 April 2025]

8.8 Conclusion

What is clear from this chapter is that there is a structured approach that must be followed to ensure successful deployment of charge points and that there are defined methodologies, standards and requirements to follow to ensure safe and effective planning, installation and commissioning. This is brought to life through five case studies which have followed the process for three local authorities in a residential setting, one destination deployment for a hotel and one en route deployment alongside the strategic road network. Prior to deployment, the CPO should ensure the business case is still relevant, that a PID is in place and that planned delivery aligns with CDM regulations including the appointment of specific roles. Once this is done, the CPO's project manager can then mobilise the project. This includes putting in place the necessary permits and permissions, sought from the DNO, local authority and, in some cases the relevant national government transport agency. This also includes completing the necessary planning activities and consulting with key stakeholders. Once planning is complete, the CPO enters into construction. Construction involves the groundworks, cabling and installation, where necessary, of an associated feeder pillar where charge point bollards are installed. The power distribution board, DNO connection to it, and site load management, monitoring and control system are also installed. The charge point itself can then be installed, tested and commissioned. Once done, the charge point and site are ready to be opened to the public.

References

[1] Health and Safety Executive. *The Construction (Design and Management) Regulations 2015* [online]. 2025. Available from https://www.hse.gov.uk/construction/cdm/2015/index.htm [Accessed 12 April 2025].

[2] HS2. *Construction Review 2022–2023* [online]. 2023. Available from https://assets.hs2.org.uk/wp-content/uploads/2023/11/HS2-Construction-Review-2022%E2%80%932023-FINAL-WEB.pdf [Accessed 12 April 2025].

[3] Health and Safety Executive. *Reporting of Injuries, Diseases and Dangerous Occurrences Regulations* [online]. 2025. Available from https://www.hse.gov.uk/riddor/ [Accessed 12 April 2025].

[4] Autel. *MaxiCHARGER 240 kW Commercial DC Fast with Payter/Nayax* [online]. 2025. Available from https://www.autel.uk/product/maxicharger-240-kw-commercial-dc-fast-with-payter-nayax/ [Accessed 12 April 2025].

[5] Autel. *MaxiCHARGER 120 kW Commercial DC Fast with Payter/Nayax* [online]. 2025. Available from https://www.autel.uk/product/maxicharger-120-kw-commercial-dc-fast-with-payter-nayax/ [Accessed 12 April 2025].

[6] Health and Safety Executive. *Electrical standards and approved codes of practice* [online]. 2023. Available from https://www.hse.gov.uk/electricity/standards.htm [Accessed 12 April 2025].

Chapter 9

Operating, maintaining and decommissioning charge points

Myles Roberts[1] and David Trousdale[2]

In Chapter 7, we outlined in detail the process of planning for charge point deployment and in Chapter 8, we provided an outline of the deployment and construction process itself. In Chapter 9, we now focus on operations, maintenance and decommissioning of charge points. Whilst the charge point infrastructure itself is crucial to operations, effective management (and improvement where possible) of the entire operational ecosystem is critical to a successful network. By ecosystem, we mean the drivers, landowners, hardware and software vendors, maintenance specialists, the end-to-end supply chain, adjacent service suppliers and interested parties in the community. Management requires a system of continuous performance monitoring, periodic reporting, review and subsequent trend analysis to identify outliers in anticipated performance. These outliers include charge point utilisation, downtime and fault analysis, maintenance costs and revenue versus the charge point operator's (CPO) commercial model and accepted operating profit, and, of course, user/driver satisfaction. Where satisfaction falls, this is often driven by factors such as cost to use a charge point, its reliability and the usefulness of the location in relation to charging. Any other supporting services, such as coffee shops and play areas, would be captured in addition when assessing satisfaction. It should be noted that operating, maintaining and decommissioning charge points is now regulated in the United Kingdom, with different geographies also adopting a similar approach. Like Chapters 7 and 8, this chapter is written from the perspective of a CPO in the United Kingdom. Key roles in the CPO include the customer support team that engage with drivers and run the helpline, the scheduler who plans field service activities related to charge point infrastructure, the field operatives team that carry out maintenance and the operational management team that manage the contracts with clients and track overall performance. The first three roles can be wholly or partly outsourced by the CPO to a different provider.

[1]Off-Street EV Charger Project Manager, char.gy, United Kingdom
[2]Principal Advisor - Climate Change and Sustainability, Newcastle City Council, United Kingdom

9.1 British government regulations

The British Government's Public Charge Point Regulations, which came into force in November 2023, have placed user-centric obligations on all CPOs operating in the United Kingdom, with penalties liable for non-compliance. Key requirements that all CPOs must consider include:

- **Pricing transparency**: the maximum price of a charging session must be displayed clearly in pence per kWh.
- **Contactless**: new public charge points of 8 kW and above and existing charge points of 50 kW and above must offer contactless payments to consumers.
- **99% reliability**: level 3 DC Rapid charge points must be 99% reliable, measured as an average across the CPO's entire rapid network over the calendar year.
- **Information on reliability compliance**: this must be published on the CPO's website. CPOs must also submit an annual reliability report to the Secretary of State and the enforcement authority.
- **Helpline**: a free-to-use 24/7 staffed telephone helpline must be available and advertised at all charge points.
- **Open data**: all data must be accurate and CPOs must use the Open Charge Point Interface (OCPI) to share their data. Government bodies, electricity network and system operators must have access to all data.
- **Roaming**: CPOs must enable consumers to pay through at least one roaming provider at their charge points [1].

From a CPO's perspective, some of these requirements are more easily complied with than others. For example, displaying the price per kWh can be done in a CPO's app, whilst the helpline contact details can be displayed on a sticker on the side of a charge point. Roaming partners and contactless payments will require commercial agreements; however, from an operations manager's perspective, the requirement focused on 99% reliability may prove challenging to achieve.

9.2 Operating a charge point and its supporting ecosystem

Once the charge point has been installed, it is registered on the CPO's charge point management system (CMS). This system is key to successful charge point management.

9.2.1 The charge point management system

The CMS (also referred to as the platform) is core to the CPO's operational business. It enables the processes outlined in this section to function, all of which are key to providing a service to customers, namely that they can find charge points and navigate to them if they are available, use them reliably to complete a charge at a fair cost and easily pay for the energy provided.

Prior to installing any charge points, the business rules must be set up on the CMS, with the client/landowner and the relevant contract details registered (as required). For example, this includes the contract start/end date, driver tariff to be applied (if relevant), maintenance budget and commercial terms. Over time, various concurrent contracts may be in force with an individual client, each covering its own subset of the CPO's charge points. The information captured in the system is used when reporting the operational and commercial performance internally within the CPO and to the client.

During installation, once a charge point has been successfully tested, it is commissioned on the CMS, with the charge point's location and related metadata (including GPS coordinates, street address and postcode, parking zone details and power rating) recorded. The charge point is assigned to the relevant client and contract. Being assigned to the correct client and contract will add each charge point's data to the CPO's operational and commercial performance reporting system. All charge point equipment that is rated at 8 kW or greater will have to offer a contactless payment terminal as per the Public Charge Point Regulations outlined in Section 9.1.

9.2.1.1 Finding a charge point

All CPOs will want to tell EV drivers where to find their charge points; both directly through their own website and mobile apps, and through websites and apps that aggregate all charge point locations and the details of multiple separate CPOs. Whether this data is served to drivers by the CPO or by an aggregator website or service, the source data will come from the CPO's CMS. The driver, via the app or mobile website, should be able to locate and check the real-time status of the charge point, navigate to it and ideally be able to start, monitor, stop and pay for a charge point session, especially if plug and charge is in place (see Chapter 11). The aggregator's website will want to offer as much of this service as possible too; at least locating and checking the real-time status before suggesting the driver navigates to the charge point. If the aggregator has been able to reach favourable commercial terms with the CPO, they will be able to offer the driver the ability to start, monitor and pay for charging sessions via the aggregator's website or app as well, likely earning the aggregator a source of revenue too.

9.2.1.2 Using a charge point

Drivers will need to create a customer account with the CPO, all controlled by the CMS. When a driver wishes to start a charge at a charge point, they log in to the CPO's app and authenticate themselves, then use the app to start the charge at the charge point they are connected to. As per the Regulations outlined in Section 9.1, the maximum price (pence per kWh) to be paid during the charging session must be clearly displayed, usually on the app itself. During the charging session, the driver can monitor the session's progress and cost of accumulated energy delivered in that session. With level 3 DC charging, the CMS can monitor and report the live state of charge. This is not available for level 1 and level 2 AC charge points. When an EV's battery is fully charged or the driver has set a lower charging limit, the charge point will advise the CMS, which in turn will inform the driver via the mobile app, that their vehicle has stopped charging. At the end of a charging session, drivers end the charge with their app (or via their vehicle) and the CMS calculates the amount of energy provided, applies the tariff relevant to that charge point at that time of day, calculates the total owed by the driver for the charging session and then bills the driver. Depending on whether the driver has an account with the CPO (which is presumably settled monthly) or the driver 'pays as you go' and the amount owed is payable immediately, the CMS will either add the amount to the driver's monthly account or will pass payment immediately to the payment gateway. Where the latter occurs, this collects the payment and reimburses the CPO, usually minus a negotiated percentage commission for the payment gateway to cover its costs. Due to stringent banking data security requirements, payment and account details are usually saved by the payment gateway and accessed securely via the CMS when the driver logs in to their account.

9.2.1.3 Driver account management

When managing their account, drivers should be able to self-serve as much as possible, enabling them to update personal and payment details, retrieve statements of accounts and records of charging history, 24/7. However, some tasks do require human interaction, especially where the driver has a complaint resulting from a poor charging experience.

9.2.1.4 CPO helpline

As per the Public Charge Point Regulations, each CPO must offer a toll-free 24/7 helpline, the number of which must be displayed on the CPO's charge point. When a driver calls the 24/7 staffed helpline, the support agent will use the CMS as the live tool to remotely investigate any issues reported by a driver. Live status and communication log details can assist the support agent to preferably resolve the issue on the call, or at least triage it, raising a support ticket in the CPO's CMS. Through the system they record as many pertinent details as possible in the support ticket, such that the operations technical team can try to understand the issue more fully to enable quicker resolution. The operations technical team will log in to the CMS to review further details and communication logs relevant to the issue to help address it. Further detail is provided in Section 9.2.2.

9.2.1.5 Energy management

Before installation a unique Meter Point Administration Number (MPAN) will have been assigned to the site based on the Distribution Network Operator (DNO) region that a charge point is located in. This will have been done following securing the DNO's permission to install the requested number and rating of charge point/s. At the end of each month, the CPO sends each DNO a list of all its charge points installed in that DNO's area and the energy provided so that the DNO can continually monitor and understand its network's performance as well as understand what spare network capacity exists.

The CPO will have an account and tariff with a chosen energy supplier. The tariff will be registered on the CMS and the charge point assigned to it. The CPO could have multiple energy supply contracts in effect at any one point in time. This is usually defined by, for example, the CPO's geographic coverage, or commercial arrangements relevant at the time. At the end of each month, a file listing each charge point per energy supply contract, with its MPAN and energy consumed per time-period on the tariff, is sent to the relevant energy supplier so that the CPO can pay for the energy that it has dispensed to its customers.

9.2.2 Driver support

CPOs may either outsource or have in-house driver support teams. They may, however, adopt a hybrid support system with first line driver support outsourced typically for the 24/7 support desk function (e.g., answering the telephone, email and social media support calls). A more experienced second line support team, with CMS access and contact with relevant CPO internal business teams, picks up unresolved and escalated tickets from the first line support team. Calls usually relate to one of the following categories:

- **User issues**: using the app, charge point or understanding how the driver's EV works.
- **Payment issues**: disputed bills and/or struggling to pay.
- **Reporting problems with the charge points**: not working when it should be available, vandalism or other issues related to the charge point itself that is not remotely monitorable.

All calls are logged and tickets raised that summarise the issue, resolution steps followed, the outcome, and time taken to resolve each ticket. Periodic evaluation is completed normally to try and establish the root cause of categories of tickets. Based on the data, CPO management explore whether clearer user instructions, better charge point hardware, software and app improvements could prevent the need for such calls and tickets in the first place.

In terms of overall driver support effectiveness and caller satisfaction, CPO management periodically review the number of tickets being raised and average time-to-resolution. They also review sentiment analysis to gain insight into the driver's satisfaction with the CPO and the quality of driver support provided. Wait times for callers phoning the helpline give an indication of whether the helpline has sufficient support agents to handle the volume and complexity of issues raised. CPO management normally put key performance indicators in place against which performance is measured and periodically reviewed. Some CPO clients, especially local authorities who offer long-term concessions to CPOs to operate charge points in their jurisdiction, will want to know that their residents are being adequately served and will want to see periodic reports concerning driver satisfaction.

9.3 Charge point monitoring and reporting

After installation, a charge point needs to be monitored to ensure correct operational performance, with the CMS playing a key role in this function. As stated in Section 9.2, the CMS continually monitors the charge point with frequent exchanges of data between the unit and the CMS to confirm the charge point's status and that they are communicating with one another. As the CMS also authorises charging sessions as requested by drivers on their mobile app, monitoring each session in real time, it is the key tool used in reporting charge point reliability and charge point utilisation.

9.3.1 Charge point reliability

9.3.1.1 Regulatory requirements

Charge point reliability is defined by the Public Charge Point Regulations as the proportion of time in the year that a charge point should be available to use. This assumes 24/7 365 days a year for public charge points, less the amount of time in the year that the charge point is either not available due to a failed supporting process (e.g., authorisation of payments) or is out of order (e.g., it has lost power). This figure is then divided by the proportion of time in the year that a charge point should be available to use (24/7 365 days a year for public charge points). Periods of planned maintenance or when the charge point is not available due to factors beyond the CPOs control are exempt from the calculation. Examples of the latter include flooding, fire or vandalism. Factors within the CPO's direct control (other than planned maintenance) will negatively affect reliability. Reliability is a useful indicator to gauge a CPO's ability to operate its network in a way that guarantees that any charge point not in use will work when required to. The Public Charge Point Regulations also state that for CPOs with level 3 DC charge points of 50 kW or more, they must maintain a reliability of 99% or face penalties. If they dip below 99% for one month but maintain 99% for a 12 month period, penalties will be waived. Penalties aside, all CPOs must submit an annual report to the Office for Product Safety and Standards.

9.3.1.2 Reputational best practice

Some CPOs do not operate rapid charge points. Those in the on-street residential segment typically operate level 1 AC 5 kW lamppost-powered charge points or level 2 single-phase 7 kW charge points. Whilst the Public Charge Point Regulations and its requirement for 99% reliability over a 12 month period do not apply to these types of charge point, it would be reputationally foolhardy to ignore such a benchmark. Issues around EV charging accessibility and reliability attract bad press. Furthermore, addressing these issues is critical to driving up EV adoption. The United Kingdom's Committee on Climate Change highlight that delivering 'a charging point network with increased availability [and] reliability . . . is a major contributor towards enabling the 2030 EV transition' [2]. With CPOs competing for the current £400 million Local Electric Vehicle Infrastructure (LEVI) capital grant fund [3], indicators such as reliability are key in demonstrating to the funders that their grant will be invested in a CPO that is capable of providing a trusted and resilient network.

9.3.1.3 Contractual requirements

In the author's experience, network availability is a key indicator that CPOs have to meet in the United Kingdom as part of the LEVI fund. Expressions of interest ask CPOs to demonstrate that they operate their networks to suitable standards, with invitations to tender creating competition between CPOs bidding to win long-term concessions. Availability and reliability are taken into consideration as part of the scoring process.

9.3.2 Charge point utilisation

Similar to reliability, charge point utilisation is a measure of how often (or how much) a charge point is actually used to charge an EV. The proportion of the year that a charge point is in use is an indicator of both the reliability of the charge point and suitability of the location to EV drivers. Utilisation is usually measured by the CPO as the power dispensed over a time period, translated into a £/hour measure or income generated per session. Many CPOs are keenly focused on utilisation as a reflection of profitability. Further details regarding utilisation can be found in Chapter 7. A CPO's operational management team regularly review individual charge point utilisation over the life of a contract. There might be the option in a contract to relocate a charge point when utilisation frequently falls below an agreed benchmark. Such an exercise is much more expensive than the cost of just the labour involved; hence, it is prudent for a compelling case to be identified earlier rather than later during a contract term. The relocated charge point must cover its relocation costs before contributing to the CPO's bottom line.

9.4 Charge point maintenance

Key to achieving 99% reliability is maintenance. Charge points can have a hard life, with internal components subjected to heat stress during charging sessions, with lower-rated charge points spending longer periods operating at or near their rated power limit when compared with higher rated charge points. This is because EV battery charging is non-linear, with charging current drawn from a level 3 DC rapid charge point peaking typically between 20% and 60% of a vehicle's state of charge and then dropping dramatically from 90%. As a result, the high-powered level 3 DC charger spends relatively less of the charging session operating at or near the rated power limit. In comparison, a level 2 7.4 kW AC charge point may regularly spend 8 –10 hours at a time charging an EV at or close to its rated limit. Thermal cycling can therefore lead to premature component failure, especially for this type of charge point.

A further source of heat in a charge point can be terminals that are not correctly torqued. Charge points located on the public highway, for example, will be subjected to constant vibration from passing traffic and over time such vibration can cause terminals to vibrate loose, thereby increasing the resistance at the terminal, leading to a hotspot that causes further thermal cycling and premature component failure. Based on the above, it is good practice when a charge point is opened during maintenance to re-torque high-current terminals and examine cable insulation for signs of overheating. Cooling systems must be properly maintained and filters and ducts regularly cleaned. If not done properly, the cooling system may fail to operate effectively, damaging the charge point in the long run.

From an external hardware perspective, tethered cable connectors and untethered sockets experience regular forceful connecting as some drivers find the charging cable heavy and cumbersome. This can lead to the pins and receptacles in the sockets and connectors becoming deformed, rendering the socket or cable useless. Based on the author's experience, level 2 AC Type-2 sockets cost in the region of £50 and a level 3 DC CCS-2 connector and tethered cable between £500 and £1,000 to replace.

Payment terminals, charge point screens and keypads are placed under strain by users and weather conditions. It is recommended that they are regularly serviced to minimise downtime. The body of a charge point can be subjected to vehicles striking it and people vandalising it. Graffiti, although an eyesore, is not likely to detract from the charge point's usability; however, collisions will result in damage ranging from cosmetic deformations to rendering the charge point unsafe. To this end, CPOs often install parking barriers around charge points to absorb the impact of an errant vehicle and protect a much more expensive asset.

The collation of these maintenance issues may indicate a wider problem. All service tickets should be reviewed on a regular basis and catalogued, with a root-cause analysis performed to try to identify any apparent trends, possibly leading to a review of work practices or components used. Where a trend is identified based on a single issue, the CPO will seek to negate this. Each charge point manufacturer provides product-specific guidance on how to maintain their charge points, and the details of the periodic inspections and tests that need to be conducted. This is key documentation that must form the central part of a CPO's maintenance schedule.

9.4.1 Planned maintenance

Planned maintenance is anticipated, scheduled and usually relates to periodic inspection and testing of charge points as required by the CPO's best practice and manufacturer's recommendations or regulatory requirements. For example, public charge points in the United Kingdom must be inspected every 6 months and tested annually. The six month test is primarily an inspection to check for any unreported cosmetic damage or aspects of the charge point that do not stop it from working but have not previously been reported. Importantly, from a health and safety perspective, the residual current device is tested to ensure correct operation in the event of a fault. The annual test will generate an Electrical Installation Condition Report which is a certificate akin to the original installation certificate, checking all the same parameters to ensure that they are still within specification as per the British

Electrical Wiring Standards (BS:7671) [4]. Another objective for planned maintenance may be the planned upgrade or modification of certain components within a particular model of charge point, arising from the manufacturer's suggestion or even the CPO's own learnings from root cause analysis performed on unplanned maintenance service tickets.

9.4.2 Unplanned maintenance

For all the reasons outlined in Section 9.5, charge points do fail, with some issues rendering charge points difficult to use, whilst others render the charge point completely inoperable. In some cases, the CMS will give the operations team an indication of a failure or issue. At other times, it is only when a user phones the CPO's helpline to report an issue that the CPO becomes aware of this at the charge point. These unanticipated issues will require unplanned maintenance to resolve and return the charge point to full availability. Once a service ticket has been created to log the issue, the CPO may request a scheduler to review tickets, plan and effectively coordinate field service activities based on traffic variability by time of day and area. Given the unexpected nature of unplanned maintenance, field service crews must be equipped with all the necessary spares, tools, product training and skills to be able to quickly and effectively repair charge points in situ, and not be located too far from the charge point to avoid excessive time being lost in travelling to the site. The variability of weather can add another dimension to field maintenance, with rain preventing on-site work unless the charge point and workplace can be effectively covered. Additionally, work will take place in the public space with the field service operative having to contend with members of the public moving in close proximity. Following health and safety protocols such as cordoning off the worksite, completing an on-site risk assessment and working to the established Risk Assessment and Method Statement will help the operative to keep themselves and the public safe. Charge point field service operatives often work alone on-site and therefore the CPO must have a lone worker policy and procedure in place that monitors their situation throughout the day so that if something untoward should happen, a responsible staff member can raise the alarm and take action in a timely manner.

An unplanned maintenance is generally triggered by a ticket, raised automatically by the CMS or operations team when certain logs or network conditions are encountered, or raised manually by the call centre answering calls. An unplanned maintenance ticket needs to be resolved within a certain timeframe as per an agreed service level agreement, usually specified in the contract with the client or with the maintenance company if independent of the CPO. Typically, issues that pose a risk of safety to people require resolving within 2 hours. The objective is not necessarily to rectify the issue but to make the site safe so that people are no longer put in harm's way. Generally, the equipment causing the issue will be isolated from its electrical supply and as appropriate, taped off with barrier tape and/or sealed to prevent any unauthorised tampering, especially if the components and wires are exposed. Tickets raised that address issues that prevent charge points from operating, typically have 2 days to be resolved. The quicker the resolution, the less impact on the availability of the charge point. The scheduler will be monitoring tickets to ensure that they do not exceed the service level agreement specified by the client.

Damage that does not affect the operation of a charge point, for example graffiti, is logged in a service ticket. These issues are less important from an operational perspective; however, they still affect the CPO's brand and image and still need to be addressed, typically within 5 or 10 days, depending on the CPO and contractual obligations with their client.

9.5 Charge point end of life

When focusing on the end of life of a charge point, there are two points to consider. The first is the location of the charge point and its ongoing ownership/viability, whereas the second is the lifespan of the equipment installed. Based on these differing perspectives, end of life is relative to each. A CPO currently operating at a location, and ceding that location to a rival CPO (either through the transfer of a contract or the acquisition of part of the CPO's charge point estate), will see that location as end of life from their perspective. To the incoming CPO, the location is treated similarly to a new deployment, as per Chapter 8; however, some initial activities such as site selection will be redundant (assuming the incoming CPO finds the location commercially advantageous). In this instance, the outgoing CPO will likely remove the charge point. A CPO may also see a site classed as end of life if the utilisation drops below what is deemed acceptable for the site. Where end of life relates to hardware becoming obsolete (anything on average over 7 years), this is removed by the CPO and replaced with a newer unit.

9.5.1 Relocation

Should a charge point be found to have, for example, an unacceptably low utilisation, or the client does not want the charge point at that location for some reason, then the same charge point could be uninstalled at its original location and reinstalled at a new location. This would likely be handled as two separate projects: first to remove the charge point from its original location and make good the area, and the second to identify and prepare a new location, and then install the original charge point. Relocating is an expensive option and the need to do so requires justification and possibly a separate business case. Contractual terms may require it, for example, where some event specified in the contract has triggered the relocation. Alternatively, if a purely commercial consideration was the key factor (utilisation, age of the charge point or cost of another charge point), a cost–benefit analysis may indicate that it is cheaper to leave the original charge point in situ and install a new and more appropriately sized charge point at the new location. This is, of course, assuming the contractual terms permit the addition of another charge point to the network in the client's area. When relocating charge points, a decision should not be hastily made. It is not something to be taken lightly and re-emphasises the need to seek viable locations during the site selection process completed as part of the original deployment stage. Case Study 9.1 outlines the practical aspects related to a relocation.

Case Study 9.1: Charge point relocation

In this case study, the local authority altered the road layout near the existing charge point. As the new road layout did not permit vehicle parking within the zig-zag lines on the approach to a pedestrian crossing, the charge point location became unviable. This highlighted the issue of effective communication and planning within a large client, such as a local authority. The Council which this example relates to has not been mentioned due to sensitivities with this charge point relocation. This particular location could possibly have been removed from the list of possible charge point locations during the planning or deployment stage if the planned road re-design had been known. Effective communication between the local authority highways team (typically responsible for roads, including pedestrian crossings) and the street lighting/EV charge point team, however, enabled the coordinated removal of the existing charge point infrastructure and relocation to another site within the authority. The key factor behind this relocation was the CPO's contract with the client determining the provision of a certain number of charge points with this being linked to revenue. Losing one charge point would reduce revenue. Similarly, the authority did not want to lose an opportunity for its residents to charge nor any benefit derived from the commercial terms associated with the charge point.

As part of any charge point design, the procedure around removal from a particular location and reinstatement of the original site should form part of the project plan. To this end, any street or other infrastructure that was modified during the initial installation, be it electrical, ground or metal works, should be reinstated to a suitable standard. In this case study, the CPO wrote a procedure specifying the secure fitment of plugs, flush with the lamppost surface to maintain water sealing and surface integrity of the post.

Source: Based on the author's experience while working for char.gy

9.5.2 Replacement

In some instances, there is a requirement to replace the current charge point with another more powerful or more reliable model, though what drove the original decision to replace it will determine the specifics. This would be handled as a mini-project, with the existing civils and electrical infrastructure examined to ensure compatibility with the proposed replacement charge point. The layout of the current foundations and placement of electrical cabling may not suit the new charge point and need modification. Similarly, the electrical capacity will need to be investigated to confirm sufficient additional power for an uprated charge point, should that be the case. To install the replacement charge point, the process outlined in Chapter 8 must be followed. Case Study 9.2 outlines the steps taken when replacing a charge point.

Case Study 9.2: Charge point replacement and upgrade

This case study refers to the scenario of changing the existing charge point at a location for another charge point. The reason behind this replacement was that the charge point had issues with spare parts availability. In this instance, the opportunity was taken to install a higher-rated charge point. Operationally, the project consisted of two parts. The first was the decommissioning and removal of the existing charge point. The second was installing the replacement charge point. Ideally, the removal of the original charge point and the installation of the new charge point should be done at the same time to avoid the site being left for an interim period in an unsafe, unfinished and unsightly state, whether on the public highway or private land. The installation of the new charge point needed to be planned in the same way as any other new installation. Existing civils and electrical infrastructure must be designed and installed to meet the requirements of the original charge point, and the new charge point will most likely require a different design of foundations and electrical power routing to enter the unit. Regulations concerning the electrical switch gear supplying a charge point will have likely evolved since the original installation, so it is quite possible that the electrical switchgear supplying the charge point will need to be replaced to comply with the current regulations.

In the event that the new charge point has the same chassis as the original, it is possible that the civils and electrical infrastructure will not need modifying; however, the issue of the electricity supply switchgear being updated would still be relevant as the entire electrical supply circuit and new charge point would need to be electrically tested and certified like any other new installation at the time, meeting the current BS:7671 electrical wiring regulations. The removed charge point might be cannibalised to provide spare parts for similar models still in use. It may, however, be scrapped in compliance with national electrical waste regulations and the CPO's procedures – a requirement increasingly dictated by the client's terms of contract.

In this case study, a further complexity was the change of CPO. Contractual issues between the client and the original CPO led to the location being withdrawn from that CPO's remit, and a new CPO engaged. The new CPO elected to install new hardware and chose not to try and integrate their CMS with the software operating the original charge point. This meant they did not need to source spare parts and technical know-how to support the original charge point until the end of its lifespan. As the location was suitable for a new dual socketed charge point rated at 22 kW, the Public Charge Point Regulations meant that the new unit had to include a contactless payment terminal.

Source: Based on the author's experience while working for char.gy

9.5.3 Retire (decommission)

Considering the two factors related to a charge point's end of life (location versus equipment), retirement is taken to mean that the location is not suitable anymore, resulting in the charge point being removed and not replaced. For example, a contract with a client may end and as per the terms of the contract, the CPO must remove its equipment. For avoidance of doubt, if the location was still required and the equipment was being retired and replaced with a newer charge point, this would be a classed as a replacement, as discussed in Section 9.6.2. In the event of a location's retirement, a project would be created to decommission and remove the original charge point making good the site and either restoring it to its original condition, or establishing a new layout, as agreed. Given the age and condition of the charge point when installed, a decision will likely be needed to decide whether to redeploy the charge point when a new opportunity becomes available, or whether to scrap the charge point, disposing of it following the CPO's waste management and national electrical waste disposal guidelines. Case Study 9.3 brings to life what happens when charge points are decommissioned.

Case Study 9.3: Charge point decommissioning

Whereas Case Study 9.2 focuses primarily on the process from the perspective of an incoming CPO replacing a charge point, this case study looks at decommissioning from the outgoing CPO's perspective. Note that the specifics of the incumbent CPO are not included to avoid reputational damage. In this case study, the relationship between the incumbent CPO ended and they were instructed by the client to remove their installed charge points or leave them for the new CPO to remove and dispose of. The incumbent CPO decided to remove the old units, leaving the site ready for the new CPO to come in and install new charge points, similar to what is noted in Case Study 9.2.

Source: Based on the author's experience while working for char.gy

9.5.4 Role of the CMS

The CMS needs to be updated when charge points reach their end of life and are decommissioned. In all cases, the CMS needs to be notified that a charge point has been removed at the respective location and that the removed charge point can no longer be controlled when a driver attempts to charge at that location. In the case of a relocated charge point, the CMS needs to be updated with all the dependencies about the new location. It must also be ensured that the updated information feeds through to the CPO's website, driver touchpoints, third-party aggregators and field services so that the charge point can be found, used and maintained. Similarly, with a charge point being replaced at a particular location, the CMS needs to transfer all the same dependencies of the changed unit to the new charge point so that testing of it can be completed at the site, the CMS can control it and field services can record the maintenance related to it.

Where a charge point is decommissioned, the CMS will need to be updated; however, there will not be a new charge point number assigned and the location will be removed from all the driver interfaces so that drivers cannot navigate to a charge point that is no longer there. It is very important that all historical data related to the site is kept for future reporting purposes covering the period that the charge point was installed at that location. Reasons for this include contract performance monitoring, driver billing and energy supplier payment reporting and field services record keeping of historical maintenance and test data.

9.6 Conclusion

Charge point operations, maintenance and, in some cases, decommissioning, is the central role of a CPO once charge points have been installed. Operations and maintenance are regulated in the United Kingdom by the Public Charge Point Regulations, which determines reliability, uptime and how EV drivers can get help. Central to managing the charge points is the CMS that holds all the key information, enabling drivers to locate, start and end a charging session. This sits alongside a customer support function that provides help to drivers when it is needed. Charge point reporting tracks reliability, utilisation and, ultimately, profitability. Planned maintenance is carried out every 6 months, with unplanned maintenance taking place when an issue develops which the CPO does not have foresight of. When a charge point reaches end of life, it can either be relocated, replaced or retired. If any of these happen, the CMS must be updated to ensure the driver has the most up-to-date information about the charge point.

References

[1] Office for Zero Emission Vehicles. *Department for Transport. Public Charge Point Regulations 2023 guidance* [online]. 2024. Available from https://www.gov.uk/government/publications/the-public-charge-point-regulations-2023-guidance/public-charge-point-regulations-2023-guidance#reliability [Accessed 05 May 2025].

[2] Willis T. *Briefing document: The UK's transition to electric vehicles* [online]. 2020. Available from https://www.theccc.org.uk/wp-content/uploads/2020/12/The-UKs-transition-to-electric-vehicles.pdf [Accessed 05 May 2025].

[3] The Energy Savings Trust. *Local electric vehicle infrastructure fund* [online]. 2025. Available from https://energysavingtrust.org.uk/grants-and-loans/local-electric-vehicle-infrastructure-scheme/ [Accessed 05 May 2025].

[4] The Institution of Engineering and Technology. *BS 7671 – 18th Edition: The IET Wiring Regulations* [online]. 2025. Available from https://electrical.theiet.org/bs-7671/ [Accessed 05 May 2025].

Chapter 10
Stakeholder perspectives

David Trousdale[1]

In Chapter 3, we introduced the key stakeholders involved in the charge point deployment process. In Chapters 7–9, we covered the activities stakeholders complete as part of the process. This chapter looks in further detail at each stakeholder and explores five key elements, roughly based on SWOT (strengths, weaknesses, opportunities, threats) analysis. The first is the role stakeholders play including what their strengths and weaknesses are in terms of fulfilling it and whether they are charge point enablers or blockers. The second and third are the opportunities they hope to unlock by being involved in a charge point deployment and the challenges or threats they might face. The fourth is where they most likely find themselves on a stakeholder matrix. Figure 10.1

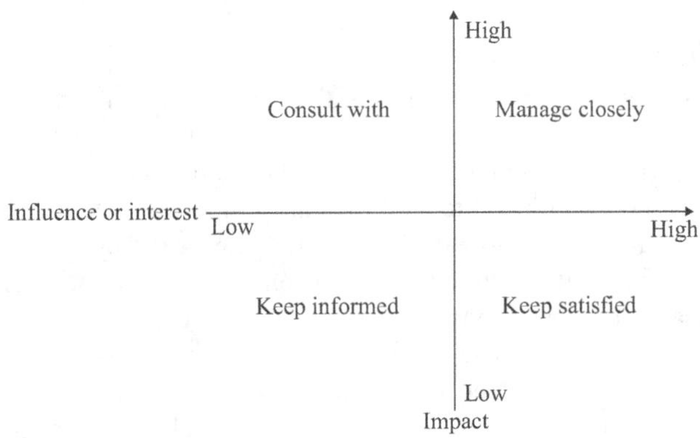

Figure 10.1 Stakeholder matrix [Source: Government Analysis Function. Stakeholder mapping https://analysisfunction.civilservice.gov.uk/policy-store/stakeholder-mapping/ [online]. 2024. Available from [Accessed 24 October 2024]].

[1]Principal Advisor - Climate Change and Sustainability, Newcastle City Council, United Kingdom

demonstrates a stakeholder matrix with influence or interest running as the x axis and impact as the y axis. Based on this model, a stakeholder should be consulted with, managed closely, kept informed or kept satisfied. The fifth and final element is how the stakeholder role might change over the duration of the deployment process. The central stakeholder around which all other stakeholders revolve is the organisation delivering or commissioning the deployment. This tends to be the charge point operator (CPO) or local/state government. National governments rarely commission charge point deployments.

This chapter will also bring to life, through various case studies, the involvement of stakeholders throughout the deployment process by focusing on one specific stakeholder in a project example. This analysis, though not exhaustive, gives an overview of the most common stakeholders and is based partly on personal experience, supplemented by research into real-world examples.

10.1 CPO

The CPO is key to all charge point deployments. Charge points can be installed without them but this is usually on a small scale and done privately. The CPO acts as the organisation delivering charge points where they are expanding their own network independently. They may also be commissioned by another organisation (e.g., a local authority).

10.1.1 Their role

The CPO plays a key role in every part of the planning, build and operation stage, either as a contractor to another entity or as the primary operator with agreement from the landowner to lease/rent land where the charge points are deployed. The CPO may also choose to own the land the charge point is installed on. Critical to their role is determining what their source of funding is. If they are commissioning their own deployment, they will need to draw on capital either privately from their reserves or via investment capital. If the latter is the case, the CPO will need to ensure that the investor backs deployment. This is why the business case as part of the planning stage is important as this includes the financial case to justify the investment. char.gy is a good example of a CPO backed by investment capital, as is demonstrated in Case Study 10.1. Zouk Capital's total investment to date is £100 million and has enhanced char.gy's ability to expand their charging networks to everyone throughout the United Kingdom over the next 5 years. If a CPO is commissioned to deliver a network by a local authority or other organisation, they may still need to draw on investment or they might be paid by a local authority to deliver it on their behalf.

Case Study 10.1: char.gy – an example of a charge point operator based in the United Kingdom

char.gy are a CPO based in the United Kingdom who were established in 2016 with the central mission to deliver society's preferred EV charging network, accelerating the transition to sustainable transport. Its vision is that by 2030 it will have made the use of EVs easy and convenient for over 1 million drivers by delivering and supporting a reliable public charging service. It currently manages 3,000 charge points. char.gy is backed by Zouk Capital, a private equity and infrastructure fund manager investing in the sustainable economy opportunities created by the global shift to EVs. Zouk's total investment to date is £100 million and has enhanced char.gy's ability to expand their charging networks to everyone throughout the United Kingdom over the next 5 years.

In 2024, char.gy has delivered various charge point deployments. In July 2024, they secured a contract with the London Boroughs of Richmond and Wandsworth to migrate and integrate existing EV infrastructure whilst also expanding the network. This included 950 existing charge points, just under 400 of which transferred over in July 2024. The remainder will transfer by November 2025. Other local authorities where char.gy is managing or deploying charge points include the London Boroughs of Brent, Redbridge, Tower Hamlets, Haringey and Southwalk, Buckinghamshire Council, Coventry City Council, Croydon Council, East Lothian Council, Harrow Council, Merton Council, Leicester City Council and Milton Keynes City Council. char.gy is also partnered with Vauxhall Motors Limited in the Electric Streets of Britain campaign. This initiative aims to support the transition to EVs for the 40% of households without off-street parking, ensuring they are not left behind in the United Kingdom's EV transition. char.gy notes that it was ranked number one as the most affordable tariff provider for the year 2023–2024, that it offers 99% network uptime and that it is 100% powered by green energy during periods of high renewable generation.

Source: char.gy. *About Us – Discover char.gy* [online]. 2024. Available from https://char.gy/about [Accessed 25 October 2024]

The strengths of a CPO are that they are agile in terms of being able to draw on resources to be able to deploy charge points. They are agnostic to technologies and charging solutions meaning they are able to deploy whichever charge point technology aligns most closely with the target use case/s. They are commercially driven meaning they will always seek to run the network efficiently and effectively, avoiding unnecessary costs. This is important to the organisation who appoints the CPO as they will want to see that they deliver impact in terms of meeting the network's objectives (e.g., enabling large-scale transition to electric vehicles (EV)). The key weakness of a CPO is that they are sometimes overly profit driven which can mean that they only deploy in locations with the highest projected demand and utilisation. This often results in targeting more affluent areas and ends up preventing a just and equitable transition as less affluent areas are ignored. CPOs may also become so focused on targeting a specific use case that they are unable to pivot towards a different one. This is often the case with CPOs specialising in en route charging, though they may choose to partner with another CPO so that multiple use cases can be satisfied. A further weakness is the CPOs' exposure to the market in terms of energy prices and needing to satisfy their investors. This can result in uncompetitive pricing, falling utilisation and possible company collapse. This is especially an issue where a CPO is trying to secure market share and has significantly undercut competitors to secure sites and win work. No one, bar a potential competitor CPO benefits from collapse, especially where a network gradually falls into disrepair as operations and maintenance resources dry up. Local users pay the price as the network they are reliant on is no longer available.

It should come as no surprise that CPOs are enablers of charge point infrastructure though it must be noted that in some cases they may act as blockers if their intention is to prevent competitors trying to establish or grow market share. They may buy up land to prevent it from becoming available to a competitor and hold that land without developing, thereby preventing a potential deployment. This is most often the case for en route charge points as competition on strategic routes can be fierce and land availability at service stations and junctions scarce.

10.1.2 What they want to achieve

The primary drivers of the CPO are to increase demand and utilisation, thereby generating a profit for the business and their shareholders. In some instances, a CPO may enter into an agreement to install charge points in a location as a 'land grab' to secure a power network connection for a future expansion or to prevent a competitor from installing charge points. During the planning stage, the focus of the CPO is to identify the sites with the best anticipated demand and utilisation. During the build stage, the CPO's efforts centre on achieving cost efficiencies and value for money by installing charge points as efficiently and as quickly and practically possible so as to start generating revenue. During the operation stage, the role of the CPO is to operate and maintain the charge points, minimising maintenance costs and downtime while pushing up utilisation by making charging as attractive as possible (this might include making changes to the price per kWh). The CPO will only decommission or upgrade a unit when it is absolutely necessary, preferring to operate the charge points until their end of life which can be anywhere between 5 and 10 years.

Some CPOs may focus their unique selling point on the social value they can deliver, specifically through supporting a portfolio approach of high- and low-demand sites in a mix of high and low income areas to enable an equitable spread of charge points. They may also deliver social value through community engagement to drive up demand. Other CPOs may focus on a technological or business, model-based selling point. Regarding the former, this might take the form of charge points that retract into the ground or rely on a lance or other piece of equipment to attach to a recessed connector in the pavement or road. Regarding the latter, this might include a focus on smart charging based on a time of use tariff offering cheaper rates at certain times or a vehicle to grid capability where the driver can make money by selling power from their vehicle back to the grid during peak times. Regardless of the uniqueness of approach, all CPOs focus on one key role, profitability through high demand and good utilisation over 15%. Case Study 10.2 highlights InstaVolt's role as a CPO providing en route public charge points in locations with good utilisation.

Case Study 10.2: InstaVolt – an example of a CPO in the United Kingdom focused on rapid public charge points

As EV adoption accelerates, drivers need confidence that they can charge conveniently wherever they travel. While slower home and on-street charging solutions are vital to a balanced infrastructure, rapid public charging plays a convenient and essential role in long distance and time-sensitive journeys, as well as for those who do not have home charging. InstaVolt, the United Kingdom's largest public rapid charging network, is focused on delivering exactly that – reliable, ultra-rapid charging in high-demand areas. A private company, founded in 2016, InstaVolt specialises in public charging, deploying 120–160 kW chargers at strategic roadside and retail locations. These chargers are designed for convenience, allowing drivers to recharge in as little as 20–30 minutes.

Networks like InstaVolt serve critical use cases: charging on the go, supporting drivers without access to off-street parking and enabling long-distance travel. Together, these charging types form a cohesive ecosystem that removes barriers to EV adoption, offering a truly nationwide infrastructure. InstaVolt's strength lies in its user friendliness and reliability, with a first to market contactless payment approach and easy to use app plus a consistently high charger uptime of over 99%. The brand seeks to build trust in the network and therefore reduces 'charging anxiety', which is frequently cited as a concern for new EV drivers.

The company's customer first approach has seen it partner with landowners and retailers across the United Kingdom, from Costa Coffee and Starbucks to McDonalds, ensuring chargers are positioned where people already go. As of 2025, InstaVolt has deployed over 2,000 chargers nationally, with ambitious targets to grow further to 11,000 by 2030 as demand for public charging continues to rise. As EV uptake grows in response to evolving government policy, the role of ultra-rapid public networks like InstaVolt will become even more important in helping the United Kingdom reach its net zero ambitions.

Source: InstaVolt. *The UK's largest rapid EV charging network* [online]. 2025. Available from https://instavolt.co.uk/ [Accessed 27 May 2025]. Image provided directly by InstaVolt with permission to publish

10.1.3 What challenges they face

CPOs face many challenges throughout the deployment process. During the planning stage, the greatest challenge is selecting locations which will be the most profitable based on projected demand and utilisation. This is entirely dependent on the strengths of the deployment model (see Chapter 5) and includes a certain degree of risk as every model is based on assumptions which, if they do not play out, can greatly impact on profitability. For example, if EV adoption rates in an area dip by a couple of percentage points this will impact on utilisation with the possibility of making a profitable site unprofitable. An additional challenge during the planning stage is selecting the most cost-effective location based on projected installation costs. This primarily relates to the distribution network operator (DNO) costs as some sites may not have capacity or may prove more difficult to connect resulting in higher costs. The CPO, when conducting the site analysis, will need to explore where best to connect charge points and whether there are alternative options which may prove more cost effective. In some instances, the CPO may consider off grid solutions like solar and battery storage, backed up with a smaller capacity grid connection to reduce reliance on the grid and minimise connection costs. A further challenge the CPO must navigate is the planning process. The CPO in some instances will need to seek planning approval and/or a traffic regulation order which impacts on the timeline of delivery. To minimise the impact of this risk, the CPO should work closely with the local authority to understand the planning rules and expectations around traffic regulations so as to ensure applications are approved, when needed, quickly.

During the build stage, the key challenges relate to ensuring the charge points are installed in a timely manner and that health and safety are adhered to. Regarding the former, any number of factors can delay an install which ultimately impacts on how quickly a charge point can go live and become profitable. This includes contractors and civils work taking longer than expected, connections to the grid being delayed, delays in delivery of charge points, delays in data connectivity upgrades and any planning or regulation issues forcing the local authority to halt work. The CPO can minimise this risk by building in some contingency into their build plan and by ensuring detailed due diligence is conducted during the plan stage to minimise foreseeable planning risks that might affect the build. Regarding the latter, the installation of charge points is a dangerous task due to dealing with high-voltage electricity cables. Installation must adhere to local health and safety regulations to ensure what is installed is both safe to those carrying out the installation and the general public. The CPO must ensure its contractors follow the necessary regulations across all its sites. This can be achieved through spot checks during installation to ensure regulations are being followed and that certification is provided, once the installation is complete, demonstrating compliance with regulations.

During the operation stage, there are several challenges including maintaining the charge points, responding to energy price fluctuations, competition from other CPOs and data connectivity challenges. Regarding maintaining charge points, there are many factors that can prevent them from operating properly. Manmade events like vandalism or collision from a vehicle can take a unit out of action. Weather is also a major factor in some climates, especially where sites are either prone to being snowed in or flooded. Nearby foliage can block or obscure a charge point and living creatures can make their home in the charge point, resulting in it failing over time. A CPO can mitigate all of these risks by considering the most suitable location to install the charge points and carrying out annually or six monthly maintenance to look after the units. Manmade events can be avoided by installing protective street furniture and fitting cameras to units. Charge points should also provide regular status update messages to the CPO so that the back office system can identify if a charge point is starting to fail. This can be identified by voltage fluctuations or a drop in utilisation. CPOs can then take proactive action to repair units, keeping them online. Regarding fluctuations in energy prices, CPOs are exposed to this risk especially if they have not locked in their energy prices for a period of time. Higher prices when compared with competitors can result in a fall in market share, especially in locations where there is a direct competitor offering a similar charging experience at a similar speed. This risk can be mitigated by agreeing a fixed energy price with an energy provider, though care must be taken to avoid locking in at a higher price if there is an anticipated fall. New competition from another CPO is also a challenge which could not have been anticipated prior to deployment. This can have a major impact on profitability by pushing down demand and utilisation, especially where the competing CPO offers a better service or is more price competitive. To mitigate this risk, the CPO must keep a close eye on its competitors in an area and use different tools like competitive pricing or other perks like shopping points or money off coupons to maintain demand and utilisation.

A further risk is data connectivity deteriorating at a site, resulting in the charge points becoming unusable. Charge point units that regularly communicate with the CPO's back office system are key to ensuring a reliable network. Disconnection can be the result of a range of events including units having their power and data cable physically severed though the most common issue is where mobile connectivity drops out. This can be mitigated by ensuring a charge point is connected by Ethernet and that the charge point has the ability to connect to multiple mobile operators. Fringe locations at the edge of mobile signal networks should opt for wired connections if possible. The CPO should also constantly monitor charge point connectivity to identify any wirelessly connected units that intermittently connect to the network and troubleshoot whether this is down to the mobile network. If it is identified that this is the case, the CPO can then work with the mobile network operator to boost the signal or replace/update the sim in the unit to connect to a different network.

10.1.4 Their importance as a stakeholder

The CPO plays a central role in deployment.

10.1.5 Changes throughout the planning, build and operations stages

Throughout all three stages of the charge point deployment process, the CPO's role is constant, driving deployment regardless of whether they have been commissioned on behalf of a local authority or landowner, or if they are delivering the network themselves. There are only three points in the overall process where the level of control the CPO has changes. This is during the planning process where they are reliant on a local or highways authority to approve their site plans and traffic regulation requests, during planning deployment at a local level where they are reliant on quotes from the DNO regarding network upgrade costs and during the build stage where they are likely to be reliant on a contractor who is installing the charge points on their behalf. In each of these steps, the CPO has to rely on actions carried out by others which could impact the timeline, cost and quality of a charge point deployment, ultimately impacting profitability.

10.2 The landowner

The involvement of a landowner is primarily related to the use case a CPO or local authority is seeking to address. Landowners are usually involved in en route and destination charge point deployments though they can be involved to a lesser extent in residential deployments where there is limited space on the public highway to install charge points. The landowner might also be passive in the deployment process and be approached by a CPO or local authority with a request to buy or rent/lease the land. In this case, they may appoint a land agent who is interested purely in selling or leasing the land for a profit. They may also be active in the process, seeking out a CPO to deploy charge points on their land. Where this is the case, the landowner is often the operator of a broader business (e.g., a service station, mall or supermarket) which wants to enhance its business and deliver profitability through new charge points at its sites. It should be noted that a local authority can also take the form of a landowner; however, detail relating to their role as a stakeholder is captured in Section 10.3. This section only focuses on landowners as businesses independent of the public sector. The landowner may also be a business with a role independent of providing charge points but with land to deploy them on. Examples include car park operators and supermarkets.

10.2.1 Their role

The landowner's primary role is to provide access to the land where the charge points will be deployed, either by selling the land to the CPO, leasing/renting it or entering into a hybrid leasing/renting/revenue share model. A landowner may appoint a land agent to conduct commercial engagements on their behalf. Throughout the installation process, the landowner is heavily involved in the planning stage. The CPO will either identify them when looking for potential sites to deploy charge points and may have initial conversations during the longlist step, or they will be approached by the landowner directly/through a commercial selection process and enter into an agreement to install a number of charge points over one or more sites. More detailed discussions around specific sites take place during shortlisting when the site lists are finalised. The CPO and landowner will enter into negotiations to sell or lease/rent the land, with the latter being done most frequently based on an annual fee, though in some cases the landowner may insist on a revenue share where an income is earned based on every kWh dispensed at the site/s. If the landowner is the operator of a broader business and wants charge points on their site/s, they will likely choose the revenue share model over a sale or ground rent/leasing model. A hybrid payment agreement mixing a fixed lease/rental fee with revenue share when utilisation achieves a certain level might be chosen as the preferred approach. Other bespoke business models are available. Case Study 10.3 demonstrates how Tesco, as an example of an operator of a broader business, entered into an agreement with Pod Point, a UK CPO, to deploy charge points across its network of supermarkets. The model adopted in this instance was a sponsorship-based business model where Pod Point installs, operates and maintains charge points. The commercial details between both parties are not publically available.

Case Study 10.3: Tesco – both an operator of supermarkets in the United Kingdom and a landowner

Tesco is a multi-national groceries and general merchandiser with a large presence in the United Kingdom. It is not only a retailer but also a significant landowner including both the land its stores and car parks sit on as well land purchased for other purposes. In 2018, Pod Point teamed up with Tesco and Volkswagen to install charge points at over 600 stores by 2022. This target was reached and Tesco continues to expand its charge points at existing stores. In 2024, Vauxhall also partnered with Tesco to further enhance the network. The partnership creates a sponsorship-based business model for EV charging where Tesco, Vauxhall and Volkswagen can advertise their services and vehicles at advertising boards next to charge points. Pod Point installs, operates and maintains the charge points through their web app. Pricing is competitive ranging from 44 p per kWh for level 2 7 kW charge points all the way up to 69 p per kWh for level 3 DC 75 kW charge points. Pod Point is now starting to offer a service called 'Energy Flex' at its charge points in a home and workplace setting, and in public charge points in the future, where they can be turned on or off in line with energy demand so as to support a resilient grid. Customers are rewarded by opting in to the service. It is likely from the overarching commercial model that Tesco will receive some income per kWh from Pod Point; however, the details are commercially sensitive and not in the public domain. Tesco also benefits from the partnership as it contributes directly to one of its corporate goals to operate in a responsible and sustainable way.

Pod Point. Tesco Vauxhall EV Charging | UK National Rollout [online]. 2024. Available from https://pod-point.com/rollout/tesco-ev-charging#:~:text=The%20Details,rapids%20which%20cost%2069p/kWh. [Accessed 27 October 2024]

As the process progresses and the land is sold, the landowner's role stops there. If a leasing/rental agreement or revenue share arrangement is put in place, their role diminishes during the build and operation stages. They fulfil a high-level monitoring role, commercially managing the agreement to ensure it is functioning as expected. Disputes between both parties may occur if rent and/or revenue share ceases or if the land is not used as agreed. The strengths of a landowner are that they have the available land to deploy charge points on and, if they are an operator of a business with a primary aim different to providing charge points, they will proactively work with the CPO to identify the most viable locations. The weaknesses are that those motivated just to sell the land will only care about the sale and not the long-term viability of the site for charge points. This might present a risk to a CPO especially where they have not done their due diligence regarding the site and where they are looking to complete a land grab to prevent another CPO establishing itself in an area. In this instance, the landowner may overemphasise the attractiveness of the site to push up the price and achieve the sale regardless of its potential profitability.

The landowner can act as both an enabler and a blocker. Where they are an operator of a broader business, they will actively work with the CPO to select the most promising sites. Where they are a land agent on behalf of a landowner, their primary goal is to generate a profit from the land so they will work closely with the CPO either to sell or lease/rent the land to them. They can, however, be a blocker if they do not want to sell or lease/rent their land for various reasons, including if they have entered into a strategic relationship with another CPO.

10.2.2 What they want to achieve

This differs based on whether the landowner is selling or renting the land with little involvement through a land agent or the operator of a business with a separate commercial role to charging. Regarding a land agent, the opportunity they are pursuing is generating a profit from the sale or rental/leasing of the land through selecting the right customer. The land agent may look for opportunities to enhance an existing sale or lease/rental by approaching a CPO to provide charge points as part of a broader commercial opportunity with an overarching client. This might be the sale of land to a retailer which is made more attractive because part of the deal involves a charge point hub on or near to the site. Depending on the drivers behind the transaction, the land agent might want to offload the land with little specification of what it can be used for, or they may spend a lengthier period of time engaging with a range of CPOs to ensure whichever is successful in buying or leasing the land is the right fit.

Regarding a business with separate commercial aims, although generating a profit is an important opportunity they will want to realise, other opportunities will be important to them including providing charge points to differentiate from competitors, pushing up footfall in locations where charge points are deployed, achieving sustainability objectives and/or using charge points to advertise other services. Case Study 10.3 demonstrates a range of opportunities that can be realised through an operator of a broader business entering into an agreement to lease land to a CPO. Tesco recognised that charge points with advertising boards presented an opportunity to advertise services beyond charge points. They also identified that charge points contribute to their broader sustainability targets. The grocery market in the United Kingdom is also incredibly competitive, so it is likely that Tesco saw an opportunity to deploy charge points to differentiate itself from competitors.

10.2.3 What challenges they face

This again is dependent on whether the landowner is seeking to sell or rent/lease the land or is the operator of a business with separate commercial aims wanting to rent/lease the land only. Regarding the former, the land agent on behalf of the landowner will face the challenge of generating the best commercials in terms of profit and sales/leasing/rental terms for its customer. This is based on ensuring there are multiple customers competing for the site including encouraging CPOs based on potential site demand. When a rental or leasing agreement is in place, the landowner faces the challenge of needing to monitor the CPO to ensure what they are being paid reflects the revenue the CPO is generating. This is particularly the case where a rental or leasing agreement includes some form of revenue share based on a per kWh payment. The landowner needs to have the mechanisms in place to monitor revenue and challenge the CPO where income is less than expected. A further challenge is if a relationship breaks down and the landowner has to enter into legal proceedings with a CPO.

The operator of a broader business with separate commercial aims will face a slightly different challenge relating to commercials. This will include both making a profit from charge points occupying their estate and whether these drive up footfall, resulting in greater demand for their services. These challenges can all be overcome by working closely with the CPO to offer the best customer service and through ensuring that the charge points are competitive in terms of price charged to consumers versus competitors. Cost competitiveness can be achieved through reducing profit margins for both the broader business and CPO or through loss leading. In Case Study 10.3, Tesco initially offered charging for free to attract customers; however, this was removed in 2022 [1]. Ensuring the best customer service is offered can be achieved by collaboratively working with the CPO to ensure charge points have up times of 99% plus and are well maintained. A potential challenge is if the relationship with the CPO deteriorates and the broader business has to terminate the agreement, causing disruption to charging services offered to customers. This risk can be mitigated by ensuring the commercial agreements are robust and include a termination process where charge points are transferred to another CPO or to the broader business.

10.2.4 Their importance as a stakeholder

The importance is based on whether the landowner is renting/leasing the land including hybrid revenue share/leasing agreements or selling it. Regarding the latter, the influence initially is high, meaning the landowner must be managed closely during the sales process to the CPO to ensure that the agreement is satisfactory to both parties. The influence then tales off once the sale is made, meaning the landowner is no longer a stakeholder. Where the CPO enters into a renting/leasing agreement or hybrid agreement, the landowner's influence and impact are both high, meaning they must initially be managed closely. What is meant by this is that the relationship should be prioritised as the contracts are being written up. Resources should be dedicated to ensure communication is regular, with information provided promptly when requested. Legal teams should work together closely and draw on key staff members in the CPO to ensure the agreement is signed. Once this is achieved, the landowner's impact starts to fall and influence remains high once the charge points are operational. This means that the landowner should be kept satisfied through regular engagements, sharing utilisation data and other data including revenue actuals and projections if required. Where payments to the landowner are required, these should be made on time and accompanied with the associated narrative to ensure the landowner understands the financial breakdown.

10.2.5 Changes throughout the planning, build and operations stages

As stated in Sections 10.2.1 and 10.2.4 the role of a landowner is dependent on whether they are selling land or entering into a leasing/rental/hybrid agreement with the CPO. Regarding the former, the landowner is involved during the end of the planning stage and start of the build stage when the land is identified by the CPO. Commercial engagement takes place and then the agreement to buy the land is achieved. After this happens, the landowner is no longer involved. Regarding the latter, the landowner is similarly involved during the end of the planning stage and start of the build stage, resulting in a leasing/rental/hybrid agreement being put in place. Once this is achieved, the landowner's involvement scales back during the build stage and becomes more active again during the operation stage. They enter into a monitoring role, ensuring they are paid by the CPO and that the charge points meet their expectations as stated in the legal agreement between both parties.

10.3 The local authority, council or state government

The local authority, council or state government is an important stakeholder in the charge point deployment process.

10.3.1 Their role

The local authority, council or state government can fulfil multiple roles. In a limited number of cases, they act as a CPO and assume the role outlined in Section 10.1.1. In some cases, they commission a CPO, appointing them to deliver a charge point network on the authority's behalf. In this instance, they also tend to act as the landowner where deployment is taking place on the public highway. In most geographies, the role they fulfil is as the planning authority during the planning stage. This role changes depending on local planning laws and may not be required in some instances. Looking first at the role of the local authority, council or state government as a commissioner, they play an active role throughout all three stages. During the planning stage, they will determine the long list, put this out to tender and then work with the appointed CPO following contract award and signature to finalise the short list. When they determine the tender, they outline the rental or revenue share agreement (or both) when deployed on their land. They then work with the CPO to plan for installation, approving the timeline, budget and designs, supporting the securing of power connections and helping prepare traffic regulations. They also play a key role as the client approving planning applications if required for installation. During the build stage, their role is reduced to one where they monitor the progress of the installs to ensure they are delivered against the delivery programme. During the operation stage, they fulfil a similar role, monitoring the demand, utilisation and associated rental/revenue share payment for the charge points to ensure they are as expected. Disputes may occur between the CPO and authority, council or state government if payments do not reflect utilisation or if the CPO fails to maintain the service level agreements (e.g., charge point reliability) as agreed in the contract.

Looking second at the local authority, council or state government's role as the planning authority alone, its involvement in all three stages is diminished. During the planning stage, providing this is required under local law, they approve or reject the planning applications for charge points. This also includes approving or rejecting traffic regulation changes (known as traffic regulation orders in the United Kingdom). During the build and operation stages, they monitor the charge point deployments to ensure they align with the original planning application and enforce the original plans and designs if the CPO deviates from what was agreed.

The strengths of a local authority, council or state government are that where they have appointed a CPO, they are also the planning body and often the highways body, meaning that approvals for deployment can be much quicker. They also can give the CPO access to a range of sites across their jurisdiction, providing CPOs with the flexibility to deploy in locations that present the best possible utilisation. The weakness of a local authority, council or state government is that they can be slow to approve planning and traffic regulation applications. They can also limit the sites where charge points can be deployed either because they want to see a just transition through charge points deployed in high and low demand areas or because they want to minimise the disruption to an area by reducing installations. Local authorities, councils or state governments may also insist on multiple CPOs operating in one area to encourage competition; however, this may put off CPOs as this will impact on demand, utilisation and profitability.

A local authority, council or state government can be both a blocker and enabler. Regarding the former, they can refuse charge point deployments for various reasons though this usually relates to planned installations not being in alignment with their strategic goals as set out in the planning stage. Regarding the latter, they can unlock multiple locations and smooth the planning and regulatory process meaning charge points can be installed relatively quickly. Where they are actively involved, this might also mean that they want to have a hands-on role in site selection.

10.3.2 What they want to achieve

What a local authority, council or state government wants to achieve from charge point deployment relates back to the initial decision to deploy. As Newcastle City Council demonstrates in Case Study 10.4, the opportunities it is seeking to realise include improving air quality across the City through encouraging the transition to EVs, ensuring a just transition by delivering a network that is spread fairly across the City and upgrading its existing network, whilst enhancing the network already provided by CPOs in the City. For Newcastle, this primarily relates to appointing a CPO to deploy charge points in a mix of residential and destination use case settings whilst recognising that CPOs have a role independent of the Council to deploy charge points to satisfy en route charging in locations close to or on the strategic and regional road network (including the A1). Most local authorities, councils or state governments will seek to realise similar opportunities to Newcastle. They may also see the opportunity to appoint a CPO to generate an income and, if so, adopt the role of a landowner. In this instance, they will do this through renting out land they own on a per charge point bay basis or via a revenue or profit share relationship, securing part of the per kWh price charged to the customer. In other instances, they may adopt a hybrid bay rental and per kWh revenue mode. In a number of more niche cases, the local authority may seek the opportunity through deployment to use the smart charging capabilities of vehicles plugged in to power buildings. Although examples of this are currently rare, going forward, this will become more common as smart charging functionality in vehicles is more established and bidirectional charge points are deployed in more locations.

Case Study 10.4: Newcastle City Council – a local authority focused on charge point deployments

Newcastle City Council, a local authority in the North of England, finalised its EV Strategy in October 2023, its aim being to prepare the ground for charge point deployment. Factors driving Newcastle's Strategy include enabling the transition to EVs to reduce air pollution in the city, enabling a just transition through ensuring charge points are deployed equitably across the city and making sure the needs of residents who have requested a charge point are met. Newcastle has entered into agreements with various CPOs including Fastned who installed a level 3 DC charging hub at the Helix site as part of the site's office and residential development. Newcastle, in its Strategy, recognises that a number of current charge points across the city are out of service and that a range of charge points will need to be deployed in the future to meet different driver needs. This includes deploying charge points at 77 target residential locations across the City in areas where there is limited or no off-street parking: 86 need to be renewed/replaced and 143 new destination level 2 and level 3 DC charge points are required in a range of locations including sites not owned by the Council. The city is also aiming to deliver an EV car club, accelerate the Council's fleet transition and deploy level 3 DC charge points for electric taxis. The Council recognises that it is an enabler of charge points across the city. This includes ensuring private developers and landowners provide EV charge points on future development sites and that the Council has an important role to play in supporting the deployment of commercial networks run by CPOs across the City. To deploy charge points, the Council is drawing on a range of funds. This includes Levelling Up funding used to install destination level 2 and level 3 DC charge points and Local Electric Vehicle Infrastructure funding, secured through the North East Combined Authority, to install charge points in residential settings.

Sources: Newcastle City Council. *Electric Vehicle Charging Infrastructure Strategy for Newcastle* [online]. 2023. Available from https://greenernewcastle.newcastle.gov.uk/assets/assets/EV%20Charging%20Strategy%20REV%20Final%20Oct%202023.2.pdf [Accessed 28 October 2024]

10.3.3 What challenges they face

A local authority, council or state government faces a number of challenges in relation to charge point deployment. A significant challenge is the limited resources available to deliver parts of the overall process. This manifests itself in terms of the available officers to support, where relevant, the site selection process including providing potential sites and validating those that have been put forward, approval of planning applications and the relevant traffic regulation orders required for deployment. This lack of resources can be mitigated by the CPO providing as much information as possible in a clearly structured way, minimising the time required by the local authority, council or state government to analyse information and approve applications. A further challenge related to the lack of resources is the time period required to gain approvals. This can be expedited by ensuring all internal local authority stakeholders (planners, legal teams, highways departments) are aware of potential deployments, are supportive of them and have been briefed throughout the process. This should result in a more efficient approvals process. In addition to this is the challenge relating to DNO connection requests where these are completed by the local authority on behalf of the CPO. Like the CPO, these can take time to complete, delaying the overall charge point deployment. One further challenge which should also not be forgotten regarding a local authority, council or state government is their political nature. This sometimes means that the political appetite for charge point deployment can wax and wane over a period of time especially where there has been a change in political leadership or a significant resident movement against charge points. The political nature can also create a challenge in terms of influencing deployments, especially where vocal communities argue for deployments which impact overall plans and result in installations which will potentially experience poor utilisation. This can be mitigated by insisting a data driven approach underpins deployment. This, however, is only effective where there is political buy-in.

10.3.4 Their importance as a stakeholder

Where the local authorities, councils or state governments is the commissioning body appointing a CPO, they should initially be managed closely. Their level of influence will decrease over time and may result in them needing to be kept satisfied as charge point deployment moves into the operations stage, especially where rental or revenue payment is received by them from the CPO. If the local authority, council or state government is acting purely as a planning body where charge points are being deployed, they need to be consulted with during the planning stage and then kept informed during the build stage, unless the deployment has resulted in some form of planning infringement. In that instance, their level of influence and impact increases, meaning they need to be managed closely to satisfy their concerns and minimise the impact on operations.

10.3.5 Changes throughout the planning, build and operations stages

As stated in Section 10.3.1, the role of a local authority, council or state government changes throughout the stages depending on whether they are commissioning a CPO or are just acting as a planning authority. Regarding the former, they are heavily involved in the planning stage, determining the vision driving the deployment, determining the area to deploy, shaping the business case and any associated procurement documentation, appointing the CPO and working with them at a local level, including approving planning and traffic regulations, and then finalising the governance and approvals process. During the build stage, they step back and monitor the deployment. During the operations stage, they monitor charge point utilisation and income either in the form of bay rent, revenue income or a hybrid mix of both. Regarding the latter, they are only involved in the planning deployment at a local level step through the approval of planning applications and traffic regulations, and during the operations stage if there is a planning infringement.

10.4 The regional or national transport body or department

The regional or national transport body/department is not always involved as a stakeholder though in the United Kingdom (specifically England) they may be involved as a combined authority or as National Highways when deployments are planned on the strategic road network.

10.4.1 Their role

The regional or national transport body's role can depend on how hands on they are in the planning, build and operation stages. In some cases, they work with the local authority, council or state government to procure a CPO on a regional or multi-regional scale, the logic being that coordinating this achieves economies of scale and with scale comes the likelihood of a better commercial deal that benefits multiple authorities. This is especially the case if these authorities lack capability and capacity to deliver a network of charge points on their own. This approach is outlined in Case Study 10.5 where Midlands Connect represents 13 local authorities across the Midlands to procure a number of CPOs or charge points. This ensures a better commercial deal and a just transition through deployments in high-demand urban and low-demand rural areas. Rarely, though, will the regional or national transport body act as the procurer, leaving this to a lead authority, council or state government. In the case of Midlands Connect, they appointed a principal local authority within their consortia to lead procurement and commission the CPOs across the region. Three consortia were formed, each with a principal lead authority. In instances where they do fulfil the commissioning role, this tends to be if local government does not have capacity or where a national transport agency needs to deploy charge points along the strategic road network, working with service station owners to procure a CPO. The Rapid Charging Fund is a good example of this. The Department for Transport and Office for Zero Emission Vehicles established this fund, administered through National Highways, to expand the charge point network along the strategic road network and future-proof electricity network capacity. Motorway service providers apply for the funding as landowners and appoint CPOs to deploy charge points [2]. In other cases, the regional or national transport body assume a more distant role, setting charge point policy direction based on national/multi-national policy and strategy guidance and determine deployment targets regionally or nationally. In this instance, the regional or national transport body plays a key role during the planning stage, specifically in terms of the CPO or local authority aligning with policy and establishing a site long list which goes towards meeting the targets. During the build and operation stages, they monitor what has been installed and align this with progress towards overall targets. In a limited number of cases, they fulfil both the procurement coordinator and charge point monitor role.

Case Study 10.5: Midlands Connect – a regional transport body deploying charge points across the Midlands

Midlands Connect is the sub-national transport body for the Midlands and transport partner of the Midlands Engine. It is a partnership that spans 22 local authorities, eight chambers of commerce, multiple business representatives and two international airports, as well as national delivery bodies: Network Rail, National Highways and HS2 Limited. One of its responsibilities is to act as a strategic body focusing on planning and coordinating transport projects across the region. In this role, Midlands Connect published 'Supercharging the Midlands', a regional EV strategy which identified that by 2030, there are expected to be 2 million EVs on the Midlands' roads. Through their strategic business case work, they identified the importance of local authorities in supporting this journey, the challenges of installing in rural areas to support the transition to EVs and the breadth of funding needed to support areas where charge point utilisation puts CPOs off deploying infrastructure. To drive forward delivery, Midlands Connect established the Midlands EV Infrastructure Consortium in 2022, comprising 13 local transport authorities across the Midlands. The focus of the consortium is to leverage the buying power of the local authorities through a collaborative procurement exercise, led by a principal authority. This is achieved through going to market with multiple deployment location across the region. The approach also supports a just transition as the packaging together of rural and urban locations across the authorities means that the deployment covers high- and low-demand areas, ensuring communities have fair access to charge points. The initial consortium of five local authorities secured £1 million in 2022 to install 300 charge points as part of the Local Electric Vehicle Infrastructure pilot. In 2023, the consortium expanded to include a further 9 local transport authorities. This larger consortium secured a further £40.3 million to install an even larger number of charge points across the region.

Sources: Midlands Connect. *The Midlands EV Infrastructure Consortium* [online]. 2022. Available from https://www.midlandsconnect.uk/projects/electric-vehicle-infrastructure/ [Accessed 31 October 2024]

The strength of a regional or national transport body or department is that it carries greater influence due to its scale. This can mean it has the ability to influence a better contractual agreement with a CPO on behalf of local authorities. It also has the ability to use its resources to support local authorities in the planning deployments and also securing larger contracts at scale. The regional or national transport body's weakness is that it can sometimes be too far removed from local deployments meaning that the best locations might not always be selected. Due to it being larger than a local authority, it can also sometimes be slow to make decisions, impacting on deployment timescales. There can also be a difference in interest or focus between the local authority and the regional body when it comes to deployment, slowing down the process.

The regional or national transport body or department is most often an enabler of charge point deployment because they are focused strategically on ensuring there is enough infrastructure in place to support increasing EV adoption rates in the future. In their enabler role, they will produce regional strategies, as has Midlands Connect, to track adoption rates and associated charge points required to meet demand. They may also act as the source of funding for local authorities or CPOs (if the regional body is appointing a CPO). They are rarely a blocker.

10.4.2 What they want to achieve

The opportunities the regional or national transport bodies or departments are seeking to realise depend on the role they are adopting. If they are acting as either the coordinator of a procurement on behalf of local authorities or as the direct procurer of a CPO on the strategic road network, they will want to achieve value for money as part of the final contract. This is achieved through economies of scale. Where they are deploying in a regional destination or residential setting, they will want to achieve a just transition. This means deployments are spread out across rural and urban areas to ensure residents have fair access to charge points. They will also want to ensure deployments support air quality improvements by encouraging the adoption of EVs over internal combustion engine vehicles and also that the charge points deployed are affordable for residents to use in terms of per kWh pricing. If they are deploying on the strategic road network, they will want to ensure that the proposed sites have good demand and projected utilisation to deliver maximum impact in terms of supporting en route charging.

Where the regional or national transport body or department is purely acting to provide strategic direction, the main opportunity they want to realise is that their strategic analysis of required charge points to support future EV numbers is used by the organisations in their region to plan deployments. This might take the form of setting targets for the local authorities in the region or involve administering funding programmes where this is awarded to local authorities to appoint CPOs to deploy charge points. They may emphasise the added benefits of deployment including air quality improvements, though their primary goal is to support the transition to EVs through new infrastructure. This may also involve identifying opportunities where grid capacity in a region can be improved to enable CPOs to deploy charge points through private funding, and also through completing regional demand analysis, provided as a report or an interactive spatial map so that CPOs can use this data for their deployment plans. Where grid capacity issues are identified, they may use central government funding to improve the grid.

10.4.3 What challenges they face

The regional or national transport body or department may struggle to coordinate multiple local authorities in the delivery of a broader charge point deployment. This can be overcome by ensuring resources are dedicated to stakeholder management and that a detailed project plan exists to coordinate activities across the various local authorities. The regional or national transport body or department may also experience differing interests to that of a local authority if both organisation do not share the same approach to deployment. This can be overcome by ensuring at the start of the programme that the aspirations and objectives of all organisations are aligned. If they are conducting a joint procurement, they may struggle to identify the correct locations for deployment without the local knowledge that local authorities have. This can be overcome by working closely with the local authorities to ensure they play a key role in the planning stage. Where they have appointed a CPO on behalf of a number of local authorities, they will face the contractual challenges of making sure the CPO delivers what is expected within the timescales and budget.

If the regional or national transport body or department is appointing a CPO to deploy charge points on the strategic road network, the challenges will often relate to technical challenges like grid capacity in some more rural areas, practical challenges regarding where to locate charge points and contractual challenges similar to those already mentioned. Regarding grid capacity, this can be overcome by first targeting locations where there is capacity. If capacity is poor, the next step is to consider solutions that bolster local capacity like solar and wind generation backed up by a battery system. If this is not sufficient, the final step is to upgrade the grid, though this is the most expensive option. Regarding locating charge points, sites should first be identified where a charging hub can be installed in existing car parks. If there is no capacity, then purchasing land should also be considered. Regarding contractual challenges in terms of managing the CPO and ensuring delivery takes place to quality, time and budget, this can be overcome by establishing a contract management function to manage the contract and prevent commercial issues from occurring.

Where the regional or national transport body or department is setting the strategic direction, the challenges tend to relate to ensuring the delivery is realised at a local level. This can be overcome as already stated by setting local charge point targets at local authority and community level. This can also be avoided by engaging closely with local authorities within the region to make sure they understand what targets should be achieved and how. Funding should also be considered to unlock charge point deployment opportunities both through channelling this through local authorities to appoint CPOs and through funding grid capacity upgrades or CPOs directly. A further challenge faced when a purely strategic role is adopted is developing a strategy which reflects local charge point demand. This can be overcome by working closely with local authorities to ensure their understanding of local area demand and potential charge point deployment locations is built into the overall strategy.

10.4.4 Their importance as a stakeholder

If the regional or national transport body or department is procuring on behalf of regional local authorities or as itself to deploy en route charge points, it should be managed closely during the planning stage. Their level of influence will decrease over time and may result in them needing to be kept satisfied as charge point deployment moves into the operations stage, either because they will receive revenue where they are procuring charge points themselves or where rental or revenue payments are channelled through them to the local authorities. They will still need to be kept satisfied throughout the deployment process if a local authority is acting as the lead procurer across a region and the procurement opportunity has been shaped by the regional or national transport body or department. This is because their influence will remain high due to their role in working with local authorities across the region but their impact will be lower because they are not the lead party procuring the CPO/s.

If the regional or national transport body or department is acting purely to set strategic direction, their influence and impact are relatively low meaning that they should be kept informed throughout the deployment process. This might mean they are informed of which locations charge points will be deployed in as well as the impact of these charge points once installed in terms of utilisation levels.

10.4.5 Changes throughout the planning, build and operations stages

Where the regional or national transport body or department is procuring on behalf of regional local authorities or as itself along the strategic road network, they will be very active in the planning stage in terms of determining the vision driving the deployment, determining the area to deploy, shaping the business case and any associated procurement documentation and appointing the CPO. At this point, their involvement will step back as planning and traffic regulations are approved at local authority level, though they will be kept closely informed of this process. They will then become more involved during the governance and approvals process as part of the planning stage. During the build stage, they step back and monitor the deployment as most of the work at this point is done by the CPO. During the operations stage, they monitor charge point utilisation and income, either receiving this revenue directly where they are the sole procurer of the CPO or channelling this back through to the local authorities. Where the regional or national transport body or department is procuring on behalf of regional local authorities and a lead authority is appointed to procure the CPO, they will be active in the planning stage. At this point, they will determine the vision behind the deployment, determine the area to deploy, shape the business case and any associated procurement documentation. Their involvement will greatly scale back during the build and operations stages as the lead local authority will carry on all activities once procurement is complete.

Where the regional or national transport body or department is acting in a strategic role, they will shape the vision driving the deployment but then play a lesser role in determining the area to deploy and shaping the business case. After that point, they will enter into a monitoring role, tracking how deployments are progressing and the impacts charge points are having in enabling EV adoption.

10.5 The distribution system operator/distribution network operator

The DNO or distribution system operator (DSO) is a key stakeholder in every charge point deployment. They are responsible for maintaining the functioning of the distribution network and acting as the gate keeper in terms of approving or rejecting connections and upgrades to the grid if required. In the United Kingdom, each DNO functions slightly differently, with most providing an online tool that enables organisations, including CPOs, to understand power capacity in their area to enable planning.

10.5.1 Their role

The DSO/DNO's role is to provide the CPO or local authority with cost estimates relating to connections and network upgrades during the planning stage, and then to complete those connections or upgrades during the build stage. They are usually engaged at a high level during the planning stage when determining the geographic area to deploy as they can give an indication of the power capacity and headroom in an area which will help determine the most suitable locations for installation. In some instances, this information can be provided by a spatial power planning platform, usually in the form of an online map-based tool showing high, medium and low current demand. Case Study 10.6 demonstrates how Northern Powergrid, a DNO based in the United Kingdom, uses spatial mapping tools to help its customers understand where there is capacity on the network. Elements of the tool are targeted at supporting charge point uptake. DNO/DSOs are also further engaged through providing upgrade cost estimates during development of the business case so as to develop the detailed costings underpinning the deployment. The DSO/DNO will then be engaged during the planning deployment at a local level step to provide finalised costs for the shortlisted sites. As part of the build stage and during the deployment of charge points step, the DNO completes the power connections and network upgrades for each installation. This can take a number of months, which means that the CPO should build their deployment schedule around the DSO/DNO's connection schedule. They may also be involved during the operations stage if a charge point or site is upgraded and additional power is required. Like during the planning stage, costings are provided to base the investment decision on.

Case Study 10.6: Northern Powergrid – the importance of spatial mapping

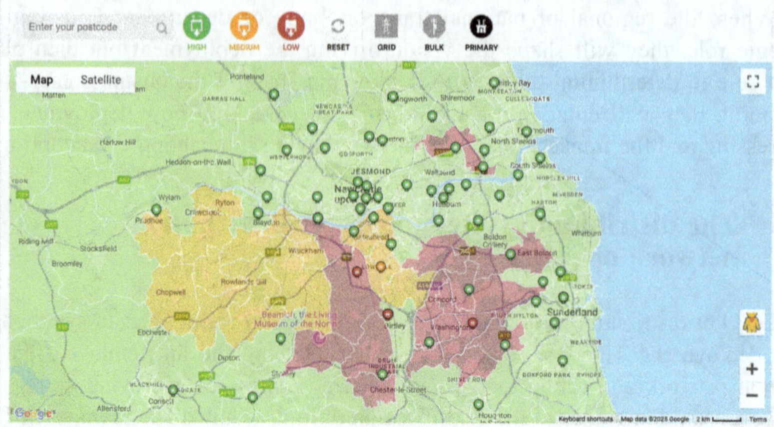

Northern Powergrid is a DNO based in the United Kingdom that owns and operates the electricity transmission and distribution network in the North East, Yorkshire and northern Lincolnshire. In their role supporting customers to help plan their projects, they provide a number of tools. This includes a demand heat map which shows high, medium and low demand at primary substations, bulk and grid level. For charge point deployment, most CPOs are interested in capacity at the nearest substation as this can influence the viability of a charge point site. Northern Powergrid also provides more EV-focused datasets as part of their distribution future energy scenarios analysis. Northern Powergrid anticipates significant growth of EVs across the region. This is determined using their reference scenario that is based on net zero compliance by the 2050 target, significant and ongoing government support, substantial investment in net-zero technologies, high consumer engagement, accelerated uptake of EVs and early adoption of heat pumps. The data can be cut in multiple ways, though one scenario is particularly useful for energy planning; that the number of EVs per substation is expected to increase every 5 years up until 2050. Using this data subset, projected EV hot spots in the region include Redcar, Norton (in the Tees Valley), Harrogate and Darlingon West. This data enables CPOs to target areas for charge points where EV growth is anticipated to be the greatest. It also enables DNOs to prepare the network ahead of demand. Northern Powergrid are planning to develop these tools further to support future network planning.

Source: Supported by Northern Powergrid Open Data. Northern Powergrid. *Demand Heat Mapping* [online]. 2025. Available from https://www.northernpowergrid.com/demand-availability-map [Accessed 25 April 2025]

Source: Supported by Northern Powergrid Open Data. Northern Powergrid. *The Distribution Future Energy Scenarios (DFES)* [online]. 2025. Available from https://northernpowergrid.opendatasoft.com/pages/dfes/ [Accessed 25 April 2025]

The strengths of a DSO/DNO are that they understand the local power network in detail and are able to draw on their network of engineers or contractors to complete upgrades for CPOs/local authorities. Their weakness is that they tend, in the United Kingdom at least, to be under resourced meaning that network upgrades often fall to the back of a backlog and can take several months to be prioritised before work can start. The DSO/DNO can reject a power upgrade or network connection and can de-prioritise a request for a connection if a more pressing request is put in place. A further weakness is inconsistency. From experience not all DSO/DNOs are as effective in planning and delivering against its backlog, meaning a CPO may have an entirely different experience in one DNO's region versus another.

The DSO/DNO is both a blocker and an enabler. Regarding the former, they can reject power upgrade applications due to lack of capacity and resources. They can also reject applications if another customer has submitted a connection request or upgrade before them. Regarding the latter, they have the ability to fast track an application in some instances and, depending on the funding available, can work with the CPO to upgrade the grid. For example, the Rapid Charging Fund noted in Section 10.4.1 included funding for the DNO to upgrade the connection to the service station to enable installation of level 3 DC charge points by the CPO.

10.5.2 What they want to achieve

The opportunities the DSO/DNO is seeking to realise all relate to supporting growth of the network through new connections and power upgrades whilst balancing this against keeping actual demand within capacity. A further opportunity includes generating revenue from charging CPOs and local authorities to connect to/upgrade the network.

10.5.3 What challenges they face

The key challenge they face also links to the opportunity of supporting growth of the network through new connections and power upgrades whilst balancing these against keeping demand within capacity. The DNO needs to ensure it understands the capacity of an area in sufficient detail to effectively support expansion within the headroom available to it. Otherwise, there is a risk that there might be a power outage due to poor planning. This risk can be overcome by ensuring there is always a small buffer on top of total demand in an area so that even if this is reached and marginally exceeded, it sits within the acceptable power range. A further challenge is prioritising different connection and upgrade demands as each customer will believe their project should be delivered as a priority. This risk can be overcome through working with customers to understand how ‘shovel ready’ their project is and also querying the size of the connection/upgrade they are seeking. The DNO may also face the challenge of resourcing the various required connections/upgrades, especially as the demand on the network is currently high due to renewable projects and charge point installations ramping up. The DNO can overcome this by building its team to support these projects and also seeking funding from central government if they are struggling to resource growth. Central government may well grant this request as the role of the DSO/DNO is recognised as critical in decarbonising transport and energy. A following challenge that might result if they are unable to support these requests is work increasingly going to independent DNOs. This, like the challenge of resourcing connection/upgrade requests, can only be overcome by identifying funds to grow the teams responsible for processing the backlog of requests and delivering the connections/upgrades.

10.5.4 Their importance as a stakeholder

During the planning stage, the DSO/DNO has medium to high influence and high impact because provision of information regarding power capacity and costings of connection/grid upgrades greatly impacts the viability of potential sites. This means they should either be managed closely when planning for sites where a connection/upgrade is critical or consulted with if connection/grid upgrades are not deemed as important in an area. This might occur, for example, if it is known there is capacity in the local area. During the build stage, their impact still remains high because a timely connection or upgrade can make or break a charge point deployment though their influence reduces because at this point the sites have been selected and the connections or upgrades are scheduled to take place. As a result, they should be consulted with throughout the build stage. During the operations stage, their influence and impact falls. They should be kept informed infrequently or may even drop off from the stakeholder matrix. They will only likely be re-engaged when there are issues with the power connection.

10.5.5 Changes throughout the planning, build and operations stages

The DSO/DNO is not involved at the same level throughout the planning, design and build stages. During the planning stage, they are involved at a high level when determining the geographic area to deploy and in shaping the business case as the costs they provide feed into the economic and financial cases. They are then more closely involved during the planning deployment at a local level step to provide finalised costs for the shortlisted sites and during the governance and approvals step when they receive the applications for connections/upgrades. Presuming these are approved, their involvement increases during the build stage as the DNO completes the power connections and network upgrades for each charge point installation. During the operations, maintenance and monitoring stage, the DNO's involvement is scaled right back. They are only engaged when there is an issue with the power network that requires repair. They may, however, become more involved if during this stage a connection upgrade is required, for example, when replacing level 2 with level 3 DC charge points. In this instance, the CPO will engage with the DNO as they did during the planning stage to get updated costs. During the build stage, the CPO will then complete an application for the upgrade and engage with the DNO once it is approved.

10.6 The national government including multi-national organisations

The national government/multi-national organisation (e.g., the European Union) is hands off throughout the charge point deployment process, setting the direction of travel in terms of the country achieving net zero and often also providing the funding underpinning deployment.

10.6.1 Their role

The core role of national government and multi-national governing organisations is to set policy and strategy direction. Case Study 10.7 demonstrates how the British Government is fulfilling this role. The national government or multi-national governing organisation also has the power to incentivise EV growth (e.g., with tax breaks), which influences demand for charge points. CPOs, local government and regional transport bodies will refer to the policy guidance set by them during the planning stage. Furthermore, national government will set building and safety regulations and standards guiding the CPOs during the build stage. They also set consumer regulations and standards that determine how the charge points are used during the operation stage. They might play a more influential role beyond setting out policy and strategy where charge points are being deployed along a strategic road network; however, they will always defer decision-making to the transport department or transport authority who engage directly with the CPO through all three stages. In some instances, the national government/multi-national organisation provides funding to part cover the cost of charge point deployments and will offer access to this through a competitive application process.

Case Study 10.7: the British Government's 2024 Autumn Budget and Taking Charge EV Strategy – framing the policy and strategy environment

The important role a national government can play in setting the direction of travel which drives charge point deployment is demonstrated through the British Government's Autumn Budget, the first for the new Labour Government following its victory in the General Election in July 2024. The budget contained several statements that relate directly to the Government's support for EVs and charge points. The Government brought forward the phase-out date of the sale of ICE vehicles from 2035 to 2030 to support transport decarbonisation, meaning that demand for charge points will continue to grow. The Government also introduced incentives to encourage EV adoption including Vehicle Excise Duty remaining low at £10 for new zero-emission vehicles and continued company car incentives in the form of benefit-in-kind tax for company cars remaining low at 2%, increasing to 9% by 2029–2030. This is substantially lower than the tax for ICE vehicles. The budget also included a commitment to invest over £200 million in 2025–2026 to increase the United Kingdom's charge point network. The funding will help expand charge point infrastructure, particularly in locations where charging access remains limited. The budget supports the Government's ongoing policies and strategies to promote EVs and charge point deployment in the United Kingdom. Although published under the previous Conservative Government, the Taking Charge: The Electric Vehicle Infrastructure Strategy is still current and aims to accelerate the rollout of Level 3 DC rapid charge points, support the charge point sector to thrive, address barriers to private sector rollout and give people confidence in the public network.

Source: Keep M., Hutton G., Harari D., Booth L., Brien P., *Autumn Budget 2024: A summary* [online] https://commonslibrary.parliament.uk/research-briefings/cbp-10124/#:~:text=Investment%20spending%3A%202024%2F25%20and,day%2Dto%2Dday%20spending, Commons Library Research Briefing Paper, 31 October 2024 [Accessed 05 November 2024] This information has been licensed under the Open Government Licence. To view this licence, visit https://www.nationalarchives.gov.uk/doc/open-government-licence/

Source: HM Government. *Taking charge: the electric vehicle infrastructure strategy* [online]. 2021. Available from https://assets.publishing.service.gov.uk/media/6245ba40e90e075f15381cf0/taking-charge-the-electric-vehicle-infrastructure-strategy.pdf [Accessed 05 November 2024] This information has been licensed under the Open Government Licence. To view this licence, visit https://www.nationalarchives.gov.uk/doc/open-government-licence/

The strengths of a national government or multi-national governing organisation are that they have a macro view of adoption rates and charge point demand, meaning that they are able to set national and regional charge point targets that drive charge point deployment at local authority/regional level. In some instances, they may provide this information in a raw format or as a national spatial planning tool which users analyse to understand demand in their specific area. The national government or multi-national governing organisation also has the ability to release funding to target specific deployments like residential charge points, as is demonstrated in Chapter 4, or focus funding on where there is a current gap in the market. A further strength is that the policy and strategy direction can have a real impact in helping make the strategic case for charge point deployment, especially where the government's objectives align with the objectives of the CPO/local authority. This can open up new funding streams and also unlock investment, especially where the alignment demonstrates projected future growth in EV drivers will increase demand, utilisation and projected profitability.

The weaknesses of a national government or multi-national governing organisation are that they do not understand localised demand and that they are politically driven. Regarding the former, the data and information they provide can only give a general indication of local demand, meaning that further analysis is required to ensure that proposed deployments are based on more accurate predictions of demand and utilisation. Regarding the latter, a change in government can have a major impact on charge point deployment, resulting in funds and support for charge points being pulled. The government also has the ability to remove the incentives driving EV growth which can result in a collapse in demand, thereby reducing the viability of charge point deployment. Based on the points above, the national government or multi-national organisation can both be an enabler and blocker. They can enable charge points by encouraging EV growth through putting in place supportive charge point policies, strategies and funding. They can also block charge point growth through removing supportive policies, strategies, and funding.

10.6.2 What they want to achieve

Provided the national government or multi-national governing organisation is supportive of EVs, the opportunities they are seeking to realise include increased EV adoption, reduced passenger vehicle and van emissions that contribute to the country's net zero targets, air quality improvements in urban areas and a just transition in terms of all citizens having fair and equal access to charge points. Regarding EV adoption, this is achieved through EV incentives and charge point deployments. A replacement of ICE vehicles on roads results in a reduction in tailpipe emissions and improvements in air quality. A just transition is achieved through high- and low-demand deployments, ensuring that there is an adequate distribution of charge points across communities and that residents are within walking distance of a charge point.

10.6.3 What challenges they face

National governments or multi-national governing organisations face a number of challenges including how to drive up EV adoption, target funding effectively and decarbonise the energy which powers charge points. Regarding EV adoption, this is closely tied to charge point deployment; however, it also directly relates to consumer behaviours in terms of buying or leasing EVs. If there is a reluctance due to various perceptions (e.g., insufficient range or charging being too expensive), the deployment of charge points will have little effect in driving EV adoption upwards. This can be overcome through a behaviour change campaign combined with further incentives in the forms of subsidising EV costs to the consumer. In relation to targeting funding, there is the challenge that funding is not sufficient to drive up deployment in low-demand areas as the deployed charge points are still perceived to be unprofitable. Regarding driving up renewables, this is a challenge because EVs cannot be called truly decarbonised transport solutions unless they are powered by 100% renewables meaning governments need to decarbonise their energy systems. Although progress is being made globally, storing renewable energy continues to be a challenge unless a country is powered primarily by hydroelectric energy or nuclear. This challenge can be overcome by adopting different forms of energy storage, including batteries, fly wheels, green hydrogen and heat storage systems.

10.6.4 Their importance as a stakeholder

The importance of the national government/multi-national organisation depends on two factors. The first is whether they are funding deployments and the second is what policy direction they are setting in a country. Where funding is available, this is sought during the planning stage and is usually secured through an application process requiring a business case. In this instance, the national government/multi-national organisation should be kept satisfied as the impact of their funding is high but their level of influence is medium. This is maintained during the build stage, especially where the funding has conditions attached. Their influence and interest drop off to low during the operations stage meaning they should be kept informed. This is primarily because the deployments need to feed in to charge point national metrics used to determine progress and demonstrate funding value for money.

If the national government/multi-national organisation is providing policy and direction alone, their influence and interest are low throughout the duration of the deployment meaning they should be kept informed, again to feed in to national charge point metrics.

10.6.5 Changes throughout the planning, build and operations stages

Changes in the role of the national government/multi-national organisation depend on whether they are setting policy and strategy direction and/or providing funding. Regarding the former, their role is constant throughout the process. The CPO should refer to the policy, strategy, regulatory and guidance documentation available to ensure what is installed is compliant and supports the government's transport decarbonisation objectives. If the national government/multi-national organisation is also providing funding, they are more involved during the planning stage in terms of setting out the funding requirements and application process which will usually include a certain degree of demand analysis, site selection and cost calculations. This is usually completed during the business case step and then submitted using data from that step and previous steps. Award of funding then comes during the governance and approvals step. Their involvement is slightly reduced during the build stage as the CPO/local authority draws down on funds to complete the installations. This may involve documentation being provided to the funding body to prove it is being spent as expected. During the operations stage, the national government/multi-national organisation's involvement scales right back. They may, however, seek information demonstrating that the charge points are active and may require these to connect to a national charge point register and/or system able to identify their location and track their operational status.

10.7 Consultancies

Consultancies are not always involved in charge point deployments but can play a role in providing capability and capacity usually to local authorities but also in some cases to CPOs where specialist skills or expertise is required as part of delivery and to national governments/multi-national organisations in shaping policies and strategies.

10.7.1 Their role

Consultancies by their very nature can be involved throughout all three stages, but only ever as advisors and only through one of the other key stakeholders such as the local authority or CPO. During the planning stage, consultancies can advise both local authorities and CPOs during the longlist exercise, utilising their modelling and analytical experience to determine demand and propose suitable locations with anticipated good utilisation. Where a local authority is procuring a CPO, they may appoint a consultancy to help develop the procurement documentation and support in the tender process. Advising on both the site selection and procurement process is demonstrated in Case Study 10.8, where Amey Consulting provided technical input as part of Telford and Wrekin's charge point strategy. A CPO may appoint a consultant to develop the business case. They also can provide advice to the CPO or local authority on the design aspects of planning including detailed site layouts, electrical schematics and civils planning. During the build stage, consultancies can project manage the installation on behalf of the CPO, commissioning contractors to complete the work if the CPO does not have that capability. During the operation stage, consultancies might advise on evaluating the utilisation of charge points and proposing options to increase this. They might advise on upgrade or expansion plans. They may also in certain cases advise the CPO on securing investment to expand or acquire another CPO. At a macro level, consultancies may also advise national and regional governments on policy and strategy development.

Case Study 10.8: Amey Consulting and Telford and Wrekin Council working closely together on a data-driven approach to EV charge point deployment

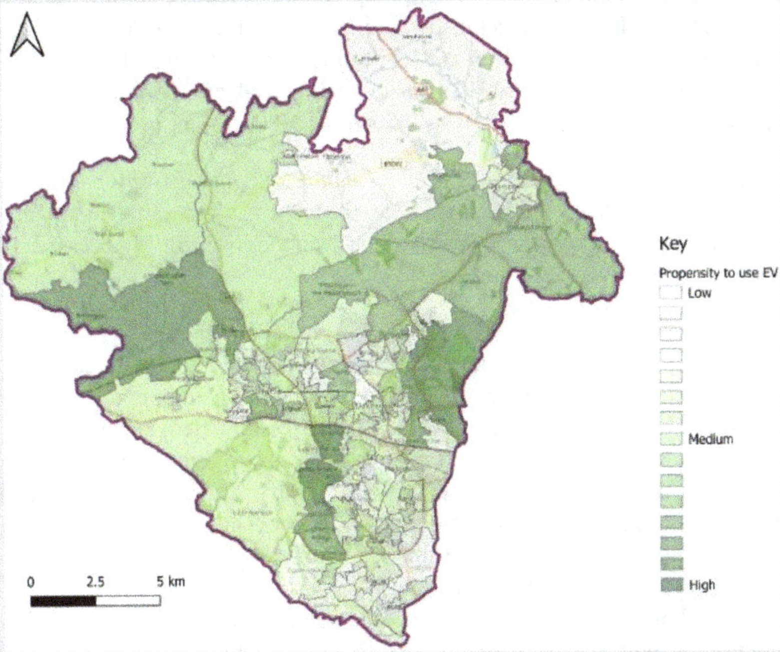

In 2020, Amey Consulting was commissioned by Telford and Wrekin Council to support the development and implementation of its Ultra-low Emission Vehicle Strategy. Amey's involvement focused on three key work streams. The first was the development of a public EV charging strategy and toolkit. The second was developing the Council fleet EV charging strategy and toolkit. The third was identifying funding and procurement options. In developing the EV charging strategy for the Council, Amey used a data-driven approach to identify optimal locations for publicly accessible EV charge points. This involved analysing various factors, including existing infrastructure, economics and human behaviour patterns, and an assessment of charge point technologies. Further analysis was also conducted to explore different commercial and procurement models. This included a policy and funding review to fully understand the policies and strategies governing charge point deployment, a technology and market review to examine the suitability of existing and emerging technologies for different charging use cases, an identification of the most suitable commercial models, including ownership models for charge point infrastructure and revenue generation opportunities from charging, parking and ancillary services. The final part of the work included a data-driven demand analysis to identify viable locations for charge point installation. Amey Consulting also supported the Council throughout the funding, procurement and planning process, helping them to secure funding for charge point deployment. This included supporting the Council in its successful bid for £683,370 of funding from the Office for Zero Emission Vehicles, delivery of a charge point trial and completion of an engagement exercise with CPOs to inform the procurement process.

Source: Amey Ltd. *Telford and Wrekin Council – a data driven approach to EV charging* [online]. 2024. Available from https://www.amey.co.uk/projects/2023/september/a-data-driven-approach-to-ev-charging/ [Accessed 06 November 2024]

The strengths of a consultancy are that they are able to draw on capacity and a range of skills and experience to provide advice and support to local authorities/CPOs. They are also often specialists in a certain field meaning they are able to provide current advice as well as access to emerging innovations. Consultancies are able to work at pace and support in developing high-quality funding applications. A further strength when they are working with a CPO during the build stage is that they are able to draw on a broader network of potential contractors to help in installations.

The weaknesses of a consultancy are that they often lack detailed local knowledge to help when determining localised sites. This means a local authority or CPO should always supplement the work done by a consultancy with their own site surveys. A further weakness is that because the consultancy is often scope based in terms of their work, they will very rarely carry out activities beneficial to a deployment that fall outside the overall scope and if they do, this will be costed as an additional deliverable. A consultancy's advice is also only as good as its consultants. This means that a CPO/local authority might appoint a consultancy based on its consultants during a procurement exercise; however, throughout the duration of delivery, these individuals may not be available or may have left, meaning the value of the consultancy to the client decreases.

A consultancy is always an enabler to charge point deployment as they see this as a growth area where they can make a profit from the sale of advisory services.

10.7.2 What they want to achieve

In the vast majority of cases, the key opportunity the consultancy wants to realise is generating a profit from the sale of consultancy/advisory services to CPOs/local authorities. In a smaller number of cases, the consultancy may carry out loss leading work or a free proof of concept project so that it develops internal capabilities which will position it well to win future work, especially where there is an existing client. Smaller more niche consultancies may seek to realise environmental, social and governance opportunities through providing consultancy services for free, specifically if they can demonstrate a deployment supports a just transition or benefits those who have been left behind by charge point deployments; however, this does not occur very often.

10.7.3 What challenges they face

The main challenges a consultancy faces in relation to supporting a local authority or CPO throughout the charge point deployment process relate to projects taking longer than planned and also gathering local knowledge to support in deployments. Regarding the former, this is because the original commission agreed with the client will have been costed to cover a certain number of months, including an overhead. If the project runs for longer than planned, the original budget becomes stretched. This can be overcome by working with the client to mothball the consultancy's involvement if there are significant time delays, with payment for activity delivered to that point. Regarding the latter, this is because the consultancy will lack a good understanding of the local area, including known issues that might affect the viability of an installation. This can be mitigated by either including costs in the commission to cover site selection or by the local authority/CPO conducting these and supplementing this with local knowledge which can then be provided to the consultancy. A further challenge for a consultancy is winning work in this space. As charge point deployments become more frequent and more consultancies offer advisory services in this area, this will mean the market becomes more competitive. This will have an impact on the win rate and likely result in each consultancy winning less work as it is spread across a broader consultancy base. This can be overcome by consultancies focusing on their unique selling point/s and adapting their offering as competition increases to secure more work.

10.7.4 Their importance as a stakeholder

Regardless of whether a consultancy is appointed by a CPO, local authority or national/regional government, they will always have high influence and impact relating to the specific work they have been commissioned to do. As such, they will always need to be managed closely by a CPO, local authority or national/regional government. For all other steps throughout the process which they are not involved in, they have no influence or impact meaning they drop off the stakeholder matrix. The only caveat is where a piece of work they have delivered impacts a later stage in the process. This might be the case, for example, when longlisting work done by them is used by the CPO/local authority for shortlisting. If this is the case and more detailed input is required beyond high-level conversations, the consultancy might be re-engaged and commissioned to carry out further work.

10.7.5 Changes throughout the planning, build and operations stages

As stated above, the involvement of a consultancy throughout the deployment process is entirely dependent on what they are commissioned to do. As a result, the change in level of involvement will be determined by their scope and how this aligns with the planning, build and operation stages, and when the work they have been commissioned to do is completed. In this sense, either they are a stakeholder who is managed closely for the work they are completing or they are not a stakeholder at all.

10.8 Resident and citizen groups

Resident and citizen groups should always be involved in the charge point deployment process in some shape or form to help determine the best location for charge points and also to gain their support if installation causes any disruption during the process.

10.8.1 Their role

Resident or citizen groups can play an influential role in the planning, build and operation stages, though this only applies to destination and residential use cases. During the planning stage, they have the ability to feed in to the governance approval process for deployment, either supporting or rejecting proposed sites. Local authorities may choose as part of the planning stage to consult with the local community, presenting them with the opportunity to make their opinions known. Community groups and residents often have strong views when on-street charge points are being installed, objecting in some cases where there is limited parking and the provision of dedicated EV bays will restrict this further. If the community is not at least consulted in some way, this might result in poor utilisation when the proposed sites become operational, especially in destination and residential charging use cases. Resident or citizen groups must be kept abreast of developments during the build stage so as to keep them on side. They have the ability to be disruptive if there is no support for charge point sites. During the operation stage, resident or citizen groups should be engaged via a communications campaign to demonstrate, specifically in the case of destination and residential charge points, the benefits of charging locally. This can ensure there is good demand and utilisation of the charge points installed. Case Study 10.9 demonstrates how a charge point deployment can successfully be delivered with community buy-in. In this example, Suffolk County Council built charge point deployment around the community, working with various community groups throughout the planning, build and operation stages to ensure deployments achieved community goals and had community support.

Case Study 10.9: Suffolk County Council's community led charge point deployment programme

As part of Suffolk County Council's Plug In Suffolk programme to make Suffolk zero emissions vehicle ready, key activities included the following:

- Delivering a county wide network that is community owned, fair, equitable, open access and easy to use.
- Providing the facilities and confidence to the community and visitors to encourage the transition to EVs
- Delivering community assets that embody the Greenest Suffolk ethos, catalysing change and developing community cohesion.

Shaping deployments around the community is a key focus of the programme. This has been achieved by engaging with 50 community groups to determine the best locations to deploy charge points. At the start of the project, the Council recognised that misconceptions of EV and charge points were driving reluctance around installations. This was overcome by the Council adopting a key role in educating and empowering communities as opposed to imposing solutions without community buy-in. The community centric approach was delivered through deploying charge points at community hubs, enabling residents and communities to transition to EVs in a visibly community focused way. Pricing is also tiered meaning that residents are offered preferential rates to encourage EV adoption and to discourage charge points being used purely by those from outside the local community.

Once installed, ownership of the asset is transferred to the local parish or community group who then operates the charge point using a management system provided by the Council. All revenues generated are kept in the community. Localised ownership means that the local community is invested in the asset and is incentivised to ensure its success.

Source: Cenex. *An electrification journey – Case studies for change – installing electric vehicle charge points in the UK* [online]. 2023. Available from https://www.cenex.co.uk/app/uploads/2023/04/CaseStudiesForChange.pdf [Accessed 06 November 2024]

The strengths of residents and citizen groups are they have a good understanding of the local community and may be able to provide important information around the viability of sites from a practical-, technical- and demand-driven perspective. They also have the ability to shape local EV adoption, thereby impacting utilisation. A further strength is their ability where they own car parks as a community group to host charge points, acting as the landowner. The weakness of resident and citizen groups is that they can sometimes be ill-informed regarding charge points and EVs as a result of misconceptions or misunderstandings. A further weakness is that there is no such thing as one single voice from the community. This means that engagements must target multiple individuals and groups to ensure the perspectives of a community are fairly represented.

Resident and citizen groups can be either enablers or blockers. If they are supportive of deployments, they can help identify sites and shape installations as well as drive up utilisation once charge points are in place. They can also block deployments and lobby their local councillor to reject a planning application, especially where deployments reduce the number of general car parking spaces.

10.8.2 What they want to achieve

Realising opportunities regarding charge point deployments all relate to when resident and community groups are enablers. The opportunities they want to realise all relate to ensuring installed charge points are accessible, easy to use and offer value for money. Achieving this makes it more convenient, cost-effective and, ultimately, attractive for communities, especially those lacking off-street parking, to adopt EVs. In a limited number of cases where communities run the charge points, as is the case in Case Study 10.9, they may also want to generate income to support the delivery of other activities.

10.8.3 What challenges they face

Resident and community group challenges primarily relate to access to and the affordability of charge points. Regarding the former, if charge points are not distributed across communities in residential areas to ensure reasonable access, communities may struggle to use them and will see this as a barrier to EV adoption, especially where there is a lack of destination charging and off-street charge points at homes. This can be overcome by ensuring that the planning stage for residential deployments explores community-wide charge point distribution, taking into consideration walking distances to nearby charge points. Regarding the latter, per kWh pricing should be comparable with domestic charging costs at home. This is important as residents are well aware, in the United Kingdom at least, of how much cheaper it is to charge domestically via a smart tariff, with some offering overnight per kWh prices as little as 6 pence. Failure to try to offer affordable community-focused tariffs for residential charge points will put residents off adopting EVs, especially if they think they are being ripped off by a CPO. This can be overcome by the CPO or local authority offering smarter resident focused tariffs which can only be accessed by local communities to improve the cost effectiveness of local charging.

10.8.4 Their importance as a stakeholder

The importance of resident and citizen groups in residential charge point deployments changes slightly throughout the deployment process as their influence fluctuates between medium and high, and their impact between high, medium and low. During the planning stage, they should be consulted with when determining the geographic area to deploy, planning deployment at a local level and during the governance and approvals step. This is because communities must be closely engaged where charge points are deployed as they will have views on suitable locations. Failure to do so can have real impact as charge points might be installed in the wrong locations, annoying communities and resulting in poor utilisation and profitability. During the build stage, their impact and influence falls to low as they cannot change the planned locations but can still influence the process, especially where support begins to fall. As a result, they should be kept informed during the build stage. This is also the case during the operations stage for similar reasons. If, for example, pricing changes and communities feel this is too expensive, they will make their voices heard, usually via their local political representatives. This is why during the operations stage communications plans should be put in place and any significant changes to how charge points operate, including pricing, should be communicated meaningfully, with communities being able to respond.

Where charge points are installed to satisfy a destination use case, the importance of residents and community groups is less than in a residential deployment; however, they should still be communicated with to avoid any risk of community resistance putting in place barriers. It is rarely the case for en route charge points that communities and citizen groups are engaged, unless there is a particularly active group. If that is the case, then they should be managed in the same way as for destination charging by being kept informed.

10.8.5 Changes throughout the planning, build and operations stages

As stated above, the level of involvement primarily relates to the charging use case. For residential deployments, resident and community groups are important during the planning stage in helping determine suitable locations. This is usually achieved through running a consultation or via an online portal where residents can request charge point locations. This is then fed into the site selection process. Communities can also be consulted on the delivery model and pricing tariff to give an indication of the likelihood that they will use the charge points. Consultation conducted earlier on in the planning stage may form a requirement to secure governance approvals, especially where a local authority is leading on the deployment. During the build stage, information should be provided to residents and communities regarding when charge points will likely go live, how to access them and anticipated pricing. During the operations stage, communities should be communicated with either via a semi-regular newsletter or via an app. Information should cover live units and down times, faulty units and known issues, changes to pricing or charge point access and planned upgrades and changes.

For destination charge points, resident and citizen group involvement is fairly constant and will take the form of ongoing communications keeping communities abreast of developments. Residents and citizen groups are rarely involved in en route deployments and if so, the approach is similar to that for destination charge points.

References

[1] Pod Point. *Price update at Tesco stores* [online]. 2022. Available from https://pod-point.com/electric-car-news/price-update-tesco-faqs-22 [Accessed 27 October 2024].

[2] National Highways. *Supporting the transition to zero emission vehicles on the strategic road network* [online]. 2024. Available from https://nationalhighways.co.uk/our-work/environment/supporting-the-transition-to-zero-emission-vehicles-on-the-strategic-road-network/ [Accessed 1 November 2024].

Chapter 11

Looking to the future: where is EV infrastructure heading

David Trousdale[1]

In previous chapters, we have focused on the here and now of charge point infrastructure, covering how in Chapter 2 the technology has developed to get to where we are today. In Chapter 3, we covered the overall process to deploy charge points and the key stakeholders, and in Chapter 4 the policy and strategy environment. In subsequent chapters, we then conducted a deep dive into the overall deployment process. This chapter switches the focus to the future of electric vehicles (EVs) and charge points, outlining the anticipated trends in the vehicles themselves and how this will impact on charge points. In addition, it gives an overview of how development in EV cars, vans and charge points is impacting on other modes including the electrification of heavy goods vehicles and buses. Although the future is unknown, the analysis conducted by various specialists in this field gives an indication of the broad direction of travel. We will draw on this analysis to bring to life these trends, helping the industry plan for the future.

11.1 Future trends in EVs which will impact on charge point infrastructure

EV adoption is anticipated to increase globally over the next decade. Major changes in battery chemistry, size and capacity are expected to occur, resulting in longer range vehicles. Changes in vehicle technologies are also expected including wireless charging, battery swapping in some geographies and even autonomous charging. Artificial intelligence (AI) will no doubt also have a major impact on how EVs function. All of these changes will heavily influence charge points.

11.1.1 Projected EV growth to 2035

When analysing projected EV growth, different countries and organisations record this in different ways. Some measure this as a percentage of the total fleet, whereas others measure this as the number of EVs in a given country. The latter can be

[1]Principal Advisor - Climate Change and Sustainability, Newcastle City Council, United Kingdom

recorded either as annual sales or by the total number of EVs on a country's roads. For the purpose of this analysis, growth is shown as the total number of EVs on a country's roads as this is more meaningful from a charge point perspective as it gives a high-level view of potential demand on current and future charge points. Using these figures, percentage growth is calculated using a baseline date of 2022, 2023 or 2024, depending on the most recent date where data is available, compared with projected numbers by 2035. Where possible we have used figures that include EV car and vans combined; however, not all geographies are able to display the data in this way.

According to the International Energy Agency, global EV car growth is projected to increase from just under 40 million in 2024 to between 390 and 440 million vehicles by 2035 [1], though not all geographies will see the same growth rate. This is growth of between 875% and 1,000%. Projected EV car and van growth in the United Kingdom varies depending on different models. The Energy Saving Trust, drawing on projections from the Committee on Climate Change's balanced pathway and Transport and Environment's charging forward report, model adoption to 2040. Projected figures for 2035 are however used as most geographies focus on this as their target date for transport decarbonisation. The Committee on Climate Change anticipate that EV car and van numbers will increase from 6 million in 2025 to 31 million vehicles in 2035. This is compared with Transport and Environment's projection which starts from a base of 2.5 million in 2025, increasing to 21 million by 2035 [2]. Taking the averages of these, we should anticipate around 26 million EVs by 2035. This is growth of between 416% and 740%, depending on the base data. The European Union's projections are that by 2035, there will be between 110 and 122 million EV cars on the roads of the 27 Member States. This is starting from a base of 9 million EV cars and hybrids in 2023 and excludes vans [3]. This is growth of between 1,122% and 1,255%. France projects that by 2035 there will be between 7 and 16 million EVs, though this does not specify the car/van split. This is growth from a starting position of 1 million EVs in 2022 [4] and amounts to between 600% and 1,500% growth. The United States projects that from its starting position of 4.5 million EV cars and light goods vehicles on United States roads at the end of 2023, this will reach 78.5 million by 2035 [5], amounting to growth of 1,644%. Canadian growth is anticipated to reach 12.4 million light-duty EVs (including cars and vans under 4,536 kg) by 2035. This is from a starting position of 248,581 in 2021 [6] and a growth of 4,888%, the largest jump in numbers when compared with the other geographies analysed above. At the time of writing, data was not available to calculate the EV car or van numbers by 2035 for Germany, Australia and New Zealand.

Although the anticipated growth figures may seem unrealistic in some geographies, what should be clear from the analysis above is that EV growth is set to continue globally and at pace over the next decade, even if the pace might be slower than some of these projections.

11.1.2 Projected developments in EV batteries

As EV adoption rates ramp up globally, so will the demand for batteries. Demand for EV batteries reached more than 750 GWh in 2023, up 40% when compared with

2022. Scale up of the mining of minerals and a standardisation in manufacturing processes resulted in a fall in battery pack prices in 2023, with lithium ferrophosphate (LFP) batteries seeing large price decreases. In addition to price, other factors are set to transform the battery market in the future. These are innovation-driven improvements in batteries and affordable chemistries. Regarding innovation-driven improvements, this relates to further development of cell-to-pack, where the design connects individual battery cells directly into a battery pack, without using modules or additional components, and advancement in cell-to-chassis, where the battery cells are built directly into the vehicle's chassis. Further innovations include improvements in battery performance, for example, through multi-layer electrodes enabling ultra-fast charging [7].

Regarding affordable chemistries, the development of sodium-ion batteries is likely to take off, though manufacturing capacity outside China is still currently at laboratory or pilot scale. In 2023, leading battery manufacturers such as BYD and CATL announced expansion plans for sodium-ion batteries. If scaled up, these batteries could cost up to 20% less than an LFP equivalent, though it is also possible that they might not improve performance or range, and increase charging times. They may not be suitable for all use cases because of this [8]. The emergence of solid-state batteries is anticipated to increase energy density and range. It will also decrease charging times. Solid-state batteries are being actively explored by Toyota [9]. Solid-state batteries also present other advantages including reduced fire risk [8]. Further improvements to the chemistry include adding graphene to EV batteries which could deliver an energy density improvement of around 20% and also increase the lifespan by 27% [9]. Other anticipated developments to battery components include changes to cathodes through reducing the nickel manganese cobalt content or replacing these altogether. Lithium iron phosphate is an alternative low-cost cathode material which is used for some lithium-ion batteries. This is commonly used by Tesla. Changes to the anode might also occur, replacing graphite with silicon to increase energy density and speed up charging, or blending both together to deliver similar improvements [8].

Battery size and capacity are anticipated to increase. The United Kingdom's Advanced Propulsion Centre anticipates that the average battery pack in an EV car or van is currently projected to increase to 65 kW by 2030, with a different scenario based on current trends suggesting this might reach 80 kW over the same period [10]. The current average battery pack size in the United Kingdom is around 50 kW [11]. Battery projections do not yet take us to 2035 for individual EVs. Most EVs have an efficiency of between 3 and 4 mi/kWh [12] meaning the current average range of an EV is between 150 and 200 mi, rising to between 195 and 260 mi in 2030 based on a 65 kW battery or 240 and 320 mi based on an 80 kW battery.

11.1.3 Developments in vehicle charging technologies

There are several emerging technologies that are likely to gain traction over the next decade. These include vehicle architecture, battery swapping, wireless charging, vehicle to grid and vehicle to load. Regarding vehicle architecture, the demand for ever faster charging is driving the development of higher voltage vehicle architectures. Most EVs currently built are based on a 400 V architecture. This is currently cheaper to manufacture than higher voltage systems. 800 V architectures are however becoming more common as they have the advantage of higher charging speeds, improved efficiency, better performance, and the use of thinner cables as well as a reduction in the weight and size of some components [13]. That being said, the challenge with higher voltage architectures is that there is the risk of greater battery degradation as faster charging speeds put greater strain on the battery, damaging it, especially when done frequently [14]. Even though this is the case, the convenience of faster charging is important to drivers in some use cases (e.g., en route) and should not have a major impact on battery health providing it is done occasionally. The trend for higher voltages is set to continue, with systems higher than 800 V also emerging. For example, Lucid and Nio are the first vehicle manufacturers to develop 900 V architectures, offering peak charging rates of up to 600 kW [15].

Focusing on battery swapping technology, this has started to take off in some geographies with Nio being the market leader in this space, having installed 2,300 battery swap stations globally. Nio are currently only breaking even in less than a fifth of stations; however, they believe this is a growing business area and is one where they have managed to secure an early advantage ahead of what they believe will be future market demand [16]. It is difficult to see that battery swapping will become part of a mass market charging solution and it feels somewhat of a step back in time (see Chapter 2). It may however have a niche use case in different geographies. Wireless charging however is likely to have a greater impact and has the potential to transform charging. There are currently two approaches to wireless charging, both of which use the same technology but one is more likely to take off than the other. The first is wireless charging on roads, where the wireless charging induction cabling is built into a road so that vehicles can charge while in motion. The second is charging pads in car parking bays or in garages where a car or van charges while parked. The former is very unlikely to become a common solution because the costs to rollout on the road network would be incredibly expensive. The latter shows much more promise with vehicle manufacturers like Tesla, Genesis and Volkswagen indicating that they plan to adopt this. Charging pads and standards are also starting to emerge [17]. A key challenge however is overcoming an adopted 'plug in and charge' behaviour. It is anticipated that some consumers will adopt the technology based on its convenience especially when charging at home privately; however, destination and residential charging may prove more challenging due to the practical aspects of needing to ensure the wireless pads are well maintained and clear of obstacles. The pads are also potentially much more susceptible to damage as a result of vehicles driving over them. Furthermore, it is also not clear whether wireless charging can deliver the same charging speeds as level 3 DC charge points. It is for these reasons that wireless charging will likely take off for home charging and also for more niche use cases like charging at depots, but it is unlikely to have mass market impact, especially when plug in charging is now established as the norm for all vehicle manufacturers.

Turning to vehicle to grid, vehicle to home and vehicle to load, this technology has the potential for mass adoption. This is because there are advantages to both the consumer and the distribution network operator (DNO)/energy system operator (ESO). For the consumer, vehicle to grid presents an opportunity to generate additional income by the user selling power back to the grid when there is high demand. Vehicle to home also enables the consumer to feed power back to their home, providing resilience especially in areas where power is at risk of failing. Vehicle to load can enable more niche use cases like providing power while camping or when working away from home (e.g., a joiner being able to power a saw). For the DNO/ESO, vehicle to grid presents an opportunity to feed energy into the grid when renewable energy generation drops, smoothing the demand peaks and troughs. Drawing on a swathe of EV batteries gives the energy system resilience and may negate the need for large scale investment in mass battery storage systems at grid level. Vehicle to grid and vehicle to home are currently supported by a handful of vehicle manufacturers. Vehicle to load is much more common. There are also relatively few charge points that support vehicle to grid; however, this is likely to change in the future, especially as the need for a more resilient renewable energy-focused power grid becomes more pressing.

11.1.4 The impact of AI on vehicles

AI is transforming multiple industries and has the potential to transform EVs. AI will no doubt transform the manufacturing process and also how EVs navigate using driving assistance systems; however, this is outside the scope of this book. Where AI will play a relevant role to charging is the battery management software and systems, energy management during charging and dynamic pricing. AI will be able to learn from previous charging cycles to perfect battery management, optimising the battery when charging and discharging, reducing battery wear, whilst training the battery to learn how to respond better in certain conditions like when cold or hot. AI will also improve charging efficiency by learning from previous charging cycles to manage energy better, resulting in real-time optimisation of energy flow, reducing the amount of energy wasted per each charging cycle. AI may also support dynamic pricing where AI-based prediction of demand is able to forecast when it is the cheapest to charge or discharge where vehicle to grid is in place, creating a charging plan which is cheapest and/or generates the greatest income for the user [18]. Further work needs to be done to explore the impacts of AI, though if its impact is anything like what is currently happening with generative AI in the creative industries, it will revolutionise the energy efficiency of EVs and greatly extend the lifespan of batteries.

11.2 Future trends in charge point infrastructure

The anticipated increase in the number of EVs over the next 10 years is expected to have a major impact on power demand, the number and type of charge points, the use cases where charging is predominantly done, the business models underpinning charging, innovations in charge point technologies, innovations in commercial models, new standards and requirements and the transformation of the whole charging process (in some cases, through AI).

11.2.1 Impact of EV uptake on power demand

Based on the anticipated number of EVs by 2035 as outlined in Section 11.1.1, it is possible to predict the demand on the grid in different geographies including the types of location where this charging will take place. Power grids and the organisations that manage them will need to adapt to this demand through managing the network differently to how it has been managed in the past.

According to the International Energy Agency, EVs of all types could account for 6–8% of total electricity demand by 2035, an increase from 0.5% today. When focusing purely on cars, vans and other light-duty vehicles, the electricity demand equated to 100 tWh in 2023 and is anticipated to increase to between 1,510 and 2,180 tWh by 2035. The range is the result of different modelling approaches adopted by the International Energy Agency [19]. This increase in demand will be met globally by a ramp-up in renewables, nuclear and energy storage. Some demand will still be met through energy generated by coal, oil and gas, though as 2035 rapidly approaches, the proportion this occupies is expected to rapidly fall. Solar photovoltaics (PV) and wind are anticipated to generate 40% of global energy by 2035 [20].

Focusing on the United Kingdom, according to the Committee on Climate Change's balanced net-zero pathway, electricity demand from all modes of electrified transport will equate to around 80 tWh in 2035, up from roughly 5 tWh in 2020. The bulk of this increase relates to an increase in EV cars and vans as per Section 11.1.1. The Committee on Climate Change states that this increased demand will be met by a grid where energy generation is entirely low carbon, 70% of which is derived from a range of renewables, with wind forming the backbone of the generation system. The remaining 30% will be generated by gas with carbon capture and storage, bioenergy with carbon capture and storage, hydrogen and nuclear. The system will be more flexible, built around energy storage, flexible energy demand pushed away from peak hours, the use of surplus energy to generate hydrogen and interconnectors between the United Kingdom and its neighbours [21]. Field Dynamics anticipates that just under 1 tWh of demand in 2035 will be derived from public charge points [22], suggesting when considered against the Committee's projections, and excluding surface transport demand that is not EV car and vans, that the bulk of charging will take place at home or at work. Turning to Europe, Enerdata anticipates that by 2035 European EV cars and vans will create a demand of 152 tWh, representing 5% of the annual electricity demand of the 27 Member States. This is an increase from a projected 72 tWh in 2030. The overall model is based on almost 100% total sales of cars and vans being EV by 2035 [23]. The data does not indicate where the demand will come from in terms of whether charging at home or via public charge points. To meet this, increased power demand will largely be dependent on each Member State decarbonising its grid in different ways; however, it is clear that renewables will play an important role across all of them. Turning to France, a report produced by RTE focused on electromobility states that by 2035 energy demand from light-duty EVs (including cars and vans) and a smaller number of heavy-duty EVs (around 10%) will be between 28 and 48 tWh a year depending on different EV adoption scenarios [4]. This is from a starting position of 1 tWh a year in 2020 [24]. France plans to meet this demand primarily through nuclear (48%), wind (22%), solar (13%) and hydroelectric (11%), with the remaining demand met by gas (4%) and biomass (2%) [4]. Similar to the United Kingdom, France also anticipates using flexible smart charging to manage demand, combined with vehicle to grid energy storage. Eighty-five per cent of charging will take place at home and work, via smart charge points either on or off-street, with the remaining 15% carried out via public charge points away from home [24].

In Germany, it is anticipated that EVs will result in an annual demand of 26 tWh by 2035. This data, however, is not published by the Federal Government and is derived from a master's theses which is 8 years out of date [25]. That being said, this information is freely available and is not too dissimilar to estimates for France, a similar sized country. Furthermore, this figure is broadly in line with that provided by Bermejo *et al.* who anticipate that EVs will result in an annual demand of 23 tWh by 2030. Under an accelerated adoption model, this demand could increase to 43 tWh by 2030. Demand in 2020 from EV cars and vans was around 1 tWh, though by 2035 a larger proportion will result from buses and heavy goods vehicles. Regarding charging locations, only 2030 figures can be found. Using these, 40% of charging is anticipated to take place at home, 14% at work, 11% en route and 5% at destination locations. The remaining 30% relates to heavy goods EVs charging at hubs. Between 55% and 60% of charging will be done via level 2 chargers [26]. Smart charging combined with vehicle to grid, like in other European nations, is key to managing demand. Germany anticipates that this demand will be met primarily by renewables by 2035 and will need to shift generation away from coal, gas and biofuels which made up roughly 51% of total generation in 2023 [27]. Energy storage will play a critical role to manage the peaks and troughs of demand.

Electricity demand from EVs of all types in the United States is expected to reach 107 tWh per annum by 2035, increasing from roughly 4.7 tWh in 2020 [28]. 2035 figures are not freely available; however, figures for 2032 derived from Pierce and Slowik's report for the International Council on Clean Transportation indicate that the bulk of charging will be done privately at home. Residential charging in flat complexes and workplace charging will also play a critical role in meeting demand [29]. The grid will meet this demand through a majority (60–80%) of energy being generated by renewables (wind and solar) with hydropower and geothermal energy also playing a role [30]. Like in Europe, energy storage and smart charging solutions will play an important role. Focusing on Canada, energy demand from all EVs (the majority of which are light-duty cars and vans) could be anywhere between 46.8 and 95.1 tWh annually by 2035 [31] with Natural Resources Canada stating EV demand of around 60 tWh. This is an increase from a base EV demand of below 1 tWh in 2020 [32]. The bulk of charging, like in the United States, will take place at home on private driveways or in multi-family flat complex car parks [33].

Canada is targeting a fully decarbonised grid by 2035. This will be achieved through expanding renewables and connecting clean power with places where it is needed. Energy storage and carbon capture will play a key part as will EV smart charging and vehicle to grid [34].

Turning to the Southern Hemisphere, modelling done by ENERGEIA for the Australian Energy Market Operator anticipates that by 2035, annual EV power demand predominantly from EV cars and vans will be roughly 7 tWh, an increase from around 2.5 gWh in 2017. Most charging will take place residentially or at work [35]. Australia is aiming to be net zero by 2035, which presents a major challenge when considering in 2023 that 65% of power was generated from burning fossil fuels [36]. Australia not only has the challenge of significantly ramping up renewables and energy storage, but it must also develop its grid to support smart charging and vehicle to grid. New Zealand is anticipated to see transport electricity demand, for which EV cars, vans and some buses/heavy goods vehicles make a substantial part, increase from 0.27 tWh annually in 2023 to around 2.5 tWh annually in 2035. Most charging will take place at home through smart charging. New Zealand already generates 87.1% of its energy via renewables and is anticipated to increase this to above 90% by 2035 [37]. Battery energy storage is anticipated to play a key role in managing demand alongside shifting the load via smart charging and providing network resilience via vehicle to grid.

What should be clear from the above geographies is that the transition to EVs will place huge additional power demands on national power networks which will be met via a shift to renewables and through encouraging charging at off-peak times. EVs will help manage demand via vehicle to grid charging and discharging. Most countries are well on their way to meeting the challenge, though Australia is perhaps the one where the most activity is required in the coming years.

11.2.2 Projected charge point numbers across different geographies

Section 11.2.1 made clear that by 2035 most charging will be done at home or at work; however, there will still be a role for public charge points in meeting residential, en route and destination needs. The focus of this section is on these projected public charge point numbers. Each geography has projected public charge point requirements out to 2035, breaking these down into the proportion which will be level 2 and level 3 DC. Focusing first on the United Kingdom, it is anticipated that by 2035 there could be between 255,000 and 490,000 charge points [2]. As of March 2024, there were just under 60,000 charge points [38]. Available projections show possible breakdowns by type up to 2030 where it is anticipated by that date that 280,000 charge points will be in place. Of this figure, about 50,000 (about 18%) will be level 1 or 2 3–7 kW, 105,000 (about 37%) will be level 2 11–22 kW, 120,000 (about 43%) will be level 3 DC 50 kW and 5,000 (about 2%) will be level 3 DC 100 kW and above [39]. Turning to Europe, 18.8 million charge points will be required by 2035 [40]. This is from a baseline of 632,000 charge points installed by the end of 2023. The European Federation of Transport and Environment provides a percentage breakdown of where charging will take place in 2030 (2035 figures were not available). Excluding home and work charging which is anticipated to make up around 70%, public charging will make up about 30%. Of this percentage, level 1 or 2 AC 3–7 kW charging will make up about 10%, level 2 AC 11–22 kW charging about 50%, level 3 DC 50 kW about 7% and level 3 DC 100 kW and above about 33% [41]. The key difference between Europe and the United Kingdom is that level 3 DC charging is anticipated to play a greater role.

When focusing specifically on France, the International Council on Clean Transportation estimates that between 436,000 and 550,000 public charge points will be required by 2035 [24]. This is from a starting position of 100,000 charge points in mid-2023 [42]. The International Council on Clean Transportation conducted a region-by-region analysis of the charge point requirements by type. Taking the Île-de-France as an example, of the 83,500 charge points required by 2035, roughly 91% (75,958) will be level 2 and the remaining 9% level 3 DC (7,565). Turning our attention to Germany, 2035 figures are difficult to identify. When focusing on the target date of 2030 between 440,000 and 843,000, public charge points are required [43]. The Federal Government Charging Infrastructure Masterplan anticipates 1 million public charge points by 2030 [44]. This is from a baseline of 130,000 charge points installed up until June 2024 [45]. It is not clear from the Masterplan or analysis done by others what the charge point mix will be in terms of level 1/2 versus level 3 DC beyond the fact that Germany acknowledges most charging will be done privately. The estimated split developed for Europe can be applied if required.

Turning our attention to the United States, it is anticipated that there will be 1.7 million charge points by 2035 [46]. This is an increase from a starting point of 192,000 charge points in August 2024 [47]. Regarding the breakdown by type, 1.2 million will be level 2 (Roughly 71%) and 500,000 will be level 3 DC (roughly 29%) [46]. To the North, Canada is anticipating that 290,000 charge points will be installed by 2035 [48], an increase from 27,000 installed up until March 2024 [49]. The breakdown by type in 2035 is anticipated to be about 260,000 level 2 charge points (roughly 90%) and 30,000 level 3 DC charge points (10%) [48]. Turning to the Southern Hemisphere, Australia is projected to install 31,500 charge points by 2033 (figures for 2035 are not available) [50]. In terms of progress to date, there are 2,000 charging sites, though it is not clear how many charge points this translates to [51]. It is also not clear what the proportion of level 2 to level 3 DC charge points will be in 2033, though if it is like other Western countries the largest proportion will be level 2 charge points. Focusing on New Zealand, projections up to 2035 are not publicly available. Using the target date of 2030, it is projected that 10,000 charge points will be required [52]. This is from a baseline of 1,248 charge points installed as of August 2024 [53]. Similar to Australia, the 2030 target is not broken down by type though current trends and comparisons with other geographies demonstrate that level 2 charge points will play a dominant role potentially at slower speeds of 11 kW and below [54]. Regarding level 1 charge points, it is anticipated that across all geographies, these will play an increasingly reduced role, satisfying a small proportion of residential on-street and off-street charging.

11.2.3 Future utilisation patterns

Section 11.2.1 and 11.2.2 highlight that across the Western world, most future EV car and van charging will take place privately at home or in workplaces (through mainly level 2 charge points and some level 1 charge points), with the remaining charging taking place publicly via level 2 charge points in residential and destination settings or via level 3 DC charge points in en route locations. Anticipating future utilisation is dependent on these differing use cases as demonstrated in Chapters 5 and 7. Basic projections of utilisation conducted by the International Energy Agency for two Western geographies using the anticipated number of EVs divided by the anticipated number of charge points indicate that by 2030, the number of EV cars and vans per public charge point in the United Kingdom will be about 35 and in the United States about 30 [55]. Although this is a helpful indicator, it has its limitations due to the assumption that EV and charge point distribution is equal across a geography. In actual fact, this is not the case and the ratio will differ in different locations. It also fails to indicate how different charge point use cases will have different usage patterns based on the EVs that tend to use them. For example, an en route charge point can cater for a larger number of EVs due to its location on a strategic road network, so will likely have a higher utilisation than a residential charge point catering for a lower number of cars in a community. The difference in use case will also have an impact on when charge points are used with a majority of charging taking place for en route and destination charge points during the day. Most residential charge points will see usage in the afternoon and in the evening, through to the morning, before people travel to work. As a result of these patterns, a more useful way to measure utilisation is time-based where usage is calculated based on the length of time a charge point is in use throughout the day as a percentage of 24 hours. Energy-based utilisation is another measure that can be used when developing an investment case but for simplicity, we will focus on time-based utilisation. This measure enables differences in different use cases to be reflected in utilisation, as demonstrated in Chapter 7. According to the Green Finance Institute, an organisation based in the United Kingdom, time-based utilisation for a level 2 on-street residential setting is the lowest when using a sample period of quarter four 2022. In this period, time-based utilisation stood at 13.7% (roughly 3.25 hours). For level 2 destination charging, this was slightly higher at 15.7% (roughly 3.75 hours). For level 3 DC 25–99 kW utilisation was 14.8% (roughly 3.5 hours). The highest utilisation over this period was level 3 DC 100 plus kW where utilisation was 16.1% (slightly under 4 hours) [56]. For residential charge points, usage was primarily in the evenings through to the mornings. For destination and en route charge points, usage was spread mainly throughout the daytime.

Projecting these utilisation figures and patterns forward to 2030 and 2035 is difficult because there are so many factors that impact utilisation. On a simplistic level, the number of EVs versus the number of charge points at a national level is a starting point; however, as previously stated, this does not reflect how utilisation differs across different charging use cases and between charge points at a local level. Other factors that will influence utilisation projections are as follows:

- **The number of homes and workplaces with private charge points**: this will influence the number of vehicles creating demand for public charge points.
- **Charge point growth in proportion to EV growth at a local level**: this will influence demand for public charge points. If either gets out of sync, this could result in higher or lower demand. CPOs deploying charge points must focus on installing ahead of the demand curve to ensure charge points do not flood the market, pushing utilisation down.
- **Emerging charging models (see Section 11.2.4)**: if community charging takes off, this could reduce demand for public charge points especially in residential or destination settings.
- **Increases in battery and vehicle efficiency**: if batteries and EVs continue to become more efficient in terms of the miles or kilometres driven per kWh, this will impact on demand for charge points as EVs will need to charge less.
- **Mass adoption of battery swapping**: this will negate the need to use public charge points, pushing down demand.

Looking at each of these factors, the number of private homes and workplaces with private charge points is likely to grow over the next decade as will the number of EVs using them. It is difficult to say that this will result in any changes to utilisation. CPOs are also becoming more aware of needing to grow their charge point base in line with EV growth, using demand analysis tools at a local level (see Chapter 5) so it is also unlikely that any changes in this will impact on current utilisation. Emerging charge point models like community charging may have an impact but it is unlikely this will occur on a mass scale to drive meaningful utilisation change. Likewise, mass battery swapping seems not to be taking off in the West, so it is unlikely to push demand for charge points down and thus impact utilisation. The one unknown is battery and vehicle efficiency advancements. This is definitely one to watch over the next decade as any improvement will push down demand and utilisation. Reflecting on the above and when looking specifically at each charging use case, we anticipate that for residential charging, utilisation might increase slightly as the number of EVs grows. Though as stated above, we also expect that charge points in this setting will grow roughly in tandem, especially as a result of the various charge point installation programmes stated in Chapter 4. In some local areas, there will likely be some local misalignments where either EVs or charge points grow out of sync.

Focusing on destination charge points, there is also nothing to suggest that current utilisation rates will change in the future, again assuming charge point and EV growth occurs in sync. One factor, however, that might drive utilisation upwards is localised schemes or programmes where destination charge point usage is incentivised through money-off coupons or other free services. These will drive up demand and utilisation but it is difficult to project this on a large scale. Turning to en route charge points, this is likely to see the greatest change in utilisation. As stated in Section 11.2.2, en route charge points are anticipated to grow over the next decade. As there are fewer projected en route charge points compared with destination and residential charge points and that charge points will be clustered on the strategic road network, this combined with anticipated rapid EV growth will likely mean that demand and utilisation at these charge points will increase. This will occur even when factoring in the uncertainty of battery and vehicle efficiency increases. This is because we anticipate that although the number of charge points will increase, the number of EVs travelling on the strategic road network requiring a charge en route will increase at a greater rate.

The future of revenue generation at charge point sites is mainly determined by utilisation. As a result, revenue is unlikely to change much for residential and destination charging use cases; however, for en route charge points, it is likely to increase based on higher future utilisation. Other factors like energy prices and operational costs are anticipated to increase with inflation, with these passed on to the customer.

11.2.4 Future business and commercial models, underpinned by innovations in charge point technologies

When looking to the future business and commercial models which will enable charge point growth, the various anticipated innovations in charge point technologies should also be considered. The future business and commercial models are focused on increasing the revenue and profit of each charge point. As stated in Chapter 7, there are three business models for CPOs, each with different commercials. A CPO can act as an owner operator, owner operator through renting or leasing the land or through a concession in the form of either owning the charge points or operating them on behalf of the owner. Across all of these business models, future potential changes will transform how charge points operate, generating revenue and profit. These are as follows:

- **The expansion of ancillary services**: as stated in Section 11.2.3, it is difficult in most use cases to increase the utilisation and resultant profit at charge point sites. One way to increase profitability is through implementing ancillary services. This might include digital advertising at sites as was demonstrated in Chapter 10 where Tesco and Pod Point entered into an agreement to provide both charge points and digital advertising boards. At large destination and en route charge point hubs, ancillary services might take the form of shops or other services similar to a traditional service station. Across the different business models, the ability to implement this approach varies. For an owner operator, they have greater freedom to implement these services, whereas renting or leasing land might limit what, other than charge points, can be installed. A concession may further limit what a charge point operator can do.
- **Community charging**: as mentioned in Section 11.2.3, community charging could provide an opportunity for CPOs or another form of provider to generate a profit through providing a platform where private owners of charge points make these available to EV drivers, charging a competitive price per kWh. The platform owner can charge a hosting fee, generating income from every kWh sold. This approach is attractive to EV owners as they are able to access

cheaper prices than public charging, for example, in the United Kingdom, where value-added tax (VAT) is 5% for domestic energy usage. The risk for a CPO in relation to community charging, however, is that this may cannibalise demand for public residential or destination charging, pushing utilisation down at charge points they provide. This is a potential reason why CPOs might enter into this market so that they are able to influence it by becoming a platform provider themselves. They may also choose to offer packages where they install the charge point under the proviso that they can generate an income from every kWh sold through community charging. Another approach might be to reward owners of community charge points through offering money off when the owner charges while out and about when using the CPO's public charge point network. A CPO operating through any of the three business models could offer community charging alongside public charging.

- **Vehicle to grid**: this innovation is perhaps one of the most talked about in terms of having a major future impact both for private residential and public charging. Vehicle to grid is where an EV is able to both charge and discharge energy back to the grid, generating an income from every kWh that is drawn from the vehicle. A growing number of vehicles support vehicle to grid; however, there are currently a small number of charge points, though this is expected to increase in the coming years. The commercial model around vehicle to grid will likely take the form where both the driver and the CPO are rewarded when power is fed back to the grid, with the driver being paid a per kWh payment and the CPO being able to take a percentage of this. Vehicle to grid presents an interesting opportunity for vehicles that are plugged in at destination and residential chargers for long periods of time throughout the day and into the early evening when the grid is at peak demand. The DNO is able to draw down power from vehicles to smoothen supply peaks and troughs. EVs give the network flexibility, providing enough are connected, to ensure a constant power source. Furthermore, vehicle to grid can provide resilience to the power network especially during periods where energy generated from renewables drops or when there is exceptional demand. Vehicle to grid is only likely to become mainstream when the financial benefits to both users and CPOs outweigh the additional cost to install compatible charge points. However, as the technology becomes more commonplace and cheaper, and the growth of renewables means the network needs to be more flexible, it is increasingly likely that CPOs offer this as an option to EV drivers. At scale, this could generate considerable income, enabling CPOs to generate additional profit from charge points.
- **Solar and battery storage**: as the cost of solar PV is anticipated to decrease over the next decade, the likelihood that these become more commonplace in a range of locations increases. To date, solar and battery solutions, in the form of a solar canopy over the charge points and battery storage nearby, have been deployed in en route and destination charging hubs. This has mainly been in locations either with limited grid capacity or no connection at all and also to demonstrate the sustainability of hubs using locally generated renewable

energy. The present challenge, however, is that the additional costs resulting from the batteries, panels and structures can impact on the profitability of sites, even when taking into consideration the fact that CPOs can sell freely generated renewable energy back to the user and also reduce the energy they need to buy from the grid. The reduction of solar PV and battery costs will make the deployment of these more affordable in a greater number of locations, including those with lower utilisation rates and not just in locations with constrained grid capacity. This approach may only be limited to owner operator and land leasing/rental business models where the CPO has full control of designing the site. This is not to say that a concession model might not support this, though it depends on who is offering the concession and whether they are focusing purely on charge points alone. It should also be noted that this infrastructure requires additional planning approvals including engagement with the DNO especially in instances where the CPO wants to sell power back to the grid. The commercials of solar and battery storage focus on two revenue-generating approaches. The first is selling the energy generated locally or charged off-peak to the batteries when energy is cheapest to the driver, resulting in a greater profit due to paying reduced energy import costs. The second is selling this energy back to the grid when demand is high. Future sites that might struggle from a projected profitability perspective might become viable due to the additional income that can be generated.

- **Valet charging**: this is where a charge point or charging hub is manned by individuals responsible for managing vehicle charging. This approach currently takes place in some locations with the sole aim being to drive up utilisation and revenue. For example, a current location where this is used is at long stay airport charge points where a valet company offers to charge the vehicle ready for when travellers arrive back at the airport to retrieve it. Valet charging is not appropriate for en route charge points due to the length of stay or residential charge points due to the number deployed and how they are spread out over larger areas. It is more suitable for destination charge points, especially in high-demand areas where vehicles are left for several hours and achieve full charge, blocking the charge point for other users. Valet charging in these settings is more likely if the growth of EVs outpaces charge point deployment, though the commercials always need to be stress-tested before this approach is deployed as the additional cost to hire someone to do this may not be viable in all locations. CPOs should consider the current and predicted utilisation and analyse the different usage patterns throughout the day to see when a valet service would have the greatest impact. In terms of business models, valet charging could be deployed across all three, though as stated this is dependent on the use case.
- **Robotic and autonomous charging**: robotic charging may become more prevalent in the future and would play a similar role to valet charging. There are multiple approaches in how it could operate. One approach is a robotic arm that is able to open a charge port and plug in a cable. This services multiple vehicles parked in EV bays, though relies on the driver leaving the vehicle or

port unlocked so that it can be accessed. The second approach is via a battery robot that services multiple vehicles parked again in specific bays by navigating to them and accessing the charging port. This robot may need to recharge throughout the day, so multiple robots may be required. Similar to valet charging, this approach is only suitable in destination charging locations. Regarding the appropriate business model, it is not suitable for the concession model as it is more infrastructure heavy, unless this is included in the concession agreement. Similar to solar and battery charging, there is a significant investment requirement which must be considered and factored into the commercial modelling of the increased utilisation and revenue versus the higher infrastructure costs. Certain destination settings like airports or busy shopping centres may be well suited to this approach though, unlike valet charging, there is still the risk of a vehicle being fully charged and then blocking a bay. This is why significant effort must be made to right size the infrastructure and sense check the commercials to satisfy the specific use case.

- **Large scale wireless charging**: this is where an enabled EV drives over a wireless pad, positions itself to enable the most efficient induction location and then starts a charge. The session ends when the vehicle drives away. In Section 11.1.3, we spoke about the role of wireless charging from a vehicle's perspective though for this to become mainstream requires the associated infrastructure to be installed. To date, wireless charging from an infrastructure perspective has not become commonplace because the industry has yet to settle on a group of standards to enable infrastructure scale up. Currently, wireless charge points are operating in niche use cases like depot charging for operational vehicles where space is limited and a large number of wireless charging pads are installed in existing parking bays to maximise the number of vehicles that can charge at the same time. In these deployments, vehicles are retrofitted to support wireless charging. It is likely, however, that as more EV cars and vans support wireless charging as standard that this is rolled out in more locations. Wireless charging presents the opportunity across all use cases to maximise utilisation, presuming the driver, once positioned over the pad, starts charging almost immediately. This is because every charging session can start and end more quickly because the driver no longer needs to get the charging cable out of their vehicle, plug it in and set up the charge via an app or pay via a debit or credit card. These time savings add up over the day, potentially enabling more charging sessions to take place and therefore greater revenue generation. Wireless charging presents unique benefits across each use case. For en route charge points, it will drive up utilisation through greater vehicle throughput due to the time savings mentioned above. For destination charging, it will enable more vehicles to charge, potentially at slower speeds, but across a larger number of parking bays as wireless charging pads have a smaller infrastructure footprint meaning they can be deployed at scale. The larger number of charging pads minimises the risk of bay blocking as it is likely there will always be a wireless-enabled EV bay spare. For residential use cases, wireless charging presents the opportunity to install bays in long stretches along residential streets without disrupting those without EVs as the installation of multiple charging pads should mean that even if some are blocked by non-EVs, a significant number will still be available to EV drivers. Wireless charging will likely play a role across all three business models, though as mentioned above, significant standardisation and rollout in vehicles will need to happen first before it becomes commonplace. This is why it is likely that this will only become the preferred way to charge as we enter into the 2030s.

- **Load management**: this is where multiple charge points are managed by a module that distributes load across the charge points. This is advantageous as this means that multiple relatively simple connection points can be installed and controlled from one module which is connected to the grid. Vehicles will likely not receive the maximum charging rate at each charge point when multiple vehicles are connected as the total load is spread across multiple bays. The solution is also smart enough to feed greater power to those vehicles who need it versus those that are topping up their batteries. The grid connection to the module can be specified to be smaller than if it was based on combining the maximum power draw of each individual charge point in a location. A smaller connection size in terms of maximum kWh drawn is usually cheaper. It is also easier to supply via solar and battery storage solutions. Load management is more applicable in destination and residential settings as drivers are likely in these locations to leave their vehicles for lengthy periods, so they are not necessarily reliant on the need for fast charging rates. Load management in en route locations may play a role across level 3 DC charge points of different rates but the value of it is less important because it is likely that in determining the connection the CPO opted for the maximum size. It is also less important because it is assumed that when using en route charge points, drivers want to charge as fast as possible, so are unlikely to want load management to reduce charging speeds. That being said, there may be an approach in the future where drivers at en route locations are happy to opt for slower load managed speeds, for example, at lunchtime if they know they will be in a location for a longer period, and that opting to do this reduces the cost per kWh to charge. Load management will increasingly play a role across all three business models though like other solutions or innovations mentioned, a concession arrangement may limit this as an option if it is not allowed in the agreement. Regarding commercials, load management can increase the profits of a site by reducing the infrastructure and connection costs. It may drive up site utilisation where more connection points can be added though utilisation at a point-by-point basis may not increase.
- **Smart charging**: this is where the charge point is intelligent enough to only charge the vehicle when electricity is cheapest (off-peak) or there is reduced demand on the grid and the driver has specified that they are happy for the vehicle charging schedule to be determined by the charge point. Smart charging is increasingly common in private residential charge points, coupled with a smart energy tariff; however, it has yet to take off in public charge point settings. The benefits of smart charging tend to result from when an EV is

plugged in for a long period of time so that charging can start and end multiple times during a session, mainly as a result of fluctuations in energy prices and demand on the power grid. Charging costs tend to be less for the driver and the CPO because they are able to charge off-peak. This may drive up utilisation at smart charge points as drivers are keen to save money as part of each session. Smart charging is likely to become more common in residential settings where drivers are acutely aware of cheaper charging costs at private residential charge points twinned with a smart domestic energy tariff. Certain geographies like the United Kingdom, for example, have made this a requirement for residential charge points. Smart charging may also play a role in destination charging where vehicles are stationary and plugged in for several hours. Smart charging will only have a limited impact at en route charge points, though it is increasingly being used by CPOs to encourage utilisation in off-peak periods through incentivising drivers to charge through lower per kWh prices. Tesla charge points have been doing this for some time. Regarding the three business models, smart charging is applicable across all of them. In terms of the commercials, smart charging should drive up utilisation and site profitability though there is the risk that this does not materialise in some residential and destination settings where EVs that have long since fully charged remain plugged in, blocking the charge point for other users. This can be discouraged with overstay fees and may become less of an issue if combined with another innovation like wireless charging, where a greater number of bays can be installed in a location.

- **Ultra-rapid charging based on 800 V plus EV systems**: as mentioned in Section 11.1.3, the emergence of 800 V and above EV architectures means that ever faster level 3 DC charge points are in development, capable of delivering 300 kW plus to the vehicle. These charge points are only relevant in en route settings and will likely only apply in owner operator and leasing/renting business models as concessions tend to be less common in en route settings. As each charging session should in theory be shorter, assuming optimal vehicle state of charge, this should also mean that each charging session is quicker and that the throughput of vehicles at en route sites is greater per hour, pushing up utilisation and profit. Countering this is the fact that installing these charge points is expensive as is the power connection cost. There are other challenges regarding site practicalities. Recognising all of these points, it is still likely that over the next decade these charge points become increasingly common to service drivers who need to rapidly charge and continue their journey. An example might be delivery van drivers. These speeds are also what will underpin the electrification of freight, which is mentioned in Section 11.3.
- **New forms of public–private relationship:** this development is difficult to predict though as the number of charge points ramp up and the demand on both public and private funding increases to enable the transition to EVs, it is likely that new public–private relationships will develop. We may see an increasing number of joint ventures between local authorities and CPOs to install charge points. We may also see strategic relationships form between CPOs so that

different use cases can be exploited in locations where each CPO has focused on a different charge point use case. Due to the criticality of the need to shift to EVs to meet global climate targets and the central role that charge points play in this space, some geographies may opt for full nationalisation of charge point infrastructure, though we anticipate that this will be the exception and not the rule in most Western countries.

Amongst all of these innovation, one that has not been mentioned and has been discounted is wireless charging integrated into the road. The cost to install this is incredibly expensive, having been estimated at £1.6 million/mi [57]. Although it is likely that this cost will fall, the advancement in battery capacities as stated in Section 11.1.2 should reduce the need for this for EV cars and vans at least. Also, as highways become increasingly designed to support autonomy, this is resulting in less infrastructure, not more. Wireless road charging would increase the infrastructure build and maintenance costs when highways authorities are trying to reduce costs.

11.2.5 Emerging charge point standards and requirements

In Chapter 4, we referred to the current charge point standards and requirements across our focus geographies and, where known, those that are emerging. This section summarises the broad emerging trends and provides further detail of their anticipated impacts. A vast majority of these focus on improving the interoperability between different CPOs and improving the customer experience. Though not exhaustive, the interoperability changes mentioned below are expected to have the greatest impact on how charge points operate:

- **Standardisation of payments**: currently, there are a myriad of ways to pay for a charging session, with level 1 and 2 charging sessions in most geographies being covered mainly by app payment and level 3 DC charging sessions being covered by a mix of app payment and card payment, where the latter can be done by chip and pin or contactless. To make it easier for drivers to pay for charging sessions, most geographies (the United Kingdom being an example) are making it a requirement that all level 3 DC charge points accept physical payments. This should prevent the need for drivers to download multiple apps and ultimately make charge points easier to access and use as drivers at level 3 DC charge points will only require a card. The emerging standards do not specify whether those who pay by card and do not have an existing account with the CPO will be able to access preferential per kWh charging rates often offered to app users though it is likely that if a growing disparity between physical and app-based payments occur, governments will legislate to change this. Standardisation of payments is likely to have the greatest impact on en route and some destination focused CPOs because these focus on installing level 3 DC charge points that are the target of legislation.
- **Plug and charge**: this is where the driver connects to the charge point and automatically pays for the charging session. This is done through the driver

having an account set up with their vehicle including associated payment details. When the driver connects to the charge point, the vehicle communicates with it, completes a 'digital handshake', starts the charge and then shares payment details at the end of the session to cover the charging cost. The driver is then subsequently billed on a session-by-session basis or monthly. Plug and charge is enabled by ISO 15118, an international standard for charging EVs. Tesla currently supports this service as standard when a Tesla uses a Tesla supercharger; however, it is not yet commonly available in other vehicles. This is because it requires both the vehicle and charge point to be enabled to support the standard. Increasingly, both CPOs and vehicles are adopting the standard. Plug and charge is set to transform the charging experience for non-Tesla drivers, making it much easier for them to charge. There is also the likelihood that plug and charge will push up utilisation as the time savings achieved through drivers not needing to start and end sessions via app or payment should enable more sessions to take place each day. Plug and charge is being rolled out in the United States in 2025, with other geographies on a similar trajectory [58]. Plug and charge will impact on residential, destination and en route focused CPOs as it is applicable to levels 1, 2 and 3 DC charge points.

- **eRoaming**: this is where an EV driver can charge at most charge points with a single account. This account is managed by their preferred electromobility service provider (EMSP). The EMSP has operational and data sharing agreements with multiple CPOs so that when a driver arrives at a charge point, they are able to use the EMSP's app or Radio Frequency Identification (RFID) card to start and end a charging session. The EMSP manages the back office interaction with the CPO, including payment, meaning the driver only needs to ensure they carry around the same EMSP RFID card or app for charging sessions. eRoaming is only as effective as the number of agreements the EMSP has with CPOs. An EMSP with multiple agreements is more attractive to a driver than one with fewer agreements. In certain geographies, eRoaming networks exist based on standardised agreements and harmonised technical integrations with multiple CPOs, enabling larger eRoaming networks. Hubject, for example, is an international eRoaming network provider that operates in this way [59]. The likely impact of eRoaming is that it increases utilisation by enabling EV drivers to use more charge points. eRoaming will impact on residential, destination and en route focused CPOs and is applicable to levels 1, 2 and 3 DC charge points.

It should be noted that with regard to the changes outlined above, most will complement each other. For example, if plug and charge becomes commonplace, this should in theory also enable eRoaming, provided the vehicle is able to interface with multiple CPOs. That being said, standardisation of payments still assumes that the driver physically pays for each transaction, meaning plug and charge and eRoaming are not applicable as these are primarily digital payment mechanisms.

Although interoperability is the key focus, other changes to standards and requirements are anticipated to make charge points open, safer, more secure to use, more accessible and better supported. The below will have the greatest impact:

- **99% up time and 24/7 customer support**: governments are increasingly legislating to improve the customer experience at charge points. One of the focus areas is improving the uptime where a charge point can be used and the time period where support is provided. As stated in Chapter 4, this is being legislated in the United Kingdom and Europe, though it is likely that other geographies will follow. This should in theory improve utilisation as greater uptime means the charge point is available to be used more often. Improved customer support should encourage drivers to charge more, knowing that they will be supported if something goes wrong. Improving the customer experience will impact on all three charge point use cases. There is some reluctance to these changes from CPOs as they argue that the 99% uptime is not achievable due to relatively routine maintenance activities and that providing customer support continuously is excessive when a vast majority of charging takes place during peak periods.
- **Accessible EV charge point standards**: various geographies are starting to design their charge points to be accessible to those with different abilities and needs. In the United Kingdom for example, PAS 1899 has been introduced, with other geographies adopting a similar standard. PAS 1899 sets out requirements related to the charge point design, placement including streetscape and public realm. Accessibility standards are applicable to residential, destination and en route use cases and should drive up utilisation by ensuring everyone can use charge points.
- **Open data protocols**: as the charge point market grows, more users expect a seamless experience across charge points. The push to standardise how data can be shared between charge points and CPOs has resulted in the Open Charge Point Protocol (OCPP). This is a system of rules that allow EV charge points and CPOs to communicate with each other, sharing data related to a number of parameters including their location, status and cost to charge. There are various versions of the OCPP, with the earlier versions widely adopted in most geographies. The latest version, OCPP 2.0, focuses on the core functionality of interoperability, remote control, support and maintenance, authorisation, configuration of the device model, transaction support and advanced security. It is yet to be rolled out more widely, though it will be adopted increasingly over the next decade. That is, unless a new standard develops. OCPP should in theory drive up utilisation by making it easier for drivers to find and use charge points. It is applicable to residential, destination and en route use cases.
- **Cyber and physical security**: as mentioned above, cybersecurity measures are critical to charge point operations. All versions of OCPP include security elements with later versions providing enhanced levels. Cybersecurity measures focus both on the digital and physical aspects of charge points. Regarding

the digital aspects, this includes when a charge point sits idle and is sharing data via OCPP and when a charging session is started via app or RFID card. When this occurs, the charge point shares data about the session, including possible personal data and payment details. If eRoaming, this data will also be shared with the EMSP. Data security ensures that data exchanged is both encrypted and anonymised and that where data is shared, digital handshakes take place, underpinned by digital signatures and certification. These handshakes ensure that only data is shared with those authorised to process and/or control it [60]. Furthermore, where communications occur with the operating system, including during firmware and software updates, these should be authenticated and encrypted.

Regarding physical aspects, this includes access to the charge point unit electrics and circuitry, operating system and communication modules. Security improvements include making sure the charge point cabinet and feeder cabinet is locked to prevent tampering, that access ports are also locked and if they can be accessed, that multi-factor authentication is in place when connecting to the unit to prevent those without authorisation from accessing and tampering with the systems. Tamper mechanisms, alarms and CCTV are also helpful in enhancing security. These security measures are being adopted in more locations as a result of current and emerging regulations and standards. It is anticipated that security measures going forward will continue to improve to counter new threats including those resulting from AI. Recent examples of where security has been breached, including the hacking of Russian charge points following the start of the Ukraine war in 2022, show that threats to charge point infrastructure are real and that they must be managed and mitigated, regardless of the charge point use case [61].

When looking at future standards relating to charge point connectors, most geographies anticipate consolidation around a single charge point socket for level 2 and level 3 DC charging, with the same socket being used for all in most cases. Public level 1 charge points will likely disappear. In the United States, charge point connections are consolidating around the North American Charging System (NACS), covering level 2 and level 3 DC, with this expected to continue over the next couple of years as more original equipment manufacturers adopt the standard. In Europe and Australasia, the combined charging system (CCS) for level 2 and level 3 DC charging is dominant and is unlikely to change going forward, though in Australia and New Zealand, NACS may become more common. As stated in Chapter 2, CHAdeMO is gradually disappearing, meaning that all future charge points targeted for deployment in the West need to provide CCS and NACS. It is unlikely that further consolidation will occur over the next decade, though refinements to each standard will deliver improvements in functionality.

11.2.6 The role of AI in charge points

AI will have an impact on charge point developments over the next decade. The five areas where it is anticipated to have the greatest impact is vehicle to grid, load management, smart charging, 24/7 customer support and cybersecurity. Regarding vehicle to grid and smart charging, AI in charge points should be able to determine when it is best in terms of grid demand and charging cost to charge. For vehicle to grid, the same will also apply to when the vehicle should discharge based on demand and potential income. This should result in a more resilient grid while rewarding charge point users with lower charging costs and/or income generation. AI in load management will enable more efficient charging as only those vehicles with the greatest need to charge at the fastest rate will achieve this. The algorithm will learn over time how to better manage load, resulting in power being drawn from the grid consistently. This should deliver improvements in terms of being able commission smaller grid connections which will cost less whilst taking less time to connect and also in terms of grid resilience as load management will ensure a consistent and predictable power draw. Focusing on 24/7 customer support, AI will enable CPOs, through chat bots, to provide support services much more cost effectively without the need for multiple call agents to be available all the time. These chat bots will triage calls meaning that only the most difficult and challenging ones end up being dealt with by call agents, meaning fewer are required. Chat bots should also reduce support costs in off-peak periods like early in the morning as they can answer queries without these being directed to call agents. AI should also transform cybersecurity through software being able to learn how to deal with incoming and emerging threats.

11.3 The impact of development in EV cars and vans on other modes

The primary focus of this book has been holistically looking at charge point infrastructure from an EV car and van perspective as this area has seen rapid growth in the past 5 years and will continue to expand in the future. It is also the area which will impact most people on a day-to-day basis as they switch from ICE vehicles to EV, requiring a behaviour change across a range of activities like charging and maintaining EV batteries. That being said, it would be remiss of us to not acknowledge the rapid electrification of other modes of transport which is being heavily influenced by the electrification of cars and vans, and the rollout of associated infrastructure. Analysing the impact of electrification on each of these modes and the requirement for charge point infrastructure would have resulted in a book 10 times its current size. There is no doubt that this change will inspire a range of journal articles and new books on the topic. That being said, it is helpful to at least outline the impact that developments in charge point infrastructure for EV cars and vans is having on other modes. These impacts are many though the main ones relate to understanding demand for charge points, right sizing grid connections and making charge points convenient and easy to use.

Regarding understanding demand for charge points, the analytical approach for EV car and van charge points is directly applicable to deploying infrastructure for heavy goods vehicles and buses, albeit in more focused areas due to the fact that the footprint of heavy goods vehicles journeys is concentrated along the strategic road network and that of buses in urban areas. That being said, even in these concentrated focus areas, demand for charge points needs to be properly understood, to ensure that the investment is matched with anticipated utilisation. The same methodology outlined in Chapter 5, albeit with some tweaks, can be applied to ensure a data-driven approach to deployment is followed. For heavy goods vehicles, this will mean charge point infrastructure at existing and new motorway truck stops, whereas for buses, charge points will be deployed in stopping locations, like bus stations and at depots, serving the most vehicles. The fact that this approach is being increasingly used for EV car and van charge point infrastructure should mean that the same approach gradually is adopted by bus operators, haulage companies and any other key stakeholder in deployment.

Turning to the right sizing of grid connections, key learnings from the installation of large-scale en route level 3 DC charge point hubs for EV cars and vans are directly applicable to heavy goods vehicle motorway hubs and depots, with the latter applying to buses. Potential locations need to be able to handle a large MW connection which will require a costly major grid upgrade. To make the case for the upgrade, good utilisation will be required, especially where energy is dispensed at truck service stations, for example. Learnings from deploying en route hubs for EV cars and vans should help in planning similar locations for heavy goods vehicles and buses. Focusing on making charge points convenient and easy to use is a key learning from the deployment of charge points for EV cars and vans as doing so drives up utilisation, strengthening the business case for deployment. Although the infrastructure for heavy goods vehicles and buses differs slightly, the approach still stands. For example, the deployment of pantograph charge points makes starting a charge for the driver much quicker and easier as all they need to do is position the vehicle below the charging pad and press a button to connect the arm to the pad. Because of its simplicity this saves time at the start and end of charging sessions, meaning the charge point becomes available much more quickly for other users. Likewise, as charge point charge rates become greater, the cabling to connect the socket to the vehicle is becoming increasingly thicker to prevent energy loss. This in turn makes it more difficult for drivers to connect their heavy goods vehicles and buses to charge points. Learnings regarding the accessibility of charge points for EV cars and vans should be applied to those for heavy goods vehicles and buses to ensure that drivers with different abilities can connect and charge. Elements of PAS1899 should be directly applicable.

11.4 Summary of future trends

Changes in EV cars and vans and associated charge point infrastructure will transform transport over the next 10 years. Looking at EV adoption across the West, this is set to grow rapidly up to 2035, reaching over 80% in most geographies. This growth will occur alongside advancements in EV batteries, increasing their size, energy density and range. Changes in charging technologies, including in architectures in excess of 800 V, will speed up charging and widespread adoption of vehicle to grid will improve grid resilience while generating an income for drivers. Battery swapping may play a role but this will primarily be for more niche use cases. AI will transform battery management, energy management and support dynamic pricing. Turning to charge points, these will rapidly grow alongside EV adoption up to 2035. Looking at utilisation, most charging will take place in the future at home or at work. We anticipate that for residential charging, this might increase slightly as the number of EVs grows. For destination charging, we expect utilisation to stay broadly on par with current rates. For en route charging, we anticipate that utilisation will grow.

Focusing on emerging trends in business and commercial models, we expect the core business models of owner operator, leasing/rental and concessions to stay the same; however, the changes will focus on how charge points operate and generate revenue and profit. We expect to see an increase in the number of sites providing ancillary services, though this will primarily relate to en route charging hubs. Community charging will also play a role, impacting mainly on utilisation at residential sites. As mentioned above, vehicle to grid from a charge point perspective should ensure grid resilience and enable those using public charge points to generate an income. Solar and battery storage solutions will increasingly play a role in locations with constrained grid connections and will also reduce the size and cost of grid connections, especially for charge point hubs. Valet and robotic/autonomous charging will play an increasingly important role in large destination and en route hubs to drive up utilisation. Wireless charging is set to transform the charge point industry, though further progress needs to be made around standardisation in vehicle and charge point connections. Load management is anticipated to make connections smaller and less costly, whilst enabling more efficient management of multiple charge points. Greater rollout of smart charging will improve the charging experience for customers using public charge points, giving them the ability to charge at cheaper times, especially in residential and destination settings. Ultra-rapid charge points supporting 800 V plus vehicle architectures will reduce the time required to charge, though per kWh charging costs are likely to increase. As the industry evolves further, we expect new public–private relationships for charge point deployment.

Focusing on emerging charge point standards and requirements, we anticipate standardisation of payments, the greater rollout of plug and charge beyond Tesla and the adoption of eRoaming, all of which are likely to increase utilisation. Other standards and requirements include the greater rollout of charge points that are much more accessible with 99% up times, 24/7 customer support, open data protocols and improved cyber and physical security measures. All of these changes should in theory drive up utilisation. AI will play a role in charge points, mainly in the areas of vehicle to grid, load management, smart charging, 24/7 customer support and cybersecurity.

We anticipate that the learnings from charge point deployment for EV cars and vans will continue to feed through to the deployment of infrastructure for heavy goods vehicles and buses. The main learnings relate to understanding demand for charge points, right sizing grid connections and making charge points convenient and easy to use. Deployment of public charge point infrastructure should help right size heavy goods vehicle and bus charge point infrastructure in terms of charge point numbers and charging rates based on demand analysis and appropriate grid connections. Making sure depot and hub charge points are easy to use, like their public charge point equivalents, should drive up utilisation and improve customer experience, especially in truck charging hubs where drivers have the option to use other locations if the experience is poor.

References

[1] International Energy Agency. *Outlook for electric mobility* [online]. 2024. Available from https://www.iea.org/reports/global-ev-outlook-2024/outlook-for-electric-mobility [Accessed 11 November 2024].

[2] Energy Saving Trust. *Forecasting public electric vehicle charging demand* [online]. 2022. Available from https://energysavingtrust.org.uk/wp-content/uploads/2022/09/EST_LGSP-Guide_Forecasting-public-electric-vehicle-charging-demand.pdf [Accessed 11 November 2024].

[3] Enerdata. *Electric vehicles' deployment: Will it reach expected targets* [online]. 2024. Available from https://www.enerdata.net/publications/executive-briefing/electric-vehicles-trends.html [Accessed 11 November 2024].

[4] RTE. *Integration of electric vehicles into the power system in France* [online]. 2019. Available from https://assets.rte-france.com/prod/public/2020-09/RTE%20electromobility%20report.pdf [Accessed 11 November 2024].

[5] Edison Electric Institute. *EEI Projects 78.5 Million EVs Will Be on U.S. Roads in 2035* [online]. 2024. Available from https://www.eei.org/en/news/news/all/eei-projects-78-million-evs-will-be-on-us-roads-in-2035#:~:text=The%20number%20of%20EVs%20on,on%20U.S.%20roads%20in%202035. [Accessed 12 November 2024].

[6] Canadian Vehicle Manufacturers Association, Canadian Automobile Dealers Association, Global Automakers of Canada. *Road to 2035 – Destination: Zero emissions* [online]. 2021. Available from https://www.roadto2035.ca/zev-timeline [Accessed 12 November 2024].

[7] International Energy Agency. *Trends in electric vehicle batteries* [online]. 2024. Available from https://www.iea.org/reports/global-ev-outlook-2024/trends-in-electric-vehicle-batteries [Accessed 13 November 2024].

[8] Crownhart C. *What's next for batteries* [online]. 2023. Available from https://www.technologyreview.com/2023/01/04/1066141/whats-next-for-batteries/ [Accessed 13 November 2024].

[9] Stokel-Walker C. *The future of EV batteries* [online]. 2024. Available from https://www.imeche.org/news/professional-engineering-digital-exclusives/the-future-of-ev-batteries [Accessed 13 November 2024].

[10] Advanced Propulsion Centre UK. *Future lithium availability for EV batteries continues to be a cause for concern* [online]. 2022. Available from

https://www.apcuk.co.uk/news-events/news/future-lithium-availability-for-ev-batteries-continues-to-be-a-cause-for-concern/#:~:text=While%20the%20APC%20sets%20a,large%20battery%20pack%20size%20scenario. [Accessed 13 November 2024].

[11] Etxandi-Santolya M., Canals Casals L., and Corchero C. 'Estimation of electric vehicle battery capacity requirements based on synthetic cycles'. *Transportation Research Part D: Transport and Environment*. [online]. 2023; 114. Available from https://www.sciencedirect.com/science/article/pii/S1361920922003716 [Accessed 13 November 2023].

[12] Blackmore K. *MPG to kWh – electric car efficiency explained* [online]. 2022. Available from https://www.fleetalliance.co.uk/driver-ev/mpg-to-kwh-electric-car-efficiency-explained/#:~:text=Most%20EVs%20will%20cover%20between,it%20will%20cost%20to%20run. [Accessed 13 November 2024].

[13] Nedelea A. *What are 400V and 800V EVs and what does that mean for charging?* [online]. 2024. Available from https://insideevs.com/features/710280/400-volt-800-volt-evs-explained/ [Accessed 15 November 2024].

[14] Institution of Mechanical Engineers. *Fast-charging can damage electric car batteries in just 25 cycles* [online]. 2020. Available from https://www.imeche.org/news/news-article/fast-charging-can-damage-electric-car-batteries-in-just-25-cycles [Accessed 15 November 2024].

[15] Johnson P. *NIO's first 900V EV drive system rolls out with 150 mile fast charge in 5 min* [online]. 2024. Available from https://electrek.co/2024/03/27/worlds-first-mass-produced-900v-ev-drive-system-rolls-out/ [Accessed 14 November 2024].

[16] Cheng E. *China's Nio to expand battery swap services to gain an edge on EV infrastructure* [online]. 2024. Available from https://www.cnbc.com/2024/04/05/chinas-nio-to-expand-battery-swap-services-to-gain-ev-infra-edge-.html [Accessed 15 November 2024].

[17] Alaka T. *Wireless charging for electric cars is inching closer to reality* [online]. 2024. Available from https://www.bloomberg.com/news/articles/2024-02-20/wireless-charging-for-electric-cars-is-inching-closer-to-reality?embedded-checkout=true [Accessed 15 November 2024].

[18] Rauf M., Kumar L., Zulkifli S. A., and Jamil A. 'Aspects of artificial intelligence in future electric vehicle technology for sustainable environmental impact'. *Environmental Challenges*. [online]. 2024; 14 Available from https://www.sciencedirect.com/science/article/pii/S2667010024000209 [Accessed 15 November 2024].

[19] International Energy Agency. *Outlook for battery and energy demand* [online]. 2024. Available from https://www.iea.org/reports/global-ev-outlook-2024/outlook-for-battery-and-energy-demand [Accessed 28 November 2024].

[20] International Energy Agency. *Pathways for the energy mix* [online]. 2024. Available from https://www.iea.org/reports/world-energy-outlook-2024/pathways-for-the-energy-mix [Accessed 28 November 2024].

[21] Climate Change Committee. *The Sixth Carbon Budget: The UK's path to Net Zero* [online]. 2020. Available from https://www.theccc.org.uk/wp-content/uploads/2020/12/The-Sixth-Carbon-Budget-The-UKs-path-to-Net-Zero.pdf [Accessed 28 November 2024].

[22] Field Dynamics. *Great Britain needs 8x more EV chargers to meet 2035 demand* [online]. 2024. Available from https://www.field-dynamics.co.uk/research/gb-charger-scenario/#:~:text=Assuming%20current%20utilisation%20and%20power,larger%20than%20it%20is%20today. [Accessed 28 November 2024].

[23] Enerdata. *Electric vehicles' deployment: Will it reach expected targets?* [online]. 2024. Available from https://www.enerdata.net/publications/executive-briefing/electric-vehicles-trends.html [Accessed 28 November 2024].

[24] Bernard R.M., Hall D., and Lutsey N. *Charging infrastructure to support the electric mobility transition in France* [online]. 2021. Available from https://theicct.org/wp-content/uploads/2021/12/france-evs-infrastructure-transition-nov21_0.pdf [Accessed 29 November 2024].

[25] Groeneweg T. *Electric vehicle adoption and its impact on 2035 German power demand* [online]. 2016. Available from https://studenttheses.uu.nl/bitstream/handle/20.500.12932/23703/160701_German-EV-adopton-2035_TG.pdf?sequence=2&isAllowed=y [Accessed 29 November 2024].

[26] Bermejo C., Geissmann T., Nägele F., Möller T., and Winter R. *The impact of electromobility on the German electric grid* [online]. 2021. Available from https://www.mckinsey.com/industries/electric-power-and-natural-gas/our-insights/the-impact-of-electromobility-on-the-german-electric-grid [Accessed 29 November 2024].

[27] International Energy Agency. *Germany* [online]. 2024. Available from https://www.iea.org/countries/germany/energy-mix [Accessed 29 November 2024].

[28] Statista. *U.S. electric vehicle energy charging demand between 2020 and 2035* [online]. 2024. Available from https://www.statista.com/statistics/952069/electric-vehicle-charging-energy-demand/ [Accessed 30 November 2024].

[29] Pierce L., and Slowik P. *Assessment of U.S. electric vehicle charging needs and announced deployments through 2032* [online]. 2023. Available from https://theicct.org/wp-content/uploads/2024/03/ID-89-%E2%80%93-Chargers-2032_final-v2.pdf [Accessed 30 November 2024].

[30] National Renewable Energy Laboratory. *100% clean electricity by 2035 study* [online]. 2022. Available from https://www.nrel.gov/analysis/100-percent-clean-electricity-by-2035-study.html#:~:text=As%20modeled%2C%20wind%20and%20solar,terawatts%20of%20wind%20and%20solar. [Accessed 30 November 2024].

[31] van Kooten C. *Ottawa's EV mandate unrealistically requires the equivalent of 10 new mega hydro dams or 13 large natural gas plants nationwide within 11 years* [Online]. 2024. Available from https://www.fraserinstitute.org/studies/electric-vehicles-and-the-demand-for-electricity [Accessed 30 November 2024].

[32] Natural Resources Canada. *ICF executive summary* [online]. 2020. https://natural-resources.canada.ca/sites/nrcan/files/Executive%20Summary%20ICF_English.pdf [Accessed 30 November 2024].

[33] Natural Resources Canada. *Electric vehicle charging infrastructure for Canada* [online]. 2024. Available from https://natural-resources.canada.ca/energy-efficiency/transportation-alternative-fuels/resource-library/electric-vehicle-charging-infrastructure-for-canada/25756 [Accessed 30 November 2024].

[34] Natural Resources Canada. *Powering Canada forward: Building a clean, affordable, and reliable electricity system for every region of Canada* [online]. 2023. Available from https://natural-resources.canada.ca/sites/nrcan/files/electricityVisionPaper/Electricity%20Paper_ENGLISH.pdf [Accessed 30 November 2024].

[35] ENERGEIA. *Electric vehicles insights* [online]. 2017. Available from https://aemo.com.au/-/media/files/electricity/nem/planning_and_forecasting/demand-forecasts/efi/2018/final—aemo-ev-insights—september-2017.pdf [Accessed 30 November 2024].

[36] Department of Climate Change, Energy, the Environment and Water. *Australian electricity generation – fuel mix* [online]. 2024. Available from https://www.energy.gov.au/energy-data/australian-energy-statistics/data-charts/australian-electricity-generation-fuel-mix [Accessed 30 November 2024].

[37] Ministry of Business, Innovation and Employment. *Electricity demand and generation scenarios: Results summary* [online]. 2024. Available from https://www.mbie.govt.nz/assets/electricity-demand-and-generation-scenarios-report-2024.pdf [Accessed 30 November 2024].

[38] The RAC. *The road to electric – in charts and data* [online]. 2024. Available from https://www.rac.co.uk/drive/electric-cars/choosing/road-to-electric/ [Accessed 5 September 2024].

[39] Deloitte. UK *EV charging infrastructure update (part 2): Show me the money* [online]. 2024 Available from https://www.deloitte.com/uk/en/Industries/energy/analysis/uk-ev-charging-infrastructure-update-show-me-the-money.html [Accessed 7 December 2024].

[40] European Automobile Manufacturers' Association. *Automotive insights – Charging ahead – accelerating the roll-out of EU electric vehicle charging infrastructure* [online] 2024. Available from https://alternative-fuels-observatory.ec.europa.eu/sites/default/files/document-files/2024-05/Charging_ahead_Accelerating_the_roll-out_of_EU_electric_vehicle_charging_infrastructure.pdf [Accessed 07 December 2024].

[41] Todts W. (ed). *Recharge EU: How many charge points will Europe and its Member States need in the 2020s* [online]. 2020. Available from https://www.transportenvironment.org/uploads/files/0120202020Draft20TE20Infrastructure20Report20Final.pdf [Accessed 07 December 2024].

[42] Colaluce L. *eMobility report: France records 37,850 EVs and already has 129,525 public charging points* [online]. 2024. Available from https://mobilityportal.eu/frances-q1-2024-emobility-insights/ [Accessed 07 December 2024].

[43] National Leitstelle Ladeinfrastrktur. *At a glance – Charging infrastructure after 2025/2030* [online].2021. Available from https://nationale-leitstelle.de/wp-content/uploads/2021/02/NLL_Factsheet_en.pdf [Accessed 07 December 2024].

[44] Federal Ministry for Digital and Transport. Charging Infrastructure Masterplan II [online]. 2022. Available from https://nationale-leitstelle.de/wp-content/uploads/2023/01/Masterplan-Ladeinfrastruktur-II-der-Bundesregierung_Englisch_DIN_A4_barrierefrei.pdf [Accessed 07 December 2024].

[45] National Leitstelle Ladeinfrastrktur. *ÖFFENTLICH ZUGÄNGLICHE LADEINFRASTRUKTUR - Report der Nationalen Leitstelle Ladeinfrastruktur – Mrz – Jul 2024* [online]. 2024. Available from https://nationale-leitstelle.de/wp-content/uploads/2024/07/oeLIS_Report_2024_Mrz-Jul.pdf [Accessed 07 December 2024].

[46] International Energy Agency. *Outlook for electric vehicle charging infrastructure* [online]. 2024. Available from https://www.iea.org/reports/global-ev-outlook-2024/outlook-for-electric-vehicle-charging-infrastructure [Accessed 08 December 2024].

[47] U.S. Department of Transportation. *INVESTING IN AMERICA: Number of publicly available electric vehicle chargers has doubled since start of Biden-Harris administration* [online]. 2024. Available from https://highways.dot.gov/newsroom/investing-america-number-publicly-available-electric-vehicle-chargers-has-doubled-start#:~:text=Since%20the%20start%20of%20the,chargers%20being%20added%20each%20week. [Accessed 08 December 2024].

[48] Natural Resources Canada. *Electric vehicle charging infrastructure for Canada* [online]. 2024. Available from https://natural-resources.canada.ca/energy-efficiency/transportation-alternative-fuels/resource-library/electric-vehicle-charging-infrastructure-for-canada/25756 [Accessed 08 December 2024].

[49] Jarratt E. and Yakub M. *2024 EV charging network report: Canada's public charger installations up 33% in 12 months* [online]. 2024. Available from https://electricautonomy.ca/news/2024-03-07/2024-canada-ev-charging-networks-report/ [Accessed 08 December 2024].

[50] McGinn E., and Garnham D. *EV charging infrastructure - The next frontier* [Online]. 2024. Available from https://www.deloitte.com/au/en/Industries/infrastructure/analysis/ev-charging-infrastructure-next-frontier.html [Accessed 08 December 2024].

[51] Cardinal E. *These are the most EV-friendly states in Australia* [online]. 2024. Available from https://www.drive.com.au/caradvice/how-many-ev-chargers-in-australia-2024/ [Accessed 08 December 2024].

[52] Energy Efficiency & Conservation Authority. *Plugging into the future: How New Zealand is electrifying its roads: An overview of public EV charging infrastructure in New Zealand* [online]. 2024. Available from https://www.eeca.govt.nz/insights/eeca-insights/plugging-into-the-future-how-new-zealand-is-electrifying-its-roads/ [Accessed 08 December 2024].

[53] Energy Efficiency & Conservation Authority. *Public EV charger dashboard* [online]. 2024. Available from https://www.eeca.govt.nz/insights/data-tools/public-ev-charger-dashboard/ [Accessed 08 December 2024].
[54] Energy Efficiency & Conservation Authority. *Electric vehicle charging technology – New Zealand residential charging perspective* [online]. 2019. Available from https://www.eeca.govt.nz/assets/EECA-Resources/Research-papers-guides/EV-Charging-NZ.pdf [Accessed 08 December 2024].
[55] International Energy Agency. *Prospects for electric vehicle deployment* [online]. 2024. Available from https://www.iea.org/reports/global-ev-outlook-2023/prospects-for-electric-vehicle-deployment#charging-infrastructure [Accessed 09 December 2024].
[56] Green Finance Institute. *Demystifying utilisation* [online]. 2023. Available from https://www.greenfinanceinstitute.com/reports/demystifying-utilisation/ [Accessed 10 December 2024].
[57] BBC. *Wireless charging: The roads where electric vehicles never need to plug in* [online]. 2024. Available from https://www.bbc.com/future/article/20240130-wireless-charging-the-roads-where-electric-vehicles-never-need-to-plug-in [Accessed 14 December 2024].
[58] Poultney L. *The EV charging nightmare could soon be over – universal Plug-and-Charge protocol due for US rollout in 2025* [online]. 2024. Available from https://www.techradar.com/vehicle-tech/hybrid-electric-vehicles/the-ev-charging-nightmare-could-soon-be-over-universal-plug-and-charge-protocol-due-for-us-rollout-in-2025 [Accessed 19 December 2024].
[59] Hubject. *Join the largest eRoaming network for borderless electric mobility* [Online]. 2024. Available from https://www.hubject.com/ [Accessed 19 December 2024].
[60] Metere R., Morisset C., Bellekens X., Neaimeh M., Czekster R.M. and Maple C. *Securing the electric vehicle charging infrastructure - state-of-the-art review and recommendations with a focus on smart charging and vehicle-to-grid* [online]. 2022. Available from https://www-users.york.ac.uk/roberto.metere/preprints/Securing-the-Electric-Vehicle-Charging-Infrastructure.pdf [Accessed 19 December 2024].
[61] Massie G. *Russian EV charging stations hacked with 'Putin is a d***head' message* [Online]. 2022. Available from https://www.independent.co.uk/news/world/europe/putin-charging-station-hacked-ukraine-russia-b2026260.html [Accessed 19 December 2024].

Chapter 12

Conclusions

David Trousdale[1]

Across the previous 11 chapters, we have been on journey, starting with the reasoning behind why we are focusing on electric vehicle (EV) cars and vans as key to decarbonising road-based transport. We outlined why the West's approach to deploying charge point infrastructure is central to this future, specifically focusing on the infrastructure supporting the decarbonisation of cars and vans. We then went on to position EVs and charge point infrastructure within the context of the broader understanding of climate change and emerging approaches to tackling it. The electrification of cars is playing a key role, one which started in the early twentieth century, stalled with the mass adoption of the internal combustion engine (ICE) and witnessed a renaissance in the early twenty-first century as the result of technology advancements, government promotion and increasingly affordable mass market EVs. Alongside this we saw the rise of charge point infrastructure including the creation of level 1, 2 and 3 DC charge points and the development of the three stage process to deploy infrastructure.

The first is the planning stage comprised of multiple steps, all focused on the 'why', 'where' and 'what' behind charge point deployment. The second stage focuses on the building of charge point infrastructure. The third and final stage focuses on operating, maintaining and decommissioning infrastructure. Across these stages, there are multiple stakeholders that are involved who must be managed. We then looked to the future in terms of how charge point infrastructure will evolve from a technological, operational, commercial and business model perspective. Across these chapters, there are several key factors that must be considered when deploying charge points:

- **EV batteries and charge points versus hydrogen**: EV batteries offer superior energy efficiency for vans and cars, while hydrogen remains viable for heavy vehicles. Plug-in charge points are essential for EV adoption, with wireless charging and battery swapping considered niche solutions.
- **Grid and energy considerations**: the transition from coal-based power to renewables supports EV growth, with smart charging strategies optimising energy demand and grid capacity.

[1]Principal Advisor - Climate Change and Sustainability, Newcastle City Council, United Kingdom

- **Utilisation and profitability**: understanding projected utilisation is key to charge point profitability and must be considered alongside the capital and development expenditure when installing charge points so that a charge point operator (CPO) can understand when a site is likely to break even.
- **Technological innovations**: advances such as vehicle to grid and vehicle to load integration, charge point planning using artificial intelligence (AI), and GIS-based site selection are becoming increasingly important in current deployments.
- **Stakeholder collaboration**: key players include landowners, governments, transport bodies and community groups, each influencing infrastructure deployment and regulations.

Looking to the future, we know that EVs are key to the decarbonisation of all forms of road-based transport and that EV charge point infrastructure is critical in enabling this revolution.

12.1 Key takeaways from each chapter

12.1.1 Introduction

In Chapter 1, we outlined why batteries are key to the future of road-based decarbonisation of cars and vans, discounting hydrogen as an alternative fuel because it is incredibly energy intensive to produce and less energy efficient than EV batteries. The infrastructure required to enable mass adoption is also not yet available in Western geographies. It will require major investment over the next 25 years; however, this is occurring against a backdrop of rapid and ever-advancing improvements in EV battery chemistries and efficiencies. This is not to say that there is not a case for hydrogen as a fuel for buses, heavy goods vehicles, trains and other heavy vehicles like refuse collection vehicles and diggers. In these use cases, hydrogen fuelling infrastructure is required on a smaller, more concentrated scale with fuelling stations able to serve multiple vehicle types at depots and strategic locations. Synthetic fuels are also discounted because they are carbon neutral, non-zero emissions at the tailpipe, effectively endorsing the current climate status quo. Synthetic fuels are less efficient when compared with battery EVs.

We then focused on why plug-in charge point infrastructure is critical to the EV revolution, discounting wireless charging (for now at least) and battery swapping as niche solutions when compared with the well-established charge point ecosystem. Plug-in hybrids are also discounted as the focus. This is because the technology, though a useful solution in enabling drivers to switch away from ICE vehicles, will not achieve meaningful reductions in emissions. Various studies have actually shown that plug-in hybrids produce more emissions than quoted in standard laboratory tests. Plug-in hybrids may be a stop gap on the road to private vehicle decarbonisation but their impact is small compared with EVs. They are also not reliant on EV infrastructure.

Once we established the focus on EV infrastructure, we then turned to establishing the geographic focus of analysis. The adoption of EVs is occurring globally in all developed and most developing countries. Variances in geographical adoption rates are down to a number of reasons ranging from incentives encouraging EV adoption to lack of infrastructure and greater average distances travelled meaning that range anxiety is putting off the transition. The deployment of charge point infrastructure is largely driven by governance systems and societal structures. These differ when comparing the West and East. As our understanding of Western governance and societal structures is greater than that in the East, our focus is on the Western approach to deployment, including the broader process and steps that must be followed. This includes the stakeholders who should be engaged with to roll out infrastructure effectively. There is no single approach across the West; however, the consistencies in approach mean that broad trends in how to deploy infrastructure can be outlined. This is not to say that elements of this book might not apply in other geographies outside its focus; however, care should always be taken, especially where those locations operate differently to the West.

12.1.2 The drive to decarbonise, a historical perspective and where EV infrastructure plays a part

This chapter provided an overview of the history of EVs as a key transport intervention to tackle climate change and the need for charge point infrastructure to enable their adoption. It outlined the origins of the scientific study of climate change and the ways to tackle it through decarbonisation and net zero. The nineteenth century witnessed the foundation of the theory behind climate change which then subsequently took shape in the early twentieth century. The advancement of our understanding was furthered through various technological innovations in the mid-twentieth century including the advancement of satellites and the use of ice cores to understand historic climate compositions and how this has changed over time. The development of both the science and theory enabled the Intergovernmental Panel on Climate Change (IPCC) to settle on the definition of climate change in 1988, describing it as 'a change in the state of the climate that can be identified... by changes in the mean and/or the variability of its properties and that persists for an extended period, typically decades or longer' [1]. The origins of the term net zero can be traced to 2009 and was defined by the IPCC in 2013 as 'achieved when anthropogenic emissions of greenhouse gases to the atmosphere are balanced by anthropogenic removals over a specified period'. In the same year, the IPCC also defined decarbonisation as 'the process by which countries, individuals or other entities aim to achieve zero fossil carbon existence' [2].

As the scientific understanding of climate change and the need to achieve net zero developed so did the efforts of international organisations and national governments to do something about it through reducing emissions. In 1972, the United Nations, the first organisation to take action, emphasised the importance of international collaboration regarding the environment. By 1988, global warming and the growing hole in the ozone layer captured the public imagination. In that year, the IPCC was established as a forum for the examination of greenhouse warming and global climate change. In 1990, a world climate conference was held which stated that climate change was a global problem of unique character for which a global response was required. Following this, the General Assembly decided to convene in 1992 in Rio de Janeiro, Brazil, the United Nations Conference on Environment and Development. This came to be known as the Earth Summit and resulted in the establishment of a new framework for seeking international agreements to protect the integrity of the global environment. The most significant event during the Conference was the opening for signature of the United Nations Framework Convention on Climate Change (UNFCCC). By the end of 1992, 158 States had signed it. In 1994, it came into force. In 1995, the Conference of the Parties (COP) to the Convention adopted the Berlin Mandate, launching talks on a protocol containing stronger commitments for developed countries and those in transition. Following this, the IPCC's Second Assessment Report was published in 1995 which led in 1997 to the Kyoto Protocol, signed by 192 parties. In 2015, the Paris Agreement established the target to limit global warming to less than 2°C above pre-industrial levels.

The progress of national governments to make tackling climate change legally binding began with the United Kingdom introducing the Climate Change Act in 2008 which set a target requiring emissions reductions of at least 100% by 2050. It also established the Committee on Climate Change, an organisation which sets out carbon budgets every 5 years and tracks progress against these. Various countries have emulated the Climate Change Act, with most basing their targets on reaching net zero by 2050, as set by the Paris Agreement. To achieve net zero, Western countries have planned their interventions around a structural approach. In 2021, the United Kingdom's Net Zero Strategy: Build Back Greener established a structure of interventions transferable to most countries and focused on seven sectors. These were power, fuel supply and hydrogen, industry, heat and buildings, transport, natural resources and waste and greenhouse gas removals. Focusing specifically on transport, using 2018 as an example year, this accounted for 21% of total emissions. When looking specifically at road-based transport, this accounted for 15% of total emissions. Interventions to decarbonise transport are a key focus of Western governments in supporting them to meet their net zero ambitions. Although modal shift towards less carbon intensive transport will play a key role in decarbonisation, the electrification of cars and vans is also critical to the future of transport decarbonisation.

The chapter then provides an overview of the history of EVs, noting that their origin can be traced back to the mid-to-late 1800s, with the early 1900s seeing advancements in EVs. With the launch of the petrol Model T in 1908, EVs would take a back seat for roughly the next 100 years until the 1990s when advancements in batteries enabled the establishment of Tesla in 2004, and what is considered the first mass market electric car, the Nissan Leaf, in 2010. The 2010s witnessed an explosion of EV makes and models which continues to this day. Charge point infrastructure also developed alongside EVs, originating in the early 1900s with basic public charge points but then rapidly advancing in the 1980s and 1990s as a result of the development of smaller standardised on-board chargers (OBCs) in EVs. As EV adoption took off, this gave rise to a demand for public level 1 and level 2 AC charge points ranging from 3.7 to 22 kW and level 3 DC charge points ranging from 50 to 350 kW. From the 1990s onwards, there have been advancements in charge point standards which have coalesced around the Mennekes type 2 connector for level 1 and 2 charging and the combined charging system (CCS) for level 3 DC charging in the United Kingdom and Europe. North America is gradually transitioning to the North American Charging Standard (NACS). Australasia is mainly CCS and type 2 focused though NACS is also being adopted.

The chapter concludes by providing an overview of the history of setting EV and charge point targets, noting that California led the way in establishing these in the 1990s. Norway adopted an incentivisation approach over the same time period making it attractive to own an EV through value-added tax exemptions and reduced road tolls among other incentives.

12.1.3 Deploying EV infrastructure: the overall process and key stakeholders

This chapter gave an overview of the process for installing EV charge points and the stakeholders involved. When looking at deploying EV charge point infrastructure, there are eight key steps that should be followed; however, prior to this, it is important to consider the charging use case or cases charge point deployment supports. There are three use cases that must be considered; en route, destination and residential/work-based charging. En route charging focuses on drivers who are travelling longer distances and require a quick (15–30 minutes) charge to enable them to continue on with their journey. Charge points are level 3 DC chargers with the ability to charge at 50 kW or more per hour. Destination charging focuses on drivers who are travelling to a location and will be stationary for several hours (1 hour 30 minutes to 4 hours) while they carry out other activities. They will not rely on the charger as the only means to charge their vehicle so expect slower charging rates. Charge points are level 2 AC with the ability to charge at between 7 and 22 kW/h. Residential/work-based charging focuses on drivers who will leave their vehicle potentially for 7 or more hours where they live or work. Residential charging is further broken down into off-street private and on-street public charging. Drivers rely on these charge points primarily to charge their vehicle but may supplement this with destination and en route charging. Charge points can be level 1 or 2 AC, typically charging at rates of between 1.5 and 7 kW/h. Work-based and private off-street residential charging is not the focus of this analysis because both are privately driven deployments, whereas en route, destination and on-street residential charging require publicly accessible infrastructure.

When looking at the process that applies to the three charging use cases the initial stage is the planning stage which is made up of multiple steps, the first of which is the decision to deploy charge point infrastructure. This involves determining the 'why' behind deployment and explores what the justification is for charge point deployment. The 'why' differs depending on the charging use case. Following this, the next step determines the 'where' in terms of the geographic target area. Key questions include whether the deployment is part of a broader plan to expand into a new geography, build on an existing network, target a new user group or rationalise/refresh a charge point network. The completion of spatial analysis creates a site long list of potential locations where charge points could be deployed depending on the use case.

The following step then determines the 'what' behind deployment via completion of a business case to further determine the rationale and secure funding for deployment. The business case is based on the Five Case Model and comprises five key parts that must be analysed. The strategic case sets out the case for change and demonstrates how a proposal provides strategic fit. The economic case sets out how the proposal will deliver value for money versus other proposals and the value to society which includes wider social and environmental effects. The commercial case establishes that the preferred option is commercially deliverable in terms of procurement and supply chain. The financial case demonstrates the affordability and funding of the preferred option, including the support of stakeholders and customers, where required. The management case determines that robust arrangements are in place for the delivery, monitoring and evaluation of the scheme, including feedback into the organisation's strategic planning cycle. Completion of this analysis should ensure that the proposal has a strong business rationale and should therefore be pursued. The business case step is complete when approval is granted to progress with the proposal, translating it into a project.

Deployment then moves into the 'how', setting out project plans for deployment at a local geographic level and the governance and approvals required to pass through the various steps to complete installation. Planning deployment at a local level builds on all previous steps and starts with the CPO finalising the site locations at a granular level, having developed a long list during the geographic areas analysis completed earlier. The CPO then completes a detailed project plan for all sites and develops the final financial cost of all elements to deliver the sites, based on capital and development costs using quotes and agreed prices. These elements plus the business case are the key gateway documents, which, once approved, and all associated stakeholders are happy, enable the CPO to move to the deployment stage. The final step of the planning stage is securing approvals and determining the governance arrangements for the next stages. In most cases, the process will involve a mix of internal and external approvals. Completion of all of these steps enables deployment to move into the build stage.

The only step in the build stage is the deployment of charge points. The step is lengthy and can take several months or even years to complete. Following the build stage is the operations stage that contains two steps. The first is operations, maintenance and monitoring and covers all the activities related to managing the charge point network. Customer management and support is the activity that takes up most of the CPO's time and involves a number of tasks including the CPO setting and changing pricing, processing charging session payments and providing a helpline. The other key activity is maintaining charge points. The second step in this stage is upgrading and decommissioning. This covers activities related to ensuring charge points do not become outdated or defective and that if they do, a plan exists to repair, replace or decommission them.

There are eight stakeholder groups involved in charge point deployment. The CPO plays a key role in every part of the planning, build and operation stage, either as a contractor to another entity (e.g., a local authority or state government) or as the primary operator with agreement from the landowner to lease/rent land where the charge points are deployed. The landowner's primary role is to provide access to the land, either by selling it to the CPO or leasing/renting it. The local authority, council or state government can fulfil multiple roles. In a limited number of cases, they act as a CPO. In some cases, they act as the client to a CPO, appointing them to deliver a charge point network on their behalf. In most geographies, the role they fulfil is as the planning authority during the planning stage. This role changes depending on local planning laws and may not be required in some instances. The power provider/distribution network operator's (DNO) role is to provide the CPO or local authority with cost estimates relating to connections and network upgrades during the planning stage, and then completing those connections or upgrades during the build stage.

The regional or national transport body's role can depend on how hands on they are in the planning, build and operation stages. In some cases, they work with the local authority, council or state government to procure a CPO on a regional or multi-regional scale. Rarely, though, will they act as the procurer, leaving this to the authority, council or state government. In other cases, they fulfil a more distant role, setting the charge point policy direction based on national/multi-national policy and strategy guidance. They also determine deployment targets regionally or nationally. The core role of national government and multi-national governing organisations is to set policy and strategy direction. Consultancies by their very nature can be involved throughout all three stages, but only ever as advisors. Resident or citizen groups can play an influential role in the planning, build and operation stages, acting as enablers or blockers.

12.1.4 The strategy, policy and funding environment

This chapter outlined the current (as of 2025) strategies, policies and funding in place from a charge point infrastructure perspective and demonstrated the difference in approaches between countries.

Focusing on the United Kingdom, since 2008 several policy and strategy documents have been produced which set the direction of travel, alongside a funding regime to enable deployment. Significant strategy documents include Powering Up Britain: The Net Zero Growth Plan, the Net Zero Strategy: Build Back Greener, Decarbonising Transport: A Better, Greener Britain, Taking Charge: The EV Infrastructure Strategy and the Transitioning to Zero Emission Cars and Vans: 2035 Delivery Plan. Building Regulations: Approved Document S sets the requirements for the installation of EV charge points in new buildings, major renovations and certain existing structures. Funding to install charge points is derived from the Rapid Charging Fund, Local Electric Vehicle Infrastructure Fund (LEVI) and On-Street Residential Chargepoint Scheme. Standards and regulations are set by the Public Charge Point Regulations 2023. This focuses on improving the consumer experience. The British Standards Institute (BSI) PAS 1899:2022 sets out the specifications that support the building of inclusive charge point infrastructure in the United Kingdom. To date, the United Kingdom has made good progress and should meet the target of 300,000 charge points required by 2030; however, significant effort will be required over the coming years to continue deployment at pace and ensure the right mix of charge points are installed. Anticipated changes in policy and strategy going forward include further funding for local authorities to roll out additional level 1 and level 2 on-street infrastructure, further investment in level 3 DC rapid charge points where there are gaps in the network, equity of public and private charging costs, major changes to how DNOs plan for and deliver upgrades and further promotion of and investment in smart charging and vehicle to grid.

Turning to the European Union, it plays an important role in setting the strategy and policy direction for charge point infrastructure for the 27 Member States. It is each state's responsibility to interpret the policy and strategy direction set by the Union and apply this in their specific geographic area. Funding also primarily comes from Member States. Key documents setting the direction of travel include the European Green Deal, Fit for 55, Alternative Fuels Infrastructure Regulation (AFIR), the Sustainable & Smart Mobility Strategy and Recharge and Refuel Strategy. Progress to date in achieving the 3.5 million charge points required across the Union by 2030 has been good, with most Member States having recently met their annual targets as set by the AFIR to stay on track in relation to the target date. The key challenge however is that growth in charge points is not consistent across Member States, nor is EV adoption rates. The European Union's push back of its EV mandate to 2035 where all vehicles should be EVs is sending the message to the market of an EV adoption slowdown, even when charge point deployment is planned to ramp up. Anticipated changes in policy and strategy for the European Union include relief for European car makers struggling to make the transition to EV, a targeted funding programme for those countries struggling to meet their charge point targets and an extension of the AFIR and mobility strategy charge point deployment targets beyond 2030, to 2035.

Focusing on France, it has a supportive policy and strategy environment which enables the installation of public charge points. Its aims and objectives are articulated through various policies, strategies, funding streams, laws and plans. The core document is the Law on Mobility. Funding to install charge points is derived from the France Recovery Plan (France Relance) and EV Infrastructure Charging Programme (ADVENIR). Further regulations on charge point installation are derived from French building control regulations. French progress to date has been good. Based on the AFIR, France needs 400,000 charge points by 2030 and has so far installed 129,000 (as of April 2024). Anticipated changes to policy and strategy include the development of a French charge point accessibility standard and the adoption of relevant policy and strategy changes derived from the European Union.

Like France, Germany has a number of policies, strategies and funding schemes that enable charge point deployment, though regarding the latter a number of these are no longer available. The key policies are the Charging Infrastructure Masterplan and Deutschlandnetz Initiative. There are various subsidies and incentives, with the key one at federal level being the charging infrastructure for EV programme in Germany (Öffentlich zugängliche Ladeinfrastruktur für Elektrofahrzeuge in Deutschland) which runs until 2025 and established a €500 million fund. Most other funding sources are at state level. Further funding for charge point deployment is available via the European Investment Bank and German state-owned bank Kreditanstalt für Wiederaufbau with the latter offering subsidies up to the value of €50 million per project and the former grants of €30 million plus for environmental protection projects of which charge point infrastructure falls into the category. German progress to date is measured against the AFIR target of 1,000,000 charge points needed by 2030. Germany is not on track, having installed 130,000 charge points as of June 2024, and is likely to significantly miss its target. Funding for deployment is constrained and Germany is currently in a period of stagnant growth paired with an unpredictable political environment. The key anticipated change to policy and strategy is an injection of subsidies to support charge point growth and EV adoption. This would sit alongside the adoption of relevant policy and strategy changes derived from the European Union.

Looking at the United States, up until fairly recently, charge point deployment has been driven by individual states championing charging infrastructure, with California being a good example of a state promoting rollout. However, in the past three years, the Federal Government began to play a major role in financing infrastructure deployment. This had an impact, with the Infrastructure Investment and Jobs Act having kicked off a period of investment and growth in 2021. As part of this Act, $7.5 billion of funding was set aside to support the rollout of a national public charge point network. The fund aimed to deploy 500,000 charge points. $5 billion was allocated to states as part of the National Electric Vehicle Infrastructure Formula Programme. Alongside this, the Inflation Reduction Act, which came into force in 2022, aimed to encourage the transition to EV. Various state initiatives complemented the previously noted schemes, setting public charge point targets. Various other federal funding programmes existed promoting the deployment of charge points. Progress to date has been strong. As of August 2024, there were over 192,000 publicly available charge points across the United States and current progress would suggest that the target will be met by 2029. With the election of President Donald Trump in 2024, various funding programmes have ended. It is likely that charge point deployment will stall, though certain states continue to support infrastructure rollout.

Turning to Canada, the Canadian Government is supportive of charge point deployments and offers a range of incentives and grant-funded projects to encourage rollout alongside the provincial governments. Key policy, strategy and standards include the Electric Vehicle Availability Standard, Electric Vehicle Charging Infrastructure for Canada Strategy, Zero-Emission Vehicle Infrastructure Programme and Canadian Electrical Code and National Building Code. Funding to deploy charge points is derived from the Electric Vehicle and Alternative Fuel Infrastructure Deployment Initiative and Canadian Infrastructure Bank funding for EV charge points. Progress to date is mixed. 679,000 level 2 and level 3 DC charge points are required by 2040. At the start of 2024 27,000 charge points have been deployed but this needs to ramp up massively. Challenges include ensuring a spread of charge points in urban and rural areas as well as upgrading power infrastructure. Charge point reliability and resilience is also a challenge as is consistency in charge point pricing across Canada. Anticipated changes in policy and strategy is unlikely following the re-election of the Liberal Government in 2025. That being said, Canada needs to address a number of factors including the inconsistency in pricing and the provision of continued funding to meet the targets.

When looking at Australia, EV adoption has been slow when compared with other Western nations. The Australian Government, state governments and local governments, however, recognise the importance of EVs and associated infrastructure in reducing emissions, and have developed a number of strategies and policies to encourage EV adoption and charge point deployment. At a national level, the key strategy document is the National Electric Vehicle Strategy that focuses on deploying level 3 DC charge points with minimum charging speeds of 75 kW when four vehicles are charging simultaneously. At a state and territory level, each has its own localised strategy. The legal framework for the installation and operation of charge points is determined by state or territory regulations. There is yet to be harmonisation or standardisation across the country. Progress to date is difficult to assess as there are no charge point targets set. Challenges include the lack of target, sluggish EV adoption rates and lack of consistency of charge point deployments on strategic highways with each state or territory opting for different intervals. The spread-out nature of Australian populations is a further challenge as is the lack of a consistent charge point technical specification or standard across Australia. With the Labor Government having just been re-elected in Australia (as of May 2025), it is not clear if there will be any policy or strategy changes. That being said, there is the need to continue installing en route level 3 DC charge points and focus efforts on residential and destination charge point infrastructure as well as setting charge point targets in numbers. There is also the need to establish standardisation and harmonisation in charge point technical specifications and requirements.

Focusing on New Zealand, EV adoption is strong. There are various policies and strategies that promote both EV up-take and charge point deployment. These include Charging Our Future: National Electric Vehicle Charging Strategy for Aotearoa New Zealand 2023–2035 and the Supercharging EV Infrastructure Work Programme. The latter set the goal to install 10,000 charge points across the country by 2030. In 2017, the New Zealand Transport Agency set the vision to build a nationwide charge point network with level 3 DC charge points deployed at 75 km intervals. Different regions of New Zealand have made progress but at different paces and overall progress to date has been slow. New Zealand will significantly miss its target by 2030 even though EV adoption continues to grow. Challenges include the mismatch between deployment and adoption, the urban focus of deployment and the resources required to deliver electrical network upgrades and upgrades to supply. Anticipated changes in policy and strategy include further co-investment in charge point deployments by the New Zealand

Government and CPOs and further government investment in upgrading the grid to enable charge point deployment and power demand management. Further government funding (in most cases without co-funding) will be required in rural and disconnected areas to deploy charge points so as to ensure equitable access.

12.1.5 Demand and supply modelling including predicting utilisation

In Chapter 5, we first focused on power demand and supply from a British perspective. The sources, locations and carbon dioxide intensities of power supply are outlined, including how this has changed over time. The distribution of different power producers is geographically determined with most wind and hydro production located in Scotland, whereas production from gas is concentrated mainly in England. However, the rapid expansion of large wind farms in the North Sea which connect to England is changing this distribution. The grid recognises different types of generation known as 'dispatchable' and 'non-dispatchable'. Dispatchable generation can be externally controlled, for example a gas power station, or a pumped hydroelectric plant. These can react quickly when in operation to changes in demand, although start-up from 'cold' is much longer for gas plants. Non-dispatchable generation covers the renewable generation of wind and solar, which, although it can be predicted to a large extent, cannot be fully controlled. Given the requirement to generate electricity producing minimum carbon dioxide, these intermittent generators are generally allowed to fluctuate to maximise their generation, and the dispatchable fossil fuel plants modulate their output accordingly.

The electrical energy generation mix has changed considerably since 1950 when coal provided more than 97% of the feedstock for electricity production. By 2010, the generation mix had not changed much, but gas was more dominant. It is only in the last decade in the United Kingdom that the grid has seen a considerable reduction in carbon intensity driven primarily by increased wind generation and an elimination of use of coal for electricity production. The modern grid is very different from the grid of only 25 years ago in terms of diversity of generation: geographically, temporally and spatially. This change has had major implications for electricity generation carbon dioxide intensity and has resulted in this falling from 500 g/kWh in 2000 to under 200 g/kWh in 2023. The reduction in emissions per unit of electricity produced when coupled with the increasing penetration of EVs is the driving force behind the decarbonisation of the road transport sector. Focusing on energy demand measured through consumption there are three major consuming groups: industry, domestic and commercial. Over the last 20 years, there has been a significant decline in electricity demand, reducing from a peak of 349 tWh in 2005 to 269 tWh in 2023. Energy efficiency initiatives have reduced demand across the board as rising costs of electricity became more important to consumers. Domestic demand has seen a 22% fall over this period. EV charging is increasingly driving demand, though primarily during off-peak periods as a result of smart tariffs. The chapter highlighted an interesting observation that energy supply is not equally distributed across the United Kingdom in terms of location and source, meaning that some locations have an abundance of spare power capacity with a low emissions intensity. Generally, at the moment, there is little thought by CPOs regarding the emissions of the electricity used to charge vehicles.

In the second part of Chapter 5, we focused on the methodologies behind modelling charge point demand, utilisation and revenue. When deploying charge points, the overarching methodology remains the same regardless of the three en route, destination and residential use cases. The only aspect that changes is the data that is used. The methodology is broken down into four parts. Demand analysis is the first step which is then used to create utilisation projections and financial models. This all then feeds into determining real-world locations for charge point deployments. The outputs of this analysis forms part of the business case and planning of deployments at a local level. Demand analysis focuses on understanding how current and future demand for charge points in a specific area translates into potential locations for deployment. For each use case, the data required to conduct this analysis differs slightly and ranges from traffic flow and EV adoption rates locally to socio-demographic and off-street charging availability, especially in residential settings. The broad approach stays the same and involves analysing the data, usually via a geospatial tool, and then identifying potential sites based on this data. Whereas demand is about the potential vehicles which might charge, utilisation analysis is about predicting how demand translates into actual and predicted use. The two main ways utilisation is measured is time based and energy based. When applying these in the real world both have their value. The outputs from demand and utilisation modelling are used to financially model the potential revenue generated at sites which then can be compared with the capital, development and operational costs of deploying and running a charge point. Unlike demand and utilisation modelling, the financial modelling methodology is identical regardless of the use case. The outputs of demand, utilisation and financial modelling can be combined with power supply data in a localised setting to determine the most viable sites based on profitability and practicality.

12.1.6 Technology considerations

In Chapter 6, we looked in detail at the various technological aspects of EVs, starting with the EV itself. EVs have a high-voltage battery, sometimes referred to as the Main Battery, operating at normally close to 400 or 800 V, and an auxiliary 12 V battery used for low-power circuits and to energise the main battery contactors that connect to the drivetrain. Batteries are usually lithium ion with variations in chemistry such as lithium ferrophosphate and nickel metal hydride. EVs usually charge via two methods once plugged in. The first uses an OBC to convert the incoming AC power supply from a level 2 charge point to the right parameters for the car battery. The second is where level 3 DC charge points supply directly to the battery via the battery management system (BMS). This is commonly used for rapid charging. Analysis of the United Kingdom EV car fleet battery sizes shows that the majority have a lithium ion traction battery with a size from 40 to 80 kWh. Nearly all cars have OBC ratings matched to a single-phase home supply which is less than 8 kW. This means that the time to charge is directly proportional to the battery size. It is only when DC charging where charge rates increase, with more expensive models able to charge at faster rates.

Focusing first on AC level 2 charge points, the incoming supply is from the grid voltage and frequency. In the United Kingdom and Europe, this is 230 V and 50 Hz, whereas in North America, this is 110 V and 60 Hz. Three-phase supply is needed for high power chargers such as AC above 8 kW. Most EVs have an OBC with a power rating of 11 kW. The AC charge point consists of a control system, contactors and communication methods with the vehicle and potentially the home mains supply. Public level 1 AC charge points can be those with standard power levels of up to 5 kW, and situated in places where the available charging time is more than 4 hours – for example at workplaces or community facilities. Level 1 charge points are also installed in homes with most charge points rated in the United Kingdom at either 3.6 kW or level 2 up to 7.6 kW. Few homes in the United Kingdom have 415 V three-phase power supply, the overwhelming majority being a single-phase supply at 230 V. Level 1 and 2 AC charge point innovations include vehicle to load, where power is used to power items locally or vehicle to grid, where power is fed back into the grid. Advantages of vehicle to load and vehicle to grid include supporting grid stability and generating an income for the driver. Level 1 and level 2 single-phase charge points tend to be used at home or work. Other charging situations where fast rates are suitable include places of interest, overnight stops at hotels and long-stay parking.

Turning to DC charge points, the system provides power directly to the BMS, which controls the battery health, makes estimation of state of charge and health, and other functions such as thermal management, switching and safety. In general, DC charging allows higher charging speeds, from 15 to 350 kW. DC charging is also more efficient since the limited power OBC stage is omitted, along with associated losses. When DC charging, the charge point will constantly communicate with the vehicle BMS to control the power transfer, taking into account the vehicle battery conditions, the incoming power supply and the vehicle charge acceptance rate. Level 3 DC charge points tend to be used to provide power during longer journeys in en route settings though lower power charge points may exist in locations where vehicles are stopped for less than an hour. Level 3 DC charge points, though advantageous in enabling rapid charging, may result in damage to the battery. As a general rule of thumb, higher power means higher cost and higher energy losses.

The chapter provided an overview of the various platforms used to manage charge points including how to find a charge point and the systems that enable payment. There are various charge point standards supporting cross network use. These include standardisation of payments, the greater rollout of plug and charge and the adoption of eRoaming. Charge point management platforms and back office systems are vital to CPOs as they enable them to effectively track utilisation and manage payment as well as providing the basis for customer relationship management. The chapter concludes by providing an overview of other technology developments linked to charging infrastructure including battery swapping, solar canopies and battery storage at charge point stations. It then summarised what the various technological aspects of charge points mean for planning charging stations with multiple charge points, including the design and layout.

12.1.7 Planning for deployment

This chapter summarised the planning for deployment process. This comprises five key steps focused on the 'why', 'where', 'what' and 'how' behind the deployment. The initial decision to deploy charge point infrastructure is the first step in understanding the 'why' behind charge point deployment. Understanding the outcomes at the start is critical to ensuring a successful deployment, as is understanding in detail the use case which a charge point deployment is seeking to satisfy. There are three use cases for public charging which are the focus of deployments. These are en route, destination and residential charging. It is critical that whoever is installing and operating the charge points understands the use cases they are seeking to satisfy from the very start of the deployment project. This determines a myriad of factors including the type and number of charge points to be installed, the types of locations to be installed in and the likely demand and utilisation. In addition to understanding the outcome/s and use case/s driving charge point deployment, it is important to understand which business model will be adopted as this can set in motion a range of different tasks further along in the planning, build and operation stages. There are three models: owner operator, owner operator leaser/renter and concession.

Determining the 'where' in relation to charge point deployment follows on from the 'why'. At this step in the process, a local authority or CPO does not necessarily need to know the exact location to deploy charge points. What they must do, however, is adopt an analytical approach that aligns with the target use case as this will determine what information is needed to be analysed to identify deployment locations. This is usually achieved through a geographic information system that gathers information from a number of sources and can then display this through different perspectives dependent on target use cases. The output of determining the geographic area to deploy step is a long list of potential sites which feeds in to the process to develop the business case and planning for deployment at a local level step. Completing the business case is primarily about setting out the 'what' behind deployment, drawing on information already gathered earlier in the process and supplementing with further analysis to build the case. There are several different approaches to writing business cases; however, a commonly used structure, in the United Kingdom at least, is the Five Case Model as outlined in Section 12.1.3.

Planning deployment at a local level takes the long list developed earlier in the process and focuses on determining specific locations at a granular street level so as to finalise the deployment locations. This requires a range of data sources to ensure analysis considers real-world opportunities and challenges impacting on the viability of a location. Once this data is layered and visualised, this enables the identification of certain locations suitable for deployment. Completing this prioritisation will result in a short list of sites. The short list is further refined through analysis of the aspects which will impact on deployment including virtual and physical site visits. This helps understand whether a site is suitable for deployment based on features at the site. The final task is the finalised short list collating the relevant data for each location. A detailed site design is developed at this stage. Alongside this, a financial model and project plan is completed covering each site, as well as combining all sites into one overarching model and plan. These documents, alongside the short list and detailed site designs for each location, are critical to the governance and approvals process.

The final step of the planning stage is securing approvals and determining the governance arrangements for the next stages. Where a CPO is driving deployment, this involves consolidating the outputs from earlier steps and finalising a business case with an associated business plan to secure investment. Once this is done, the CPO can then set in motion the various approvals needed to be put in place for the build stage. This includes finalised quotes for power connections and upgrades if required and completed civils plans and associated quotes to carry out the work at each site. Planning applications may also be required for certain deployments that are submitted to the local authority. Traffic regulation approval, if applicable, is required in on-street charge point deployments where bays need to be designated specifically for EV charging and is sought from the local transport authority. Planning to purchase, lease or rent land, including associated costs, should start at this point. If any of these changes result in significant changes to projected costs, the business case may need to be re-visited and adjusted to ensure the deployment still makes financial sense. Once these final tasks have been completed and funding secured for deployment, the CPO can move into the build stage. The CPO should then put in place the management structure required for the build stage. To ensure all governance steps are taken into consideration, this involves mapping out the decision-making points. Planning for these ensures, there are no surprises in the build and operation stages.

12.1.8 Deploying charge point infrastructure

In Chapter 8, we outlined the key elements of charge point deployment. At the start of deployment, the project manager must ensure that they have as much detail to hand for the proposed installation, with sufficient information about the location of the deployment and power supply to progress the project. This includes a detailed understanding of the nominated site, with its given location, size and transport accessibility, the proposed project design and that the civils work has been factored in to the project. The project manager should also establish the key stakeholders, including the charge point supplier/other suppliers, local or state authority or landowner, the electrical contractors, the DNO/s and the civils contractors, and make sure initial engagement with them has begun.

The chapter outlines the project management methodologies that should be followed as part of deployment. This includes project planning and the production of an associated project initiation document (PID) as well as establishing the contract management approach to ensure delivery and the consequences if things do not go to plan. Change management is key and sets out the process if something changes during deployment. Risk management assesses and mitigates identified risks. Further methodologies include resource management, budgeting and cost control, quality management, progress tracking and stakeholder management. As charge point deployment is classed as construction, the Construction (Design and Management) (CDM) regulations must be adhered to. This involves defining the roles of the client, principal designer and principal contractor and ensuring each fully understands the responsibilities and activities linked to these. Key deliverables completed by the principal contractor include the construction phase plan, a method and risk assessment, a site works plan and an accident reporting mechanism. Principal designer and designer responsibilities include ensuring the client knows their responsibilities throughout the deployment. Principal contractor and contractor responsibilities include managing the work they are delivering, complying with direction given by the principal contractor and preventing unauthorised access to the site.

When beginning deployment, the project manager must initially consider the charge point type and numbers planned as this directly affects the costs related to electrics. Site geology and earthworks required to enable the design to be built are also relevant. Public access roads may need to be built or modified, and proposed on-site facilities built. The CPO should have a good sense of what the costs will be from experience. All of these factors should be detailed in the business case and reviewed during mobilisation and deployment. The business case should give the project manager enough information to draft a PID, outlining all the high-level details such as budget timelines, project objectives and other relevant information the project manager needs to consider when working out the project plan. The project manager will review the PID with the project stakeholders to ensure alignment with the overall business objectives and will repeat this throughout the project.

Once the project's objectives are clearly understood by the project manager, they can then put in place the necessary project plans and infrastructure to deliver the project. They can mobilise the installation, deliver the construction and commission the charge points. The first part in mobilisation is localised planning, gaining permissions and securing permits. DNOs must be informed before any EV charge point is installed. Data requirements in terms of LAN cabling, land ownership in terms of confirming who owns the land and wayleaves for power and access all play a key part in mobilisation. Other elements including ensuring the relevant local government permissions are in place (including those from the national government transport agency if relevant|) and that stakeholder consultation has been considered. Once all of these steps have been completed, and any subsequent changes to the design and/or business case have been made, the project manager then seeks approval from their own company's management to commit to the project by establishing contracts with suppliers to provide the required services. Achieving this completes localised planning, resulting in the transition to the construction step.

During the construction step, the principal contractor takes over delivery at this point. They refine the construction phase plan as defined by information given over by the principal designer. The principal contractor will henceforth manage the build step, ensuring that materials and skills are delivered to site as and when required to enable the construction plan to proceed as scheduled. Construction will involve the groundworks, cabling and installation of associated feeder pillar. The power distribution board, site load management, monitoring and control system will also have been installed as part of the construction step. This includes the cabling to provide power from the DNO connection point to the power distribution board. Foundations will be put in place for each charge point unit in preparation for installation. The charge point can then be lifted onto the foundations and then connected to the site power and data cables. With installation complete and electrical power provided to the charge points and on-site control systems, the charge points are then powered up in sequence.

Once all the electrical equipment on-site is powered and certified as safe to use, commissioning involves testing all functionality and is completed by following the manufacturer's instructions. At the end of the construction step, the site is ready to be made available for the public to use and operation of the site is handed over from the project deployment team to the CPO's network management team, who will operate, maintain and periodically test all equipment throughout the operational life of the site. One of the last steps in making the site ready for the public is ensuring the site's details are added to CPO's website and charge point finder tools as well as publishing links to third-party websites and apps to enable EV drivers to establish availability of and use the charge points.

12.1.9 Operating, maintaining and decommissioning infrastructure

In Chapter 9, we focused on the role of the CPO in operating, maintaining and decommissioning charge points. Once a charge point goes live, effective management of the entire operational ecosystem is critical to a successful network. This requires a system of continuous performance monitoring, periodic reporting, review and subsequent trend analysis to identify outliers in anticipated performance. Key roles in the CPO include the customer support team who engage with drivers and run the help line, the scheduler who plans field service activities, the field operatives team who carryout maintenance and the operational management team who manage the contracts with clients and track overall performance. The first three roles can be wholly or partly outsourced to a different provider who provides a service to the CPO.

In the United Kingdom, the operations of charge points are regulated by the Public Charge Point Regulations. Central to managing the charge points is the customer management system (CMS) which holds all the key information recorded including the relevant client and contract the charge point is assigned to. The CMS enables drivers to locate, start and end a charging session. This sits alongside a customer support function that provides help to drivers when it is needed. With level 3 DC charging, the CMS can monitor and report the live state of charge. This is unfortunately not available for level 1 and level 2 AC charging. At the end of a charging session, drivers end this with their app (or via their vehicle) and the CMS calculates the amount of energy provided during the charging session, applies the tariff relevant to that charge point at that time of day, calculates the total owed by the driver for the charging session and then bills the driver. Energy management is critical to this and ensures the driver is billed the right amount of kWhs used. Each charge point has a Meter Point Administration Number which measures energy usage. At the end of each month, the CPO sends each DNO a list of all its charge points installed in that DNO's area and energy provided so that the DNO can continually monitor and understand its network's performance and consider when future requests are made to connect new electrical loads to the electricity grid.

When managing their account, drivers should be able to self serve as much as possible, enabling them to update personal and payment details, retrieve statements of accounts and records of charging history, 24/7. However, some tasks do require human interaction, especially where the driver has a complaint due to various factors including incorrect charges or a poor charging experience. If they require help, drivers can call the 24/7 staffed helpline. The support agent will likely use the CMS as the live tool to remotely investigate any issues reported by a driver. Issues are triaged and support tickets raised where a greater deal of support is required. Calls usually relate to user issues, payment issues or the reporting of problems with charge points.

The CPO monitors various elements core to operations, including reliability. Level 1 AC charge points are technically not regulated by the Public Charge Point Regulations, though it is reputational best practice to ensure 99% uptime for these and level 2/level 3 DC charge points, especially where CPOs are assessed based on their network averages as part of tender exercises. Monitoring also includes tracking contractual requirements, with network availability and reliability both being key performance indicators. Utilisation is monitored by the CPO and is a measure of how often a charge point is used, and the energy dispensed. A CPO's operational management team regularly reviews individual charge point utilisation over the life of a contract.

Charge point maintenance is key to reliability. Charge points can have a hard life, with internal components subjected to heat stress during charging sessions. The body of a charge point can be subjected to vehicles striking it, people vandalising it and graffiti. Each charge point manufacturer provides product-specific guidance on how to maintain their charge point, and the details of the periodic inspections and tests that need to be conducted. Planned maintenance is anticipated, scheduled and usually relates to periodic inspection and testing of charge points as required by the CPO's best practice, manufacturer's recommendations or regulatory requirements. Another objective for planned maintenance may be the planned upgrade or modification of certain components within a particular model of charge point.

Unplanned maintenance is where the charge point fails, with some issues rendering the charge point difficult to use whilst others render the charge point completely inoperable. In some cases, the CMS will give the operations team an indication of a failure or issue. At other times, it is only when a user phones the

CPO's helpline to report an issue that the CPO becomes aware of this at the charge point. Once a service ticket has been created to log the issue, the CPO may request a scheduler to review tickets and plan and coordinate field service activities in the most effective manner. Given the unexpected nature of unplanned maintenance, field service crews must be equipped with all the necessary spares, tools, product training and skills to be able to quickly and effectively repair charge points in situ. An unplanned maintenance ticket needs to be resolved within a certain timeframe as per an agreed service level agreement, usually specified in the contract with the client or with the maintenance company. Typically, issues that pose a risk of safety to people require resolving within 2 hours. Tickets raised that address issues that prevent charge points from operating typically have 2 days to be resolved. The quicker the resolution, the less impact on the availability of the charge point.

When a charge point reaches end of life, this is usually down to two factors. The first is that the charge point location is no longer suitable due to ownership or viability, whereas the second is when the equipment itself reaches the end of its lifespan. In both instances, slightly different approaches are required. When a charge point location is no longer suitable due to ownership or viability, removal and reinstallation somewhere else is run as two separate projects. Relocating is an expensive option and the need to do so requires justification and possibly a separate business case, though contractual terms may require it as may commercial requirements. When charge point hardware reaches the end of its lifespan, it can either be replaced with a more powerful or more reliable model, this being treated as a mini project, or it can be retired, either due to age or because the CPO is exiting the location. In all end-of-life scenarios, the CMS must be updated to ensure the data held is accurate.

12.1.10 Stakeholder perspectives

This chapter looked in detail at each stakeholder and analysed five key elements, roughly based on SWOT (strengths, weaknesses, opportunities, threats) analysis. The central stakeholder around which all other stakeholders revolve is the organisation delivering or commissioning the deployment. This tends to be the CPO or local/state government. The CPO is key to any charge point deployment. Charge points can be installed without them but this is usually on a small scale and done privately. The CPO acts as the organisation delivering charge points where they are expanding their own network independently. They may also be commissioned by another organisation (e.g., a local authority). A key strength is that they are agile in terms of being able to draw on resources to be able to deploy charge points. The main weakness is that they are sometimes overly profit driven which can mean that they only deploy in locations with the highest projected demand and utilisation. CPOs are key enablers in charge point deployments. Their primary aim is to increase demand and utilisation, thereby generating a profit for the business and, if applicable, their shareholders.

Turning our attention to the landowner, their role primarily relates to the use case a CPO or local authority is seeking to address. Landowners are usually involved for en route and destination charge point deployments though they can be involved to a lesser extent in residential charge point deployments where there is limited space on the public highway to install charge points. The landowner might also be passive in the deployment process and be approached by a CPO or local authority with a request to buy or rent/lease the land. In this case, they may appoint a land agent interested purely in selling or leasing the land for a profit. They may also be active in the process, seeking out a CPO to deploy charge points on their land. A local authority can also take the form of a landowner. The landowner may sell the land to the CPO, lease/rent it or enter into a hybrid leasing/renting/revenue share model. The local authority, council or state government is an important stakeholder in the charge point deployment process. The local authority, council or state government can fulfil multiple roles. In a limited number of cases, they act as a CPO. In some cases, they commission a CPO, appointing them to deliver a charge point network on the authority's behalf. In this instance, they also tend to act as the landowner where deployment is taking place on the public highway or public car parks they own. In most geographies, the role they fulfil is as the planning authority during the planning stage. This role changes depending on local planning laws and may not be required in some instances.

Turning to regional or national transport body/department, they are not always involved as a stakeholder. In the United Kingdom, their role can depend on how hands on they are in the planning, build and operation stages. In some cases, they work with the local authority, council or state government to procure a CPO on a regional or multi-regional scale, the logic being that coordinating this achieves scale and with scale comes the likelihood of a better commercial deal that benefits multiple authorities. Rarely, though, will they act as the procurer, leaving this to a lead authority, council or state government. In other cases, they assume a more distant role, setting charge point policy direction based on national/multi-national policy and strategy guidance and determine deployment targets regionally or nationally. If they are acting as either the coordinator of a procurement on behalf of local authorities or as the direct procurer of a CPO on the strategic road network, they will want to achieve value for money through the final contract. Where they are deploying in a regional destination or residential setting, they will want to achieve a just transition. If they are deploying on the strategic road network, they will want to ensure the proposed sites have good demand and projected utilisation as they will want their investment to deliver maximum impact in terms of supporting en route charging. Where the regional or national transport body or department is purely acting to provide strategic direction, the main opportunity they want to realise is driving up EV adoption.

The distribution system operator (DSO) is a key stakeholder in every charge point deployment. They are responsible for maintaining the functioning of the distribution network and acting as the gate keeper in terms of approving or rejecting connections and upgrades to the grid connection if required. In the United Kingdom, each DNO functions slightly differently, with most providing an online tool that enables organisations, including CPOs, to understand power capacity in their area to enable planning. The DSO/DNO's role is to provide the CPO or local authority with cost estimates relating to connections and network upgrades during the planning stage, and then complete those connections or upgrades during the build stage. Focusing on national governments including multi-national organisations, their role is hands off throughout the charge point deployment process, setting the direction of travel in terms of the country achieving net zero and often also providing the funding underpinning deployment. The core role of national governments and multi-national governing organisations is to set policy and strategy direction. The national government or multi-national governing organisation also has the power to incentivise EV growth and also sets building safety and consumer regulations and standards.

Turning to consultancies, they are not always involved in charge point deployments but can play a role in providing capability and capacity usually to local authorities. They also, in some cases, provide this to CPOs where specialist skills or expertise is required as part of delivery and to national governments/multinational organisations in shaping policies and strategies. Consultancies will always have high influence and impact relating to the specific work they have been commissioned to do. Their involvement is entirely dependent on what work they need to complete. Focusing finally on resident and citizen groups, they should always be involved in the charge point deployment process in some shape or form to help determine the best location for charge points and also to gain their support if installation causes any disruption during the process. Resident or citizen groups can play an influential role in the planning, build and operation stages, though this only applies to destination and residential use cases. If the community is not at least consulted in some way, this might result in poor utilisation when the proposed sites become operational, especially in destination and residential charging use cases.

12.1.11 Looking to the future: where is EV infrastructure heading

This chapter focused on the future of EVs and charge points, outlining the trends in the vehicles themselves and how this will impact on charge points. It gave an overview of how developments in EV cars and vans and charge points is influencing the electrification of other modes, including heavy goods vehicles and buses. EV adoption is anticipated to increase globally over the next decade. Major changes in battery chemistry, size and capacity are expected to occur, resulting in longer range vehicles. Changes in vehicle technologies are also expected, including wireless charging, battery swapping in some geographies and even autonomous charging. AI will no doubt also have a major impact on how EVs function. All of these changes will heavily influence charge points.

According to the International Energy Agency, global car EV growth is projected to increase from just under 40 million in 2024 to between 390 and 440 million vehicles by 2035 [3]. Europe including the United Kingdom, North America and Australasia are all anticipated to see large increases, albeit the projections may prove unrealistic and the pace of growth slower than anticipated. As EV adoption rates ramp up globally, so will the demand for batteries. Price per kW will fall further. In addition to price, other factors are set to transform the battery market in the future. These are innovation-driven improvements in batteries and affordable chemistries. Battery size and capacity are anticipated to increase. There are several emerging technologies that are likely to gain traction over the next decade. These include vehicle architecture advancements, battery swapping, wireless charging and vehicle to grid and vehicle to load. Where AI will play a relevant role to charging is the battery management software and systems, energy management during charging and dynamic pricing.

The anticipated increase in the number of EVs over the next 10 years is expected to have a major impact on the power demand, the number and type of charge points, the use cases where charging is predominantly done and the business models underpinning charging. It will also impact on innovations in charge point technologies, innovations in commercial models, new standards and requirements and the transformation of the whole charging process through AI. By 2035, most charging will be done at home or at work; however, there will still be a role for public charge points in meeting residential, en route and destination needs. Each geography has projected public charge point requirements out to 2035. All predict rapid growth over the next decade.

When analysing future utilisation patterns, most future EV car and van charging will take place privately at home or in workplaces, with the remaining charging taking place publicly via level 1 or 2 charge points in residential and destination settings or via level 3 DC charge points in en route locations. When looking to the future of business and commercial models that will enable charge point growth, the various anticipated innovations in charge point technologies should also be considered. The future business and commercial models are focused on increasing the revenue and profit of each charge point. Potential changes include the expansion of ancillary services, community charging, vehicle to grid, solar and battery storage. Others include valet charging, robotic and autonomous charging, large-scale wireless charging, load management, smart charging, ultra-fast charging based on 800 V plus EV systems and new forms of public–private relationship. Amongst all of these innovation, the one that has not been mentioned but has been discounted is wireless charging integrated into the road.

The emerging trends in charge point standards and requirements mainly focus on improving the interoperability between different CPOs and improving the customer experience. Though not exhaustive, interoperability changes are expected to have the greatest impact on how charge points operate. These are the standardisation of payments, plug and charge, and eRoaming. Although interoperability is the key focus, other changes to standards and requirements are anticipated to make charge points open, safer, more secure to users, more accessible and better supported. These include 99% up time and 24/7 customer support, accessible EV charge point standards, open data protocols, cyber and physical security. When looking at future standards relating to charge point connectors, most geographies anticipate consolidation around a single charge point socket for level 2 and level 3 DC charging, with the same socket being used for both in most cases. Level 1 charge points in public settings will increasingly disappear. In the United States, charge point connections are consolidating around the North American Charging

System (NACS) covering both level 2 and level 3 DC, with this expected to continue over the next couple of years as more original equipment manufacturers adopt the standard. In Europe and Australasia, the CCS system for level 2 and level 3 DC charging is dominant and is unlikely to change going forward, though in the latter geography NACS may become more common. CHAdeMO is gradually disappearing. AI will have an impact on charge point developments over the next decade. The five areas where it is anticipated to have the greatest impact is vehicle to grid, load management, smart charging, 24/7 customer support and cybersecurity.

Although the focus of this book is on charge point infrastructure for EV cars and vans, the development of charge points for these modes is influencing the infrastructure for other modes. The impacts are many though the main ones relate to understanding demand for charge points, right sizing grid connections and making charge points convenient and easy to use. In conclusion, changes in EV cars and vans and associated charge point infrastructure will transform transport over the next 10 years. We anticipate that the learnings from charge point deployments for EV cars and vans will continue to feed through to the deployment of infrastructure for heavy goods vehicles and buses.

12.2 Key factors common across the chapters

There are several key factors common across the chapters which must be considered when deploying charge points. These are that the use case must be understood in detail at the start of the process, utilisation is key to everything and properly costing the capital, development and operational costs is central to profitability. Furthermore, construction and deployment of charge points require detailed scheduling. Maintenance and reliability are key to the success of a CPO and the stakeholders must be managed effectively throughout the process. Plugging in to charge will continue to play a key role in the future. Geographies are entering into a period of intense deployment to meet their 2030 and 2035 targets. Perhaps the most important of all is that EVs are the future for private vehicles, regardless of what the critics say.

Turning first to understanding the use case in detail, what should be clear throughout the planning, build and operations stages is that en route, destination and residential charging determines everything. This includes the demand modelling, the technologies required to meet charging needs, the funding available, the costs and timeline to deliver including the grid upgrades, the return on investment and the ongoing operations and maintenance activities and costs. When deploying charge points, a location can satisfy one, two or all three use cases; however, it should be clear when planning for sites which charge points satisfy which use case. The CPO must ensure all the subsequent aspects mentioned above are considered in the planning, build and operations plans. Turning to utilisation, this plays a central role in terms of determining the profitability of a site. Failure to predict this may result in installing a poorly used charge point which not only loses money for the

CPO but also fails to support other objectives behind deployment, including enabling the EV transition and ensuring this is equitable and accessible to everyone. Utilisation must be considered by the CPO, landowner and local/state government to ensure the success of a site. Utilisation should also be analysed through the lens of the different use cases as each use case will have a different utilisation profile.

Properly costing the capital, development and operational costs is key to site profitability. Capital costs can fluctuate based on the charge points installed, primarily due to the rate of charge. Development costs are primarily related to connection costs, whereas operational costs are primarily determined by power import and charge point maintenance costs. Development costs, for example, can fluctuate hugely based on quotes from the DSO/DNO. This is mainly based on the size of the connection. Wrong sizing a connection and then installing the wrong technology, combined with installing in an area with poor anticipated utilisation, will mean a site may take several years to become profitable and, in some cases, this may never be achieved. Focusing on stakeholder management, there are multiple moving parts behind a charge point deployment and throughout the entire process each stakeholder can influence the cost and quality of installation and the time it might take. The CPO must put in place the correct mechanisms to manage the stakeholders to ensure success.

Construction and deployment of charge points require detailed planning. The proposed sites need to have been selected based on utilisation and demand viability. The project to install the charge points must be managed and time sequenced to factor in permitting and approvals. Civils and groundworks can be lengthy depending on the site. The actual deployment of the charge point is often the quickest part of the process and is closely followed by the testing and commissioning of charge points, including addition to the CMS. Once this is completed, it can be marked as live and opened to the public.

Reliability and maintenance are key to the success of a CPO once charge points are in operation. Reliability is not only driven by regulations in some geographies; it is also a key part of a CPO's reputation and can impact on profitability, especially where poor reliability impacts utilisation. Maintenance is key to reliability, with planned maintenance needing to take place every 6 months and unplanned maintenance needing to occur as soon as possible after an issue is identified which is stopping a charge point from operating. Service-level agreements often state this should be done in 2 hours; however, the severity of the issue will influence this.

Not only is managing the process and stakeholders important. So is the need to consider the right technology and not become influenced by possible alternatives to the status quo. Wireless charging, for example, is promoted as the future for charging; however, the entire industry, not to mention drivers themselves, are behind traditional plug-in charge points. It will take a major shift in wireless technology adoption, standards and driver behaviours for this to properly take off. This is all the more important in the context of rapid projected charge point growth over the coming decade. What should be clear from the targets set across North

America, the United Kingdom, the European Union and Australasia is that significant resources will need to be invested to achieve these targets and that deployment of traditional charge points is critical to accelerated adoption rates. The size of this challenge should not be underestimated but that does not mean it is impossible or that alternatives are available. The case for hydrogen will continue to be made but without mass green hydrogen manufacture paired with the associated infrastructure and vehicles to use it, EVs will continue to be king. This is even more so because the technological advancements over the coming years, paired with huge anticipated investment, will solidify the position of EVs further.

12.3 Towards the future

The trend going forward across the globe is growth in both EV adoption rates and charge points deployed. As EVs become more commonplace, so will the advancement in EV and charge point technologies, with the former expected to see major changes in battery chemistries and vehicle architectures, increasing efficiency and range, whereas the latter will see improvements in ease of use, security and widespread adoption of vehicle to grid. As EV cars and vans become more common, so will electric equivalents of other modes, including buses and heavy goods vehicles. With this will come an expansion of charging infrastructure to meet their slightly different charging needs, though learnings and experiences derived from EV car and van charge point deployments will ensure the deployment process will be much more structured. A major challenge that should not be underestimated is the improvements and advancements required in the power grid to meet the additional energy demands from mass EV adoption. Major investment in renewables, combined with energy storage and improvements to the grid infrastructure itself, is required. Without this, our transition to EVs and the installation of charge points may slow or even grind to a halt.

12.4 Final remarks

What I hope is clear from this book is that climate change is one of the greatest, if not the greatest, challenges we face as a species and that transport emissions are a major contributing factor to global warming. Although transport emissions cover all modes including aviation and shipping, with both of these requiring significant investment to decarbonise, road-based transport makes up a significant chunk of these emissions. Whereas bus and heavy goods vehicle electrification has significant progress to make, car and van electrification is playing and will continue to play a key role in decarbonising transport, with EV adoption rates anticipated to increase at pace over the next decade up to 2035. Although the largest proportion of charging sessions take place at home or work, and this is anticipated to grow, public charge points in residential, destination and en route locations will continue to play an important role. Demand analysis must be conducted to ensure these are deployed in the right locations to have the greatest impact through good utilisation, whilst also being profitable. As EV adoption grows, charge point deployment across the West will need to ramp up in tandem. Although there will be changes over the coming decade in terms of charge point standards and business models, with wireless charging and battery swapping playing a role in niche use cases, the nature of public charging will stay the same, relying on plugging in and charging. As the decade advances and innovations in vehicle battery technology, architecture and ranges increase, the demand for public charge points may be impacted, but the growth in the number of EVs should mean that utilisation remains good. To enable this EV revolution, governments have an important role to play in working with the private sector to fund charge points in locations to enable a just transition and also in rural locations where demand and utilisation are difficult to justify investment. Significant investment is also required in grid infrastructure and renewable power generation and energy storage to enable the transition. We are on a critical journey which, if we do it right, should mean we will reach net zero for EV cars and vans by 2035. Failure is not an option.

References

[1] UK Research and Innovation, *A brief history of climate change discoveries* [online]. 2025. Available from https://www.discover.ukri.org/a-brief-history-of-climate-change-discoveries/index.html#article [Accessed 08 January 2025].

[2] Intergovernmental Panel on Climate Change. Glossary [online]. 2018. Available from https://www.ipcc.ch/sr15/chapter/glossary/ [Accessed 08 January 2025].

[3] International Energy Agency. *Outlook for electric mobility* [online]. 2024. Available from https://www.iea.org/reports/global-ev-outlook-2024/outlook-for-electric-mobility [Accessed 16 January 2025].

Index

www.ingramcontent.com/pod-product-compliance
Lightning Source LLC
LaVergne TN
LVHW011841200926
848484LV00002B/385